VOLUME ONE HUNDRED AND SIXTY SEVEN

PROGRESS IN MOLECULAR BIOLOGY AND TRANSLATIONAL SCIENCE

Molecular Basis of Neuropsychiatric Disorders: from Bench to Bedside

VOLUME ONE HUNDRED AND SIXTY SEVEN

PROGRESS IN MOLECULAR BIOLOGY AND TRANSLATIONAL SCIENCE

Molecular Basis of Neuropsychiatric Disorders: from Bench to Bedside

Edited by

SHAFIQUR RAHMAN
Department of Pharmaceutical Sciences, South Dakota State University, Brookings, SD, United States

ELSEVIER

ACADEMIC PRESS
An imprint of Elsevier

Academic Press is an imprint of Elsevier
50 Hampshire Street, 5th Floor, Cambridge, MA 02139, United States
525 B Street, Suite 1650, San Diego, CA 92101, United States
The Boulevard, Langford Lane, Kidlington, Oxford OX5 1GB, United Kingdom
125 London Wall, London EC2Y 5AS, United Kingdom

First edition 2019

ISBN: 978-0-12-818855-2
ISSN: 1877-1173

For information on all Academic Press publications
visit our website at https://www.elsevier.com/books-and-journals

Publisher: Zoe Kruze
Acquisition Editor: Ashlie M. Jackman
Editorial Project Manager: Joanna Collett
Production Project Manager: James Selvam
Cover Designer: Matthew Limbert

Typeset by SPi Global, India

Contents

Contributors

Sami Alzarea
Department of Pharmaceutical Sciences, College of Pharmacy, South Dakota State University, Brookings, SD, United States

Megumi Andoh
Laboratory of Chemical Pharmacology, Graduate School of Pharmaceutical Sciences, The University of Tokyo, Tokyo, Japan

Fatin Atrooz
Department of Pharmacological and Pharmaceutical Sciences, College of Pharmacy, University of Houston, Houston, TX, United States

Naren L. Banik
Department of Microbiology and Immunology, Hollings Cancer Center; Department of Neurosurgery, Medical University of South Carolina; Ralph H. Johnson Veterans Administration Medical Center, Charleston, SC, United States

Thomas P. Beresford
Laboratory for Clinical and Translational Research in Psychiatry, Research Service and Psychiatry, Rocky Mountain Regional VA Medical Center (116); Department of Psychiatry, University of Colorado Denver School of Medicine, Anschutz Medical Campus, Aurora, CO, United States

Robert L. Copeland
Department of Pharmacology, Howard University College of Medicine, Washington, DC, United States

Antonei B. Csoka
Department of Anatomy, Howard University College of Medicine, Washington, DC, United States

Sonja Elsaid
Translational Addiction Research Laboratory, Centre for Addiction and Mental Health; Institute of Medical Science, Faculty of Medicine, University of Toronto, Toronto, ON, Canada

Sarah A. Flynn
Department of Psychiatry and Division of Basic Biomedical Sciences, Sanford USD School of Medicine, Sioux Falls, SD, United States

Bruk Getachew
Department of Pharmacology, Howard University College of Medicine, Washington, DC, United States

Azizul Haque
Department of Microbiology and Immunology, Hollings Cancer Center, Medical University of South Carolina, Charleston, SC, United States

Yuji Ikegaya
Laboratory of Chemical Pharmacology, Graduate School of Pharmaceutical Sciences, The University of Tokyo, Tokyo, Japan

Stefan Kloiber
Institute of Medical Science, Faculty of Medicine; Department of Psychiatry, Faculty of Medicine, University of Toronto; General Adult Psychiatry and Health Systems Division; Campbell Family Mental Health Research Institute, Centre for Addiction and Mental Health, Toronto, ON, Canada

Ryuta Koyama
Laboratory of Chemical Pharmacology, Graduate School of Pharmaceutical Sciences, The University of Tokyo, Tokyo, Japan

Bernard Le Foll
Translational Addiction Research Laboratory; Acute Care Program, Centre for Addiction and Mental Health; Institute of Medical Science, Faculty of Medicine; Department of Psychiatry, Faculty of Medicine; Department of Family and Community Medicine; Department of Pharmacology and Toxicology, University of Toronto, Toronto, ON, Canada

Hesong Liu
Baylor College of Medicine, Houston, TX, United States

Kabirullah Lutfy
College of Pharmacy, Western University of Health Sciences, Pomona, CA, United States

Kebreten F. Manaye
Department of Physiology and Biophysics, Howard University College of Medicine, Washington, DC, United States

Jennifer K. Melbourne
The University of Texas at Austin, College of Pharmacy, Division of Pharmacology & Toxicology, Austin, TX, United States

Kimberly Nixon
The University of Texas at Austin, College of Pharmacy, Division of Pharmacology & Toxicology, Austin, TX, United States

Hui Peng
University of Kentucky, College of Pharmacy, Department of Pharmaceutical Sciences, Lexington, KY, United States

Shafiqur Rahman
Department of Pharmaceutical Sciences, College of Pharmacy, South Dakota State University, Brookings, SD, United States

Patrick J. Ronan
Sioux Falls VA Research Service; Department of Psychiatry and Division of Basic Biomedical Sciences, Sanford USD School of Medicine, Sioux Falls, SD; Laboratory for Clinical and Translational Research in Psychiatry, Research Service and Psychiatry, Rocky Mountain Regional VA Medical Center (116), Aurora, CO, United States

Samina Salim
Department of Pharmacological and Pharmaceutical Sciences, College of Pharmacy, University of Houston, Houston, TX, United States

Ramsha Shams
Department of Microbiology and Immunology, Hollings Cancer Center, Medical University of South Carolina, Charleston, SC, United States

Gollapudi Shankar
Outpatient Clinic, VA Affairs-VCB, Laredo, TX, United States

K. Ryan Thompson
The University of Texas at Austin, College of Pharmacy, Division of Pharmacology & Toxicology, Austin, TX, United States

Yousef Tizabi
Department of Pharmacology, Howard University College of Medicine, Washington, DC, United States

Preface

The need for new and effective treatments for neuropsychiatric disorders has never been greater. At this critical juncture of human history, although our current knowledge and understanding on biology of brain disorders have significantly advanced, we face many challenges and limitations to treating many neuropsychiatric disorders. Several critical issues may have contributed to the less effective therapeutics for these neuropsychiatric disorders including: (1) limited understanding of the underlying biology of the disorders; (2) complex etiologies of the disorders; (3) lack of appropriate molecular targets and biomarkers for the disorders; and (4) limitations of animal models for the disorders. Therefore, there is a continuing interest to translate the knowledge obtained from preclinical research on novel targets, appropriate biomarkers, and molecular mechanisms of neuropsychiatric disorders into effective therapeutics.

This volume, *Molecular Basis of Neuropsychiatric Disorders: from Bench to Bedsides*, consists of nine chapters written by eminent scientists in the field. The volume covers new molecular targets, biomarkers, complex mechanisms, and important animal models associated with several common neuropsychiatric disorders including major depression, Parkinson's disease, schizophrenia, stress and anxiety disorders, traumatic brain injury, alcohol addiction, and autism spectrum disorders. The chapters describe emerging breakthrough findings and discuss many innovative strategies for the discovery of new treatments of these neuropsychiatric disorders. The first chapter presents and discusses the evidence of shared biological substrates between Parkinson's disease and depression, focusing on several novel targets and treatment strategies. The second chapter reviews and provides a detailed overview on the role of cannabidiol in a number of neuropsychiatric disorders, including schizophrenia, mood disorders, and addiction. The third chapter presents a comprehensive account on the biochemical basis of the link between stress, and stress-induced neuronal and behavioral deficits, and discusses potential novel therapeutic drug targets. The fourth chapter focuses on calcium-induced activation of calpain and the cellular processing of α-synuclein associated with Parkinson's disease (PD) and discusses emerging therapeutic targets for PD. The fifth chapter presents the role of neuroinflammation and calcineurin signaling in alcohol use disorder and traumatic brain injury and discusses the therapeutic effect of immunosuppressant in preclinical models. The sixth chapter reviews emerging evidence on pituitary adenylyl cyclase-activating polypeptide in stress-mediated neuropsychiatric disorders and examines potential therapeutic strategies by targeting the pituitary peptide receptors. The seventh chapter concentrates on recent advances on glial mechanisms and inflammatory targets in major depressive disorder and evaluates the glial nicotinic receptor-mediated mechanisms in animal models as potential therapeutic strategy for the disorder. The eighth chapter presents the role of microglia and neuroimmune system in alcohol use disorders (AUDs) and discusses several potential therapeutic drug targets for AUDs. The ninth chapter reviews the role of microglial mechanisms in autism spectrum disorders and discusses several microglial targets as potential therapeutic approach for the disorders.

Together, these nine comprehensive chapters provide extensive insight into our current knowledge on novel cellular and molecular mechanisms for these neuropsychiatric disorders and indicate the gaps in our understanding from preclinical research to therapeutic drug discovery and development. I hope that the research summarized in this volume will foster new ideas and stimulate translational investigations for greater understanding, as well as provide increased insight into effective strategies for the discovery and development of novel or improved pharmacotherapies in the management of these crippling neuropsychiatric disorders.

I would like to thank all authors for their excellent contributions to this volume. I also thank Dr. David Teplow, the Serial Editor of the Progress in Molecular Biology and Translational Science, for his invitation and guidance. Finally, I also thank Ms. Ashlie Jackman-Juler, Acquisitions Editor, and Ms. Joanna Collett, Editorial Project Manager and all of the people at Elsevier, for their assistance and support in bringing this volume to completion.

Shafiqur Rahman
Editor
Department of Pharmaceutical Sciences
South Dakota State University
Brookings, SD
United States

CHAPTER ONE

Novel targets for parkinsonism-depression comorbidity

Yousef Tizabi[a,*], Bruk Getachew[a], Antonei B. Csoka[b], Kebreten F. Manaye[c], Robert L. Copeland[a]

[a]Department of Pharmacology, Howard University College of Medicine, Washington, DC, United States
[b]Department of Anatomy, Howard University College of Medicine, Washington, DC, United States
[c]Department of Physiology and Biophysics, Howard University College of Medicine, Washington, DC, United States
*Corresponding author: e-mail address: ytizabi@howard.edu

Contents

Abstract

With the aging population growing and the incidence of neurodegenerative diseases on the rise, the researchers in the field are yet more urgently challenged to slow and/or

Progress in Molecular Biology and Translational Science, Volume 167
ISSN 1877-1173
https://doi.org/10.1016/bs.pmbts.2019.06.004

reverse the devastating consequences of such progression. The challenge is further enforced by psychiatric co-morbid conditions, particularly the feeling of despair in these population. Fortunately, as our understanding of the neurobiological substrates of maladies affecting the central nervous system increases, more therapeutic options are also presented. In this short review while providing evidence of shared biological substrates between Parkinson's disease and depression, novel therapeutic targets and drugs are suggested. The emphasis will be on neuroplasticity underscored by roles of neurotrophic and inflammatory factors. Examples of few therapeutic drugs as well as future directions are also touched upon.

1. Introduction

Although various treatments affecting the symptomatology of the Parkinson's disease (PD), the second most common age-related neurodegenerative disease following Alzheimer's disease (AD), are available, none is proven to retard the progressive nature of the disease. Depression, a common neuropsychiatric disorder also takes a substantial toll on the individual and the society. In this case also, treatments options are not universally effective, and discovery of novel targets and/or drugs are rigorously pursued. Moreover, there is a high co-morbidity of PD and depression. Indeed, it is suggested that depressive disorders, especially in elderly people, may be an indication of latent neurodegeneration, including PD. Recent evidence signifies key roles played by the inflammatory and neurotrophic factors in the etiology of both PD and depression. Hence, understanding specific functions of these factors can suggest potential novel targets for treatment of these diseases. Here, following a brief description of PD and depression, utility of various drugs including nicotine, resveratrol, curcumin and potentially ketamine that affect both the aforementioned underlying mechanisms (i.e., inflammation and neuroplasticity) in co-morbid PD-depression is discussed.

2. Parkinson's disease

It is well recognized that PD or parkinsonism is primarily associated with loss of dopaminergic neurons in substantia nigra pars compacta (SNpc), which leads to the striatal dopamine (DA) deficiency, the target of SNpc DA neurons. PD is characterized by motor deficits such as akinesia, rigidity, resting tremor and postural instability.[1,2] The most common therapy, L-Dopa, a DA precursor, loses its full efficacy in a few years and can induce severe dyskinesia.[3,4] To delay this scenario, the patients are usually started with or are concurrently given one of the following: a dopaminergic agonist such as pramipexole; a monoamine oxidase inhibitor such as selegiline or most recently the FDA approved sulfonamide; or a catechol-methyltransferase inhibitor such as tolcapone.[1,5] However, the progression of the neurodegeneration is not affected by any of these treatments and hence drugs with neuroprotective properties that may stop or at least delay the neuronal loss are urgently needed.

2.1 Etiology

The interaction between genes and environment are believed to define an individual and his or her susceptibility to diseases including neurological and/or neuropsychiatric disorders.[6–9] Juxtaposed on these are the epigenetic modifications, which can modify the gene expression without affecting the genome sequence.[10,11] Collectively, abnormalities within the neurobiological substrates, e.g., the neurotransmitter systems, receptor functions, transduction signals, underscored by inflammatory mediators and impairment in neuronal plasticity can manifest the disease symptoms. Thus, eventual alteration in circuit connections can precipitate neurodegenerative as well as neuropsychiatric diseases.[2,12]

Although the exact etiology of PD, which affects more than 10 million people worldwide and approximately 60,000 annually in the United States remains elusive, involvement of various neurotoxicological agents as well as excess accumulation of trace elements such as manganese and iron have been suggested.[13,14] Indeed, some of these compounds have been used in cellular or preclinical models of PD. Thus, neurotoxins that selectively target catecholaminergic neurons (such as salsilinol, 6-hydroxydopamine, 1-methyl-1,2,3,6-tetrahydropiridine = MPTP, agricultural pesticides such as rotenone and paraquat, manganese, iron, etc.), as well as genetic manipulations that introduce mutations in genes involving specific proteins implicated in PD such as alpha-synuclein, Parkin or Nurr1 are routinely used to better understand the mechanisms involved in neurodegeneration and/or to develop novel therapies.[15–26]

2.2 Role of inflammation (immune system)

Immune activation is necessary to help protect the body against toxic agents, injury, stress and various diseases. In a normal functioning system, the main action of the immune system is mediated by a delicate balance between pro- and anti-inflammatory cytokines. Many cytokines are referred to as interleukins (ILs), indicating that they are secreted by some leukocytes and act on other similar cell types.[27,28] However, breakdown of this delicate balance by insults such as stress or diseases can tip the balance in favor of pro-inflammatory cytokines, which when persistent, can be quite harmful and damaging.[10,29] Whereas peripheral immune defense is performed primarily by lymphocytes, monocytes and macrophages of the hematopoietic system,[30,31] in the brain the main task is carried out by microglia. Thus, inflammation induced by systemic administration of lipopolysaccharide (LPS), found in the outer membrane of Gram-negative bacteria, can cause long-term increases in tumor necrosis factor (TNF-α), a pro-inflammatory cytokine by brain microglia.[32] This increased pro-inflammatory response can induce a delayed but progressive loss in dopaminergic neurons in the substantia nigra, similar to that seen in Parkinson's disease, suggesting that unregulated neuroinflammation could lead to neurodegeneration.[32,33] It was verified recently that the immune response in the brain changes after an infection and that the dead immune cells are replaced by immune cells from elsewhere in the body.[34] However, the new cells may have an altered response to subsequent challenges and new infections.[34]

2.3 Role of neurotrophic factors (neuroplasticity)

Neurogenesis, by which new neurons are generated from neural stem cells (NSCs) or neural progenitor cells (NPCs), is a rather complex neurobiological process, whereby all dividing cells possess some ability to generate different neural units. Thus, it is estimated that approximately 700 new neurons are added to the adult human hippocampus daily.[35–37] In both humans and rodents, the two neurogenic zones are the subventricular zone (SVZ) of the lateral ventricles and the subgranular zone (SGZ) of the dentate gyrus (DG) in the hippocampus. Although the trajectories of newborn neurons in SGZ are similar in humans and rodents, i.e., in both cases neurons migrate into the DG and become dentate granule cells, the trajectory of SVZ in rodents is the olfactory bulb, whereas in humans the destination is the striatum.[35] Unfortunately, these newly generated neurons are not sufficient to repair the damages brought about by various insults. Hence, pathological perturbation such as infection or injury may provoke a cascade of molecular and cellular events leading to microglia's release of various pro-inflammatory cytokines that can be detrimental to the neurogenesis process in NSCs and can lead to neuronal death.[35–37] Nonetheless, extensive effort is expended in understanding the mechanisms involved in neurogenesis with the hope of identifying potential novel interventions in neurodegenerative as well as neuropsychiatric disorders (discussed below).

Neurotrophic factors play an essential role in mediating the development, survival and maintenance of peripheral and central nervous system.[38,39] These varied protein molecules are also intimately involved in neural plasticity during the adulthood.[40,41] Currently, neurotrophic factors may be grouped in three major categories: brain-derived (BDNF), glial-derived (GDNF) and neurokines.[39,42] Preclinical studies strongly support potential manipulation of neurotrophic factors as therapeutic intervention in PD, however, their effectiveness in clinical studies remains to be evaluated.[43,44] In this review we will be concentrating mainly on the role of BDNF as it has been most extensively investigated and a connection between inflammatory process and this peptide is also well documented.[45,46]

3. Depression

Major depressive disorder (MDD) characterized by a despondent feeling, loss of interest in pleasurable activities, guilt, worthlessness, and trouble concentrating is a serious mental illness affecting the individual and the society with tremendous loss of productivity.[47–49] The patient may also suffer from abnormalities in appetite and sleep. It may ultimately lead to suicidal ideation and actual suicide, estimated to be about 1 million people per year worldwide.[50] The incidence of depression is relatively high as in 2013 in the United States alone, an estimated 15.7 million adults aged 18 or over had reported at least one major depressive episode in previous year.[51]

3.1 Etiology

Similar to PD, genetics plays a very important role in manifestation of depression.[52] Environmental factors, particularly stress, have been identified as culprits in precipitation of this condition.[53–55] Indeed, administration of chronic mild stress to animals can result in depressive-like symptoms and is commonly used to get an insight into the neurobiological substrates of this disease with the hope of developing novel therapies.[7,53,56,57]

The current antidepressants which heralded a new chapter in treatment of mental disorders follow the general hypothesis of "Biogenic Amine Depletion" which suggests that low levels of serotonin, norepinephrine, or dopamine in the brain are responsible for the symptom manifestation. Based on this hypothesis, monoamine reuptake inhibitors were developed as antidepressants which are still the mainstay of therapy in this disease. However, these drugs have been shown to have a limited efficacy and despite enormous effort exerted to improve these monoaminergic drugs, they still fail to produce a rapid and sustained antidepressant response in a substantial proportion of depressed patients.[48,58] This, together with the slow onset and various side effects of the current drugs, has shifted the field toward deeper understanding of the biological bases of depression and development of more effective drugs. Here, we will review several of the strategies that are not only applicable to depression in general, but also to the co-morbid condition of depression and PD.

3.2 Role of inflammatory mediators

It is now well established that inflammatory mediators play a very important role in pathophysiology of a variety of neuropsychiatric disorders including MDD.[59–62] This contention is supported by numerous studies indicating elevated levels of the pro-inflammatory cytokines such as TNF-α and interleukin (IL)-1β in plasma of depressed patients.[63–65] Moreover, depression induced by chronic stress may also be due to an increase in the release of pro-inflammatory cytokines. In this regard, Maes et al.[66] were the first to show that psychological stress in humans induces an inflammatory response through production of pro-inflammatory cytokines, such as TNF-α and interferon-γ (IFN-γ). Subsequent studies verified that mice subjected to chronic mild stress developed depressive-like behavior and increased IL-1β in the hippocampus, whereas IL-1 receptor deficient mice did not show such behavioral changes.[67–69] Moreover, direct administration of TNF-α induces a depressive-like behavior that can be blocked with an antibody against TNF-α.[70] Thus, it may be suggested that at least some of the action of antidepressants may be also mediated via suppression of inflammation. Interestingly, it was shown recently that various anti-inflammatory agents improved antidepressant treatment effects in clinical trials in humans.[62]

3.3 Role of neurotrophins

It is now apparent that neurodegenerative processes as well as impairments in cellular plasticity are important contributory factors to mood disorders.[71,72] It is believed that neuroplasticity mediated by neurotrophic factors may constitute an adaptive defensive behavior.[49] Thus, neurotrophins such as BDNF and its receptor TrkB have been shown to be critically involved in regulation of synaptic plasticity and in the pathophysiology of mood disorders.[73,74] Moreover, polymorphism in the BDNF gene may contribute to depression and/or antidepressant therapies.[75,76] Indeed, adult neurogenesis is required to maintain hippocampal volume and reduced BDNF levels are observed in postmortem brain samples and in the blood of depressed patients.[77,78] Hippocampal volume reduction, signifying a reduction in neurogenesis, has been detected in both animal models of depression,[79,80] as well as in human postmortem studies.[77,81,82] Interestingly, a reduction in hippocampal BDNF was associated with hippocampal volume reduction in an animal model of depression.[83] These reductions in hippocampal volume and BDNF levels, however, were reversible by antidepressant treatment.[80–82] Hence, antidepressants through enhanced BDNF signaling may improve the ability of critical brain circuits and help recovery from depression.[54,74,84–86] In the same vein, chronic mild stress that is frequently used to induce depressive-like behavior in animal models is also associated with a reduction in neurogenesis.[78,87,88]

4. Co-morbid PD and depression

It is now well recognized that PD is also accompanied by a range of various non-motor symptoms, including gastrointestinal disturbance (e.g., constipation), loss of smell, pain and anxiety and depression. These non-motor symptoms usually appear at early stages of the disease, sometimes even before the first motor symptoms, and have a dramatic impact on the quality of life of the patients.[89–91] Depression occurs in various neurodegenerative diseases such as Alzheimer's disease,[92] Lewy body disease,[93] multiple sclerosis,[94] Huntington's disease,[95] and in particular, in PD, where it is associated with a poorer prognosis of the latter.[59,96–98] The co-morbid PD-depression poses a further challenge in treatment options as depression itself may be an indication of latent neurodegeneration, particularly in late life.[59,99,100]

Since neuroinflammation and dysregulation of neuroplasticity are common neurobiological substrates of PD and depression, drugs that may affect these systems simultaneously would offer effective interventions in not only symptom alleviation, but also in slowing or halting the progression of the PD. It is also of relevance to note that a bidirectional relationship between cytokines and neurotrophic factors, i.e., between mediators of inflammation and neuroplasticity is amply documented.[59,100–103] Moreover, it appears that drugs with neuroprotective effects would also likely possess antidepressant effects and vice versa.[104] Thus, in the following section we discuss drugs that might have such dual action and therefore might offer novel intervention in PD-depression co-morbidity.

5. Novel targets and/or drugs based on empirical evidence

5.1 Nicotine as an antidepressant

An antidepressant effect of nicotine has been amply verified by a number of preclinical,[80,105–110] as well as limited clinical studies.[111–114] It is now believed that the high incidence of smoking among depressed patients may be an attempt at self-medication with nicotine.[114–117] Thus, sufficient evidence for applicability of nicotine and involvement of specific nicotinic receptors, particularly alpha4–beta2 and alpha7 subtypes in its antidepressant effect is currently available.[118–121]

5.2 Nicotine as a neuroprotectant

Neuroprotective effects of nicotine in general,[122] and its applicability to PD, in particular, has been amply documented.[123] Thus, in cellular models of dopaminergic cell damage, nicotine has been shown to protect against salsolinol and aminochrome that selectively damage dopaminergic cells.[15,16,124–126] The utility of aminochrome model of PD has been recently reviewed.[127] We have also observed nicotine protection against toxicity induced by iron and manganese, excess of which has been implicated in PD.[23] Similarly, nicotine has been shown to protect PD-like symptoms induced by MPTP in various animal studies, including non-human primates.[17] Nicotine's action in this regard may involve inhibition of astrocyte activation and inflammatory suppression.[128] In addition to its well documented anti-inflammatory effects,[129,130] nicotine may also modulate synaptic plasticity via its interaction with the neurotrophic mediators, particularly BDNF, which may be considered as a mechanism for its antidepressant and neuroprotective effects.[118,123,131–134] It may be therefore suggested that nicotine or nicotinic agonists via their interaction with neurotrophic and/or inflammatory mediators could offer novel intervention in depressive and/or neurodegenerative diseases as well as in PD-depression co-morbidity.

5.3 Curcumin as an antidepressant

The antidepressant effects of curcumin, the active ingredient in turmeric (*Curcuma longa*), were initially observed in stress-induced depression models,[135,136] and later in WKY rats, a non-induced animal model of depression.[137] These studies were extended to include LPS-induced depressive-like behavior in mice,[138] and in unpredictable, mild stress-induced depressive-like behavior in rats.[139] Clinical trials have also shown antidepressant properties of curcumin alone or in combination with other currently used antidepressants.[140–142]

Curcumin antidepressant activities may be mediated via diverse mechanisms involving neurotransmitters, transcription pathways, neurogenesis, the hypothalamic–pituitary–adrenal axis, and inflammatory and immune pathways, as demonstrated in various animal and human studies.[143,144] Thus, curcumin has been shown to increase biogenic amines (e.g., dopamine, serotonin, and NE) in the cortex and hippocampus,[135,145] to up-regulate hippocampal BDNF[137,146,147] and reduce pro-inflammatory cytokines such as TNF-α and IL-1β.[148–151] Thus, multiple mechanisms including interactions with the neurotrophic and inflammatory mediators are likely to contribute to the antidepressant-like effects of curcumin.

5.4 Curcumin as a neuroprotectant

Interestingly, it has been observed that societies that widely use curcumin show reduced incidence of inflammation-influenced diseases including cognitive deficits.[152–155] This is likely due to the antioxidant,[156,157] anti-inflammatory,[157,158] neuroprotectant[20,159] and autophagy modulating effects of curcumin.[160] Thus, curcumin has been shown to suppress IL-1β, tumor necrosis factor TNF-α,[161,162] and the nuclear factor NF-κB, considered a prototypical pro-inflammatory signaling pathway as well as intracellular components of glial fibrillary acidic protein (GFAP), considered a marker of gliosis.[163,164] In addition, curcumin may exert a neuroprotective effect against the 6-OHDA-induced rat model of PD[164] and prevent axonal degeneration.[165] These anti-inflammatory effects of curcumin together with its neurotrophic promoting effects discussed above, make this compound of significant interest in both neuropsychiatric, particularly MDD, and neurodegenerative diseases, such as PD and AD.[143,166,167]

An important point, however, in relation to therapeutic use of curcumin is its poor bioavailability,[167–169] which might be due to insolubility in water, poor absorption as well as rapid metabolism. Although addition of piperine could enhance curcumin's absorption[168] and improve its effectiveness in treatment of depression,[170] more research in structural modification as well delivery system via nanoparticles could provide optimal therapeutic application for this naturally occurring polyphenol.[169]

5.5 Resveratrol as an antidepressant

Like curcumin, resveratrol (3,4′,5-trihydroxy-*trans*-stilbene), is also a natural non-flavonoid polyphenol that may be extracted from the skin of red grapes, e.g., in the processing of wine, or from other fruits such as blueberries and cranberries and in some nuts such as peanuts and pistachios, and even cocoa and dark chocolate. It is believed that the plants from which these foods come make resveratrol to fight microbial and fungal infection, ultra-violet radiation, stress, and injury.[171] Resveratrol is also a potent antioxidant and anti-inflammatory compound.[171,172] Its anti-inflammatory effects are evident in its suppression of NF-κB, TNF-α, IL-1β, and nitric oxide (another proinflammatory factor), as well as inhibition of LPS-induced microglial activation.[171–174]

Antidepressant effects of resveratrol have been observed in various rodent models of depression, including depression induced by chronic unpredictable mild stress in rats,[175–177] in Wistar-Kyoto rat model,[178] and corticosterone-induced depression in mice.[179] Interestingly, in all cases, the antidepressant effect of resveratrol was associated with an increase in hippocampal BDNF.[176,178–180]

5.6 Resveratrol as a neuroprotectant

The anti-inflammatory and other properties of resveratrol (e.g., antioxidant) are likely responsible for its various beneficial effects in neurodegenerative diseases. Thus, a number of reports have highlighted possible application of resveratrol in ischemic and traumatic CNS injury,[181–183] in AD,[184–186] in improving cognitive functions[187] as well as in PD.[188–192]

Here also, similar to discussion on curcumin the important point to be borne in mind is that naturally occurring forms of resveratrol have a very limited half-life in plasma, and therefore it is necessary to develop more potent analogs with increased bioavailability.[193] An added advantage in polyphenols is that in addition to their neurotrophic and inflammatory/anti-oxidant properties, they also possess other anti-degenerative properties (e.g., through epigenetic and long chain non-coding RNA) as well as interaction with steroid receptors, that can further enhance their therapeutic potential in variety of diseases such as age-related eye diseases,[194] cardiovascular and cardio-oncology,[195–197] diabetes mellitus,[198] cancers, particularly of breast and prostate[199–201] and osteoporosis.[202,203]

5.7 Ketamine as an antidepressant

Ketamine, a derivative of phencyclidine (PCP), is a non-competitive glutamate NMDA receptor antagonist.[204] Earlier clinical studies had shown significant and prolonged reduction of depressive symptoms following an acute single sub-anesthetic dose of ketamine in treatment-resistant patients.[205–209] Ketamine's antidepressant effect appears to be actually mediated by post synaptic α-amino-3-hydroxy-5-methyl-4-isoxazolepropionic acid (AMPA) receptors, as NBQX, an AMPA receptor antagonist, can block ketamine's effects.[209,210] This contention is further supported by both clinical studies showing that expression of AMPA receptors is altered in patients with depression,[211] and preclinical studies demonstrating that AMPA itself could exert antidepressant effect in an animal model of depression.[212] Activation of post-synaptic AMPA receptors in turn appear to cause an increase in BDNF via m-TOR.[212,213] It is believed that BDNF is the actual factor in inducing the antidepressant effect of ketamine.[213–215] Recently it has been demonstrated that GABA alterations may also contribute to the ketamine response.[54] Thus, identifying the role of specific AMPA and/or GABA receptor subtypes as well as the neurotrophic pathway in depression may offer novel pharmacological interventions.[54,212] It is of relevance to note that both ketamine and NBQX are also effective in attenuating alcohol-induced depression[216] and that very recently esketamine nasal spray was approved by FDA as a fast-acting antidepressant (FDA News Release, March 5, 2019). Moreover, just-published report indicates that ketamine's long lasting effect may be due to reversal of dendritic spine losses that underlie depression-related behaviors.[217]

In addition to its effects on the neurotropic factors including BDNF, ketamine may also suppress pro-inflammatory cytokines such as TNF-α and IL-6[218,219] and promote the activity of anti-inflammatory cytokines such as IL-10.[220] Indeed, peripheral and central anti-inflammatory effects of ketamine should be considered in relation to its therapeutic potential as well as identification of novel targets for treatment of depression.[59,221–223]

5.8 Ketamine as a neuroprotectant

Neuroprotective effects of ketamine have been demonstrated both in vitro and in vivo. Thus, protective effects of ketamine against damage induced by: ischemia, oxygen and/or glucose deprivation,[224,225] spinal ischemia,[226] as well as in acute neurological injury[227] have been reported. Moreover, a combination of ketamine and atropine may offer neuroprotection against toxic (organophosphate)-induced status epilepticus in mice.[228,229] More recently, it was shown that ketamine may also act at non-neuronal cells such as inhibition of osteoclast formation[230] as well as limiting pathogen expansion in vitro.[231] We have also reported on ketamine's effect on gut microbiota, relevant to inflammatory and depressive-like behavior.[232] Thus, it may be concluded that ketamine through a variety of mechanisms may provide antidepressant as well as neuroprotective effects and hence may be of specific utility in PD-depression co-morbidity. In this regard, some evidence on utility of ketamine in non-motor aspects of PD, including depression has been provided.[97] Further investigation of theses mechanisms and the signal transduction pathways such as neuron-specific P-CREB (phosphorylated CAMP response element-binding protein) and mTOR may provide further therapeutic targets in co-morbid condition of PD and depression.

6. Concluding remarks and future direction

Although to date, a true neuroprotective drug that could substantially prevent the progression of the neurodegenerative diseases in general, and PD in particular, is not available, current research offers tremendous hope as potential novel targets become apparent. Moreover, the co-morbid condition of depression with PD could also be benefited by such advances as it appears that a neuroprotective drug would have an antidepressant effect and vice versa.[104]

In addition to the established roles of inflammation and neuroplasticity in neurodegenerative and neuropsychiatric diseases, few other fundamental factors that may have a profound effect in such diseases are being currently explored. These include involvement of genetics and epigenetics,[11,233] microRNAs, a class of small non-coding RNAs,[234] microbiota,[232,235,236] and long non-coding RNAs.[200,237] Moreover, the promise of more effective intervention including gene therapy and/or combination therapies is also on the horizon.[104,238]

Acknowledgment

Supported by NIH/NIAAA R03AA022479 (Y.T.); NIH/NIA 1R25AG047843-01 (K.F. M. and A.B.C.); 1R03G049288-01A1 (K.F.M.).

Disclaimers

The authors indicate that the views expressed in the submitted article are their own and not an official position of the institution or the funding agency.

Financial disclosure/conflict of interest

None

References

1. Cacabelos R. Parkinson's disease: from pathogenesis to pharmacogenomics. *Int J Mol Sci*. 2017;18(3):551.
2. McGregor MM, Nelson AB. Circuit mechanisms of Parkinson's disease. *Neuron*. 2019;101(6):1042–1056.
3. Picconi B, Hernández LF, Obeso JA, Calabresi P. Motor complications in Parkinson's disease: striatal molecular and electrophysiological mechanisms of dyskinesias. *Mov Disord*. 2018;33(6):867–876.
4. Martinez B, Peplow PV. Neuroprotection by immunomodulatory agents in animal models of Parkinson's disease. *Neural Regen Res*. 2018;13(9):1493–1506.
5. Schapira AH, Fox SH, Hauser RA, et al. Assessment of safety and efficacy of safinamide as a levodopa adjunct in patients with Parkinson disease and motor fluctuations: a randomized clinical trial. *JAMA Neurol*. 2017;74(2):216–224.
6. Roy M, Tapadia MG, Joshi S, Koch B. Molecular and genetic basis of depression. *J Genet*. 2014;93(3):879–892.
7. Lopizzo N, Bocchio Chiavetto L, Cattane N, et al. Gene-environment interaction in major depression: focus on experience-dependent biological systems. *Front Psychiatry*. 2015;6:68.
8. L'Episcopo F, Tirolo C, Serapide MF, et al. Microglia Polarization, Gene-environment interactions and wnt/β-catenin signaling: emerging roles of glia-neuron and glia-stem/neuroprogenitor crosstalk for dopaminergic neurorestoration in aged parkinsonian brain. *Front Aging Neurosci*. 2018;10:12.
9. Caggiu E, Arru G, Hosseini S, et al. Inflammation, infectious triggers, and Parkinson's disease. *Front Neurol*. 2019;10:122.
10. Hegarty SV, Sullivan AM, O'Keeffe GW. The epigenome as a therapeutic target for Parkinson's disease. *Neural Regen Res*. 2016;11(11):1735–1738.
11. van Heesbeen HJ, Smidt MP. Entanglement of genetics and epigenetics in Parkinson's disease. *Front Neurosci*. 2019;13:277.
12. Bredt DS, Furey ML, Chen G, Lovenberg T, Drevets WC, Manji HK. Translating depression biomarkers for improved targeted therapies. *Neurosci Biobehav Rev*. 2015;59:1–15.
13. Andrade VM, Aschner M, Marreilha Dos Santos AP. Neurotoxicity of metal mixtures. *Adv Neurobiol*. 2017;18:227–265.
14. Vaccari C, El Dib R, de Camargo JLV. Paraquat and Parkinson's disease: a systematic review protocol according to the OHAT approach for hazard identification. *Syst Rev*. 2017;6(1):98.
15. Copeland Jr RL, Leggett YA, Kanaan YM, Taylor RE, Tizabi Y. Neuroprotective effects of nicotine against salsolinol-induced cytotoxicity: implications for Parkinson's disease. *Neurotox Res*. 2005;8(3-4):289–293.
16. Copeland Jr RL, Das JR, Kanaan YM, Taylor RE, Tizabi Y. Antiapoptotic effects of nicotine in its protection against salsolinol-induced cytotoxicity. *Neurotox Res*. 2007;12:61–69.

17. Quik M, Parameswaran N, McCallum SE, et al. Chronic oral nicotine treatment protects against striatal degeneration in MPTP-treated primates. *J Neurochem.* 2006;98:1866–1875.
18. Quik M, Zhang D, Perez XA, Bordia T. Role for the nicotinic cholinergic system in movement disorders; therapeutic implications. *Pharmacol Ther.* 2014;144(1):50–59.
19. Brown D, Tamas A, Reglödi D, Tizabi Y. PACAP protects against salsolinol-induced toxicity in dopaminergic SH-SY5Y cells: implication for Parkinson's disease. *J Mol Neurosci.* 2013;50(3):600–607.
20. Qualls Z, Brown D, Ramlochansingh C, Hurley LL, Tizabi YY. Protective effects of curcumin against rotenone and salsolinol-induced toxicity: implications for Parkinson's disease. *Neurotox Res.* 2014;25:81–89.
21. Hedya SA, Safar MM, Bahgat AK. Cilostazol mediated Nurr1 and autophagy enhancement: neuroprotective activity in rat rotenone PD model. *Mol Neurobiol.* 2018;55(9): 7579–7587.
22. Garretti F, Agalliu D, Lindestam Arlehamn CS, Sette A, Sulzer D. Autoimmunity in Parkinson's disease: the role of α-synuclein-specific T cells. *Front Immunol.* 2019;10:303.
23. Getachew B, Csoka AB, Aschner M, Tizabi Y. Nicotine protects against manganese and iron-induced toxicity in SH-SY5Y cells: implication for Parkinson's disease. *Neurochem Int.* 2019;124:19–24.
24. Kluss JH, Mamais A, Cookson MR. LRRK2 links genetic and sporadic Parkinson's disease. *Biochem Soc Trans.* 2019;47:651–661.
25. Patricio F, Parra I, Martínez I, et al. Effectiveness of fragment C domain of tetanus toxin and pramipexole in an animal model of Parkinson's disease. *Neurotox Res.* 2019;35(3): 699–710.
26. Sassone J, Valtorta F, Ciammola A. Early dyskinesias in Parkinson's disease patients with parkin mutation: a primary corticostriatal synaptopathy? *Front Neurosci.* 2019;13:273.
27. Dinarello CA. Proinflammatory cytokines. *Chest.* 2000;118:503–508.
28. Brocker C, Thompson D, Matsumoto A, Nebert DW, Vasiliou V. Evolutionary divergence and functions of the human interleukin (IL) gene family. *Hum Genomics.* 2010;5(1):30–55.
29. Gao HM, Hong JS. Why neurodegenerative diseases are progressive: uncontrolled inflammation drives disease progression. *Trends Immunol.* 2008;29:357–365.
30. Streit WJ, Walter SA, Pennell NA. Reactive microgliosis. *Prog Neurobiol.* 1999;57: 563–581.
31. Stoll G, Jander S. The role of microglia and macrophages in the pathophysiology of the CNS. *Prog Neurobiol.* 1999;58:233–247.
32. Qin L, Wu X, Block ML, et al. Systemic LPS causes chronic neuroinflammation and progressive neurodegeneration. *Glia.* 2007;55:453–462.
33. Chitnis T, Weiner HL. CNS inflammation and neurodegeneration. *J Clin Invest.* 2017;127(10):3577–3587.
34. Rua R, Lee JY, Silva AB, et al. Infection drives meningeal engraftment by inflammatory monocytes that impairs CNS immunity. *Nat Immunol.* 2019;20(4):407–419.
35. Borsini A, Zunszain PA, Thuret S, Pariante CM. The role of inflammatory cytokines as key modulators of neurogenesis. *Trends Neurosci.* 2015;38(3):145–157.
36. Das S, Basu A. Inflammation: a new candidate in modulating adult neurogenesis. *J Neurosci Res.* 2008;86:1199–1208.
37. Shigemoto-Mogami Y, Hoshikawa K, Goldman JE, Sekino Y, Sato K. Microglia enhance neurogenesis and oligodendrogenesis in the early postnatal subventricular zone. *J Neurosci.* 2014;34(6):2231–2243.
38. Bothwell M. Recent advances in understanding neurotrophin signaling. *F1000Res.* 2016;5. https://doi.org/10.12688/f1000research.8434.1. pii: F1000 Faculty Rev-1885. eCollection 2016. Review.

39. Ibanez CF, Andressoo JO. Biology of GDNF and its receptors—relevance for disorders of the central nervous system. *Neurobiol Dis*. 2017;97:80–89.
40. Vilar M, Mira H. Regulation of neurogenesis by neurotrophins during adulthood: expected and unexpected roles. *Front Neurosci*. 2016;10:26.
41. Ledda F, Paratcha G. Assembly of neuronal connectivity by neurotrophic factors and leucine-rich repeat proteins. *Front Cell Neurosci*. 2016;10:199.
42. Bothwell M. NGF, BDNF, NT3, and NT4. *Handb Exp Pharmacol*. 2014;220:3–15.
43. Pramanik S, Sulistio YA, Heese K. Neurotrophin signaling and stem cells—implications for neurodegenerative diseases and stem cell therapy. *Mol Neurobiol*. 2017;54(9):7401–7459.
44. Sullivan AM, O'Keeffe GW. Neurotrophic factor therapy for Parkinson's disease: past, present and future. *Neural Regen Res*. 2016;11(2):205–207.
45. Singhal G, Baune BT. Microglia: an interface between the loss of neuroplasticity and depression. *Front Cell Neurosci*. 2017;11:270.
46. Giacobbo BL, Doorduin J, Klein HC, Dierckx RAJO, Bromberg E, de Vries EFJ. Brain-derived neurotrophic factor in brain disorders: focus on neuroinflammation. *Mol Neurobiol*. 2018;56:3295–3312.
47. Greenberg PE, Fournier AA, Sisitsky T, Pike CT, Kessler RC. The economic burden of adults with major depressive disorder in the United States (2005 and 2010). *J Clin Psychiatry*. 2015;76(2):155–162.
48. Perez-Caballero L, Torres-Sanchez S, Romero-López-Alberca C, González-Saiz F, Mico JA, Berrocoso E. Monoaminergic system and depression. *Cell Tissue Res*. 2019;377:107–113.
49. Villas Boas GR, de Boerngen Lacerda R, Paes MM, et al. Molecular aspects of depression: a review from neurobiology to treatment. *Eur J Pharmacol*. 2019;851:99–121.
50. World Federation for Mental Health. 2012. www.who.int/mental_health/.../wfmh_paper_depression_wmhd_2012.
51. National Institute of Mental Health. *Awareness month: by the numbers*. 2015. www.nimh.nih.gov/.../2015/mentalhealth-awareness-month-by-the-numbers.
52. Shadrina M, Bondarenko EA, Slominsky PA. Genetics factors in major depression disease. *Front Psych*. 2018;9:334.
53. Belleau EL, Treadway MT, Pizzagalli DA. The impact of stress and major depressive disorder on hippocampal and medial prefrontal cortex morphology. *Biol Psychiatry*. 2019;85(6):443–453.
54. Fogaça MV, Duman RS. Cortical GABAergic dysfunction in stress and depression: new insights for therapeutic interventions. *Front Cell Neurosci*. 2019;13:87.
55. Maguire J. Neuroactive steroids and GABAergic involvement in the neuroendocrine dysfunction associated with major depressive disorder and postpartum depression. *Front Cell Neurosci*. 2019;13:83.
56. Hill MN, Hellemans KG, Verma P, Gorzalka BB, Weinberg J. Neurobiology of chronic mild stress: parallels to major depression. *Neurosci Biobehav Rev*. 2012;36(9):2085–2117.
57. Wiborg O. Chronic mild stress for modeling anhedonia. *Cell Tissue Res*. 2013;354(1):155–169.
58. Ben-Sheetrit J, Aizenberg D, Csoka AB, et al. Post-SSRI sexual dysfunction: clinical characterization and preliminary assessment of contributory factors and dose-response relationship. *J Clin Psychopharmacol*. 2015;35:273–278.
59. Hurley LL, Tizabi Y. Neuroinflammation, neurodegeneration, and depression. *Neurotox Res*. 2013;23(2):131–144.
60. Culmsee C, Michels S, Scheu S, Arolt V, Dannlowski U, Alferink J. Mitochondria, microglia, and the immune system-how are they linked in affective disorders? *Front Psych*. 2019;9:739.

61. Innes S, Pariante CM, Borsini A. Microglial-driven changes in synaptic plasticity: a possible role in major depressive disorder. *Psychoneuroendocrinology*. 2019;102:236–247.
62. Köhler-Forsberg O, Lydholm C N, Hjorthøj C, Nordentoft M, Mors O, Benros ME. Efficacy of anti-inflammatory treatment on major depressive disorder or depressive symptoms: meta-analysis of clinical trials. *Acta Psychiatr Scand*. 2019;139:404–419.
63. Gold SM, Irwin MR. Depression and immunity: inflammation and depressive symptoms in multiple sclerosis. *Immunol Allergy Clin North Am*. 2009;29:309–320.
64. Felger JC, Lotrich FE. Inflammatory cytokines in depression: neurobiological mechanisms and therapeutic implications. *Neuroscience*. 2013;246:199–229.
65. Toben C, Baune BT. An act of balance between adaptive and maladaptive immunity in depression: a role for T lymphocytes. *J Neuroimmune Pharmacol*. 2015;10(4):595–609.
66. Maes M, Song C, Lin A, et al. The effects of psychological stress on humans: increased production of pro-inflammatory cytokines and a Th1-like response in stress- induced anxiety. *Cytokine*. 1998;10:313–318.
67. Steptoe A, Hamer M, Chida Y. The effects of acute psychological stress on circulating inflammatory factors in humans: a review and meta-analysis. *Brain Behav Immun*. 2007;21:901–912.
68. Goshen I, Kreisel T, Ben-Menachem-Zidon O, et al. Brain interleukin-1 mediates chronic stress-induced depression in mice via adrenocortical activation and hippocampal neurogenesis suppression. *Mol Psychiatry*. 2008;13:717–728.
69. Shapira-Lichter I, Beilin B, Ofek K, et al. Cytokines and cholinergic signals co-modulate surgical stress-induced changes in mood and memory. *Brain Behav Immun*. 2008;22:388–398.
70. Kaster MP, Gadotti VM, Calixto JB, Santos AR, Rodrigues AL. Depressive-like behavior induced by tumor necrosis factor-alpha in mice. *Neuropharmacology*. 2012;62:419–426.
71. Haase J, Brown E. Integrating the monoamine, neurotrophin and cytokine hypotheses of depression—a central role for the serotonin transporter? *Pharmacol Ther*. 2015;147: 1–11.
72. Duman RS, Aghajanian GK, Sanacora G, Krystal JH. Synaptic plasticity and depression: new insights from stress and rapid-acting antidepressants. *Nat Med*. 2016;22(3): 238–249.
73. Manji HK, Quiroz JA, Sporn J, et al. Enhancing neuronal plasticity and cellular resilience to develop novel, improved therapeutics for difficult-to-treat depression. *Biol Psychiatry*. 2003;53(8):707–742.
74. Caviedes A, Lafourcade C, Soto C, Wyneken U. BDNF/NF-κB signaling in the neurobiology of depression. *Curr Pharm Des*. 2017;23:3154.
75. Colle R, Deflesselle E, Martin S, et al. BDNF/TRKB/P75NTR polymorphisms and their consequences on antidepressant efficacy in depressed patients. *Pharmacogenomics*. 2015;16(9):997–1013.
76. Lisiecka DM, O'Hanlon E, Fagan AJ, et al. BDNF Val66Met polymorphism in patterns of neural activation in individuals with MDD and healthy controls. *J Affect Disord*. 2015;184:239–244.
77. Sheldrick A, Camara S, Ilieva M, Riederer P, Michel TM. Brain-derived neurotrophic factor (BDNF) and neurotrophin 3 (NT3) levels in post-mortem brain tissue from patients with depression compared to healthy individuals—a proof of concept study. *Eur Psychiatry*. 2017;46:65–71.
78. Schoenfeld TJ, McCausland HC, Morris HD, Padmanaban V, Cameron HA. Stress and loss of adult neurogenesis differentially reduce hippocampal volume. *Biol Psychiatry*. 2017;82(12):914–923.
79. Zhao C, Deng W, Gage FH. Mechanisms and functional implications of adult neurogenesis. *Cell*. 2008;132:645–660.

80. Tizabi Y, Hauser SR, Tyler KY, et al. Effects of nicotine on depressive-like behavior and hippocampal volume of female WKY rats. *Prog Neuropsychopharmacol Biol Psychiatry*. 2010;34(1):62–69.
81. Sheline YI, Gado MH, Kraemer HC. Untreated depression and hippocampal volume loss. *Am J Psychiatry*. 2003;160:1516–1518.
82. Czeh B, Lucassen PJ. What causes the hippocampal volume decrease in depression? Are neurogenesis, glial changes and apoptosis implicated? *Eur Arch Psychiatry Clin Neurosci*. 2007;257:250–260.
83. Hauser SR, Getachew B, Taylor RE, Tizabi Y. Alcohol induced depressive-like behavior is associated with a reduction in hippocampal BDNF. *Pharmacol Biochem Behav*. 2011;100:253–258.
84. Castrén E, Rantamäki T. Neurotrophins in depression and antidepressant effects. *Novartis Found Symp*. 2008;289:43–52. discussion 53–59, 87–93.
85. Mendez-David I, Hen R, Gardier AM, David DJ. Adult hippocampal neurogenesis: an actor in the antidepressant-like action. *Ann Pharm Fr*. 2013;71(3):143–149.
86. Duman RS. Pathophysiology of depression and innovative treatments: remodeling glutamatergic synaptic connections. *Dialogues Clin Neurosci*. 2014;16(1):11–27.
87. Sapolsky RM. Is impaired neurogenesis relevant to the affective symptoms of depression? *Biol Psychiatry*. 2004;56:137–139.
88. Toth E, Gersner R, Wilf-Yarkoni A, et al. Age-dependent effects of chronic stress on brain plasticity and depressive behavior. *J Neurochem*. 2008;107:522–532.
89. Zapiec B, Dieriks BV, Tan S, Faull RLM, Mombaerts P, Curtis MA. A ventral glomerular deficit in Parkinson's disease revealed by whole olfactory bulb reconstruction. *Brain*. 2017;140(10):2722–2736.
90. Faivre F, Joshi A, Bezard E, Barrot M. The hidden side of Parkinson's disease: studying pain, anxiety and depression in animal models. *Neurosci Biobehav Rev*. 2019;96: 335–352.
91. Knudsen K, Fedorova TD, Hansen AK, et al. Objective intestinal function in patients with idiopathic REM sleep behavior disorder. *Parkinsonism Relat Disord*. 2019;58: 28–34.
92. de Souza MB, de Lemos RR, da Cunha JE, de Lima Filho JL, de Oliveira JR. Searching for new genetic risk factors for neuropsychiatric disorders in expression databases. *J Mol Neurosci*. 2010;41:193–197.
93. Fritze F, Ehrt U, Hortobagyi T, Ballard C, Aarsland D. Depressive symptoms in Alzheimer's disease and Lewy body dementia: a one-year follow-up study. *Dement Geriatr Cogn Disord*. 2011;32:143–149.
94. Feinstein A. Multiple sclerosis and depression. *Mult Scler*. 2011;17(11):1276–1281.
95. Perlis RH, Smoller JW, Mysore J, et al. Prevalence of incompletely penetrant Huntington's disease alleles among individuals with major depressive disorder. *Am J Psychiatry*. 2010;167:574–579.
96. Gómez-Esteban JC, Tijero B, Somme J, et al. Application of depression criteria (DSM-IV) in patients with Parkinson's disease. *Clin Neurol Neurosurg*. 2009;111(8):665–669.
97. Vanle B, Olcott W, Jimenez J, Bashmi L, Danovitch I, WW IH. NMDA antagonists for treating the non-motor symptoms in Parkinson's disease. *Transl Psychiatry*. 2018;8(1):117.
98. Wang S, Mao S, Xiang D, Fang C. Association between depression and the subsequent risk of Parkinson's disease: a meta-analysis. *Prog Neuropsychopharmacol Biol Psychiatry*. 2018;86:186–192.
99. Burgut FT, Benaur M, Hencliffe C. Late-life depression: a neuropsychiatric approach. *Expert Rev Neurother*. 2006;6:65–72.
100. Baquero M, Martín N. Depressive symptoms in neurodegenerative diseases. *World J Clin Cases*. 2015;3(8):682–693.

101. Hayley S. The neuroimmune-neuroplasticity interface and brain pathology. *Front Cell Neurosci*. 2014;8:419.
102. Mancini A, Gaetani L, Di Gregorio M, et al. Hippocampal neuroplasticity and inflammation: relevance for multiple sclerosis. *Mult Scler Demyelinating Disord*. 2017;2:2.
103. Calabrese V, Santoro A, Monti D, et al. Aging and Parkinson's disease: inflammaging, neuroinflammation and biological remodeling as key factors in pathogenesis. *Free Radic Biol Med*. 2018;115:80–91.
104. Tizabi Y. Duality of antidepressants and neuroprotectants. *Neurotox Res*. 2016;30(1): 1–13.
105. Semba J, Mataki C, Yamada S, Nankai M, Toru M. Antidepressant like effects of chronic nicotine on learned helplessness paradigm in rats. *Biol Psychiatry*. 1998;43: 389–391.
106. Djuric VJ, Dunn E, Overstreet DH, Dragomir A, Steiner M. Antidepressant effect of ingested nicotine in female rats of Flinders resistant and sensitive lines. *Physiol Behav*. 1999;67:533–537.
107. Tizabi Y, Overstreet DH, Rezvani AH, et al. Antidepressant effects of nicotine in an animal model of depression. *Psychopharmacology (Berl)*. 1999;142:193–199.
108. Tizabi Y, Rezvani AH, Russell LT, Tyler KY, Overstreet DH. Depressive characteristics of FSL rats: involvement of central nicotinic receptors. *Pharmacol Biochem Behav*. 2000;66(1):73–77.
109. Tizabi Y, Getachew B, Rezvani AH, Hauser SR, Overstreet DH. Antidepressant-like effects of nicotine and reduced nicotinic receptor binding in the Fawn-Hooded rat, an animal model of co-morbid depression and alcoholism. *Prog Neuropsychopharmacol Biol Psychiatry*. 2009;33(3):398–402.
110. Kalejaiye O, Bhatti BH, Taylor RE, Tizabi Y. Nicotine blocks the depressogenic effects of alcohol: implications for drinking-smoking co-morbidity. *J Drug Alcohol Res*. 2013;2:235709.
111. Pomerleau CS, Pomerleau OF. Euphoriant effects of nicotine in smokers. *Psychopharmacology (Berl)*. 1992;108:460–465.
112. Salin-Pascual RJ, de la Fuente JR, Galicia-Polo L, Drucker-Colin R. Effects of transderman nicotine on mood and sleep in nonsmoking major depressed patients. *Psychopharmacology (Berl)*. 1995;121:476–479.
113. McClernon FJ, Hiott FB, Westman EC, Rose JE, Levin ED. Transdermal nicotine attenuates depression symptoms in nonsmokers: a double-blind, placebo-controlled trial. *Psychopharmacology (Berl)*. 2006;189:125–133.
114. Cook JW, Spring B, McChargue D. Influence of nicotine on positive affect in anhedonic smokers. *Psychopharmacology (Berl)*. 2007;192:87–95.
115. Moreno-Coutino A, Calderon-Ezquerro C, Drucker-Colin R. Long-term changes in sleep and depressive symptoms of smokers in abstinence. *Nicotine Tob Res*. 2007;9:389–396.
116. Spring B, Cook JW, Appelhans B, et al. Nicotine effects on affective response in depression-prone smokers. *Psychopharmacology (Berl)*. 2008;196:461–471.
117. Edwards AC, Kendler KS. Nicotine withdrawal-induced negative affect is a function of nicotine dependence and not liability to depression or anxiety. *Nicotine Tob Res*. 2011;13:677–685.
118. Barreto GE, Iarkov A, Moran VE. Beneficial effects of nicotine, cotinine and its metabolites as potential agents for Parkinson's disease. *Front Aging Neurosci*. 2015;6:340.
119. Picciotto MR, Lewis AS, van Schalkwyk GI, Mineur YS. Mood and anxiety regulation by nicotinic acetylcholine receptors: a potential pathway to modulate aggression and related behavioral states. *Neuropharmacology*. 2015;96:235–243.
120. Rahman S. Targeting brain nicotinic acetylcholine receptors to treat major depression and co-morbid alcohol or nicotine addiction. *CNS Neurol Disord Drug Targets*. 2015;14(5):647–653.

121. Alzarea S, Rahman S. Alpha-7 nicotinic receptor allosteric modulator PNU120596 prevents lipopolysaccharide- induced anxiety, cognitive deficit and depression-like behaviors in mice. *Behav Brain Res*. 2019;366:19–28.
122. Hoskin JL, Al-Hasan Y, Sabbagh MN. Nicotinic acetylcholine receptor agonists for the treatment of Alzheimer's dementia: an update. *Nicotine Tob Res*. 2019;21(3): 370–376.
123. Tizabi Y, Getachew B. Nicotinic receptor intervention in Parkinson's disease: future directions. *Clin Pharmacol Transl Med*. 2017;1(1):14–19.
124. Das JR, Tizabi Y. Additive protective effects of donepezil and nicotine against salsolinol-induced cytotoxicity in SH-SY5Y cells. *Neurotox Res*. 2009;16:194–204.
125. Ramlochansingh C, Taylor RE, Tizabi Y. Toxic effects of low alcohol and nicotine combinations in SH-SY5Y cells are apoptotically mediated. *Neurotox Res*. 2011;20: 263–269.
126. Munoz P, Huenchuguala S, Paris I, et al. Protective effects of nicotine against aminochrome-induced toxicity in substantia nigra derived cells: implications for Parkinson's disease. *Neurotox Res*. 2012;22:177–180.
127. Segura-Aguilar J. On the role of aminochrome in mitochondrial dysfunction and endoplasmic reticulum stress in Parkinson's disease. *Front Neurosci*. 2019;13:271.
128. Liu Y, Hu J, Wu J, et al. Alpha7 nicotinic acetylcholine receptor-mediated neuroprotection against dopaminergic neuron loss in an MPTP mouse model via inhibition of astrocyte activation. *J Neuroinflammation*. 2012;9:98.
129. Patel H, McIntire J, Ryan S, Dunah A, Loring R. Anti-inflammatory effects of astroglial α7 nicotinic acetylcholine receptors are mediated by inhibition of the NF-κB pathway and activation of the Nrf2 pathway. *J Neuroinflammation*. 2017;14(1):192.
130. Bagdas D, Gurun MS, Flood P, Papke RL, Damaj MI. New insights on neuronal nicotinic acetylcholine receptors as targets for pain and inflammation: a focus on α7 nAChRs. *Curr Neuropharmacol*. 2018;16(4):415–425.
131. Belluardo N, Mudò G, Blum M, Fuxe K. Central nicotinic receptors, neurotrophic factors and neuroprotection. *Behav Brain Res*. 2000;113:21–34.
132. Posadas I, López-Hernández B, Ceña V. Nicotinic receptors in neurodegeneration. *Curr Neuropharmacol*. 2013;11(3):298–314.
133. Yakel JL. Cholinergic receptors: functional role of nicotinic ACh receptors in brain circuits and disease. *Pflugers Arch*. 2013;465(4):441–450.
134. Wei P, Zheng Q, Liu H, et al. Nicotine-induced neuroprotection against cognitive dysfunction after partial hepatectomy involves activation of BDNF/TRKb signaling pathway and inhibition of NF-κb signaling pathway in aged rats. *Nicotine Tob Res*. 2018;20(4):515–522.
135. Kulkarni SK, Bhutani MK, Bishnoi M. Antidepressant activity of curcumin: involvement of serotonin and dopamine system. *Psychopharmacology (Berl)*. 2008;201: 435–442.
136. Bhutani MK, Bishnoi M, Kulkarni SK. Anti-depressant like effect of curcumin and its combination with piperine in unpredictable chronic stress-induced behavioral, biochemical and neurochemical changes. *Pharmacol Biochem Behav*. 2009;92:39–43.
137. Hurley LL, Akinfiresoye L, Nwulia E, Kamiya A, Kulkarni AA, Tizabi Y. Antidepressant-like effects of curcumin in WKY rat model of depression is associated with an increase in hippocampal BDNF. *Behav Brain Res*. 2013;239:27–30.
138. Wang Z, Zhang Q, Yuan L, et al. The effects of curcumin on depressive-like behavior in mice after lipopolysaccharide administration. *Behav Brain Res*. 2014;274: 282–290.
139. Zhang L, Luo J, Zhang M, Yao W, Ma X, Yu SY. Effects of curcumin on chronic, unpredictable, mild, stress-induced depressive-like behaviour and structural plasticity in the lateral amygdala of rats. *Int J Neuropsychopharmacol*. 2014;17(5):793–806.

140. Yu JJ, Pei LB, Zhang Y, Wen ZY, Yang JL. Chronic supplementation of curcumin enhances the efficacy of antidepressants in major depressive disorder: a randomized, double-blind, placebo-controlled pilot study. *J Clin Psychopharmacol.* 2015;35(4): 406–410.
141. Lopresti AL, Drummond PD. Efficacy of curcumin, and a saffron/curcumin combination for the treatment of major depression: a randomised, double-blind, placebo-controlled study. *J Affect Disord.* 2017;207:188–196.
142. Ng QX, Koh SSH, Chan HW, Ho CYX. Clinical use of curcumin in depression: a meta-analysis. *J Am Med Dir Assoc.* 2017;18(6):503–508.
143. Tizabi Y, Hurley LL, Qualls Z, Akinfiresoye L. Relevance of the anti-inflammatory properties of curcumin in neurodegenerative diseases and depression. *Molecules.* 2014;19(12):20864–20879.
144. Seo HJ, Wang SM, Han C, et al. Curcumin as a putative antidepressant. *Expert Rev Neurother.* 2015;15(3):269–280.
145. Arora V, Kuhad A, Tiwari V, Chopra K. Curcumin ameliorates reserpine-induced pain-depression dyad: behavioural, biochemical, neurochemical and molecular evidences. *Psychoneuroendocrinology.* 2011;36:1570–1581.
146. Nam SM, Choi JH, Yoo DY, et al. Effects of curcumin (Curcuma longa) on learning and spatial memory as well as cell proliferation and neuroblast differentiation in adult and aged mice by upregulating brain-derived neurotrophic factor and CREB signaling. *J Med Food.* 2014;17(6):641–649.
147. Srivastava P, Dhuriya YK, Gupta R, et al. Protective effect of curcumin by modulating BDNF/DARPP32/CREB in arsenic-induced alterations in dopaminergic signaling in rat corpus striatum. *Mol Neurobiol.* 2018;55(1):445–461.
148. Aggarwal BB, Gupta SC, Sung B. Curcumin: an orally bioavailable blocker of TNF and other pro-inflammatory biomarkers. *Br J Pharmacol.* 2013;169(8):1672–1692.
149. Sun Y, Liu W, Zhang H, et al. Curcumin prevents osteoarthritis by inhibiting the activation of inflammasome NLRP3. *J Interferon Cytokine Res.* 2017;37(10):449–455.
150. Baj T, Seth R. Role of curcumin in regulation of TNF-a mediated brain inflammatory responses. *Recent Pat Inflamm Allergy Drug Discov.* 2018;12(1):69–77.
151. Kaufmann FN, Gazal M, Bastos CR, Kaster MP, Ghisleni G. Curcumin in depressive disorders: an overview of potential mechanisms, preclinical and clinical findings. *Eur J Pharmacol.* 2016;784:192–198.
152. Chandra V, Pandav R, Dodge HH, et al. Incidence of Alzheimer's disease in a rural community in India: the Indo-US study. *Neurology.* 2001;57:985–989.
153. Vas CJ, Pinto C, Panikker D, et al. Prevalence of dementia in an urban Indian population. *Int Psychogeriatr.* 2001;13:439–450.
154. Ng TP, Chiam PC, Lee T, Chua HC, Lim L, Kua EH. Curry consumption and cognitive function in the elderly. *Am J Epidemiol.* 2006;164(9):898–906. Epub 2006 Jul 26.
155. Aggarwal BB, Sundaram C, Malani N, Ichikawa H. Curcumin: the Indian solid gold. *Adv Exp Med Biol.* 2007;595:1–75.
156. Abrahams S, Haylett WL, Johnson G, Carr JA, Bardien S. Antioxidant effects of curcumin in models of neurodegeneration, aging, oxidative and nitrosative stress: a review. *Neuroscience.* 2019;406:1–21.
157. Li C, Miao X, Li F, et al. Curcuminoids: implication for inflammation and oxidative stress in cardiovascular diseases. *Phytother Res.* 2019;33:1302–1317.
158. Aggarwal BB, Harikumar KB. Potential therapeutic effects of curcumin, the anti-inflammatory agent, against neurodegenerative, cardiovascular, pulmonary, metabolic, autoimmune and neoplastic diseases. *Int J Biochem Cell Biol.* 2009;41:40–59.
159. Singla N, Dhawan DK. N-Methyl N-nitrosourea induced functional and structural alterations in mice brain-role of curcumin. *Neurotox Res.* 2012;22(2):115–126.

160. Shakeri A, Cicero AFG, Panahi Y, Mohajeri M, Sahebkar A. Curcumin: a naturally occurring autophagy modulator. *J Cell Physiol.* 2019;234(5):5643–5654.
161. Gupta SC, Tyagi AK, Deshmukh-Taskar P, Hinojosa M, Prasad S, Aggarwal BB. Downregulation of tumor necrosis factor and other proinflammatory biomarkers by polyphenols. *Arch Biochem Biophys.* 2014;559:91–99.
162. Yuan J, Zou M, Xiang X, et al. Curcumin improves neural function after spinal cord injury by the joint inhibition of the intracellular and extracellular components of glial scar. *J Surg Res.* 2015;195(1):235–245.
163. Yang Z, Zhao T, Zou Y, Zhang JH, Feng H. Curcumin inhibits microglia inflammation and confers neuroprotection in intracerebral hemorrhage. *Immunol Lett.* 2014;160(1):89–95.
164. Yang J, Song S, Li J, Liang T. Neuroprotective effect of curcumin on hippocampal injury in 6-OHDA-induced Parkinson's disease rat. *Pathol Res Pract.* 2014;210(6): 357–362.
165. Tegenge MA, Rajbhandari L, Shrestha S, Mithal A, Hosmane S, Venkatesan A. Curcumin protects axons from degeneration in the setting of local neuroinflammation. *Exp Neurol.* 2014;253:102–110.
166. Darvesh AS, Carroll RT, Bishayee A, Novotny NA, Geldenhuys WJ, Van der Schyf CJ. Curcumin and neurodegenerative diseases: a perspective. *Expert Opin Investig Drugs.* 2012;21:1123–1140.
167. Di Martino RMC, Bisi A, Rampa A, Gobbi S, Belluti F. Recent progress on curcumin-based therapeutics: a patent review (2012-2016). Part II: curcumin derivatives in cancer and neurodegeneration. *Expert Opin Ther Pat.* 2017;27(8):953–965.
168. Hamaguchi T, Ono K, Yamada M. Review: curcumin and Alzheimer's disease. *CNS Neurosci Ther.* 2010;16:285–297.
169. Del Prado-Audelo ML, Caballero-Florán IH, Meza-Toledo JA, et al. Formulations of curcumin nanoparticles for brain diseases. *Biomolecules.* 2019;9(2). pii: E56.
170. Panahi Y, Badeli R, Karami GR, Sahebkar A. Investigation of the efficacy of adjunctive therapy with bioavailability-boosted curcuminoids in major depressive disorder. *Phytother Res.* 2015;29(1):17–21. Biomolecules. 2019;9(2).
171. Berman AY, Motechin RA, Wiesenfeld MY, Holz MK. The therapeutic potential of resveratrol: a review of clinical trials. *NPJ Precis Oncol.* 2017;1:pii: 35.
172. Reglodi D, Renaud J, Tamas A, et al. Novel tactics for neuroprotection in Parkinson's disease: role of antibiotics, polyphenols and neuropeptides. *Prog Neurobiol.* 2017;155: 120–148.
173. Zhang F, Wang H, Wu Q, et al. Resveratrol protects cortical neurons against microglia-mediated neuroinflammation. *Phytother Res.* 2012;27:344–349.
174. Pallàs M, Porquet D, Vicente A, Sanfeliu C. Resveratrol: new avenues for a natural compound in neuroprotection. *Curr Pharm Des.* 2013;19(38):6726–6731.
175. Ge JF, Peng L, Cheng JQ, et al. Antidepressant-like effect of resveratrol: involvement of antioxidant effect and peripheral regulation on HPA axis. *Pharmacol Biochem Behav.* 2013;114–115:64–69.
176. Liu D, Wang Z, Gao Z, et al. Effects of curcumin on learning and memory deficits, BDNF, and ERK protein expression in rats exposed to chronic unpredictable stress. *Behav Brain Res.* 2014;271:116–121.
177. Gu Z, Chu L, Han Y. Therapeutic effect of resveratrol on mice with depression. *Exp Ther Med.* 2019;17(4):3061–3064.
178. Hurley LL, Akinfiresoye L, Kalejaiye O, Tizabi Y. Antidepressant effects of resveratrol in an animal model of depression. *Behav Brain Res.* 2014;268:1–7.
179. Ali SH, Madhana RM, Athira KV, et al. Resveratrol ameliorates depressive-like behavior in repeated corticosterone-induced depression in mice. *Steroids.* 2015;101: 37–42.

180. Moore A, Beidler J, Hong MY. Resveratrol and depression in animal models: a systematic review of the biological mechanisms. *Molecules*. 2018;23(9). pii: E2197.
181. Sinha K, Chaudhary G, Gupta YK. Protective effect of resveratrol against oxidative stress in middle cerebral artery occlusion model of stroke in rats. *Life Sci*. 2002;71: 655–665.
182. Wang Z, Huang Y, Zou J, Cao K, Xu Y, Wu JM. Effects of red wine and wine polyphenol resveratrol on platelet aggregation in vivo and in vitro. *Int J Mol Med*. 2002;9:77–79.
183. Lopez MS, Dempsey RJ, Vemuganti R. Resveratrol neuroprotection in stroke and traumatic CNS injury. *Neurochem Int*. 2015;89:75–82.
184. Vingtdeux V, DresesWerringloer U, Zhao H, Davies P, Marambaud P. Therapeutic potential of resveratrol in Alzheimer's disease. *BMC Neurosci*. 2008;9(Suppl 2):S6.
185. Rege SD, Geetha T, Griffin GD, Broderick TL, Babu JR. Neuroprotective effects of resveratrol in Alzheimer disease pathology. *Front Aging Neurosci*. 2014;11(6):218.
186. Moussa C, Hebron M, Huang X, et al. Resveratrol regulates neuro-inflammation and induces adaptive immunity in Alzheimer's disease. *J Neuroinflammation*. 2017;14(1):1.
187. Harada N, Zhao J, Kurihara H, Nakagata N, Okajima K. Resveratrol improves cognitive function in mice by increasing production of insulin-like growth factor-I in the hippocampus. *J Nutr Biochem*. 2011;22:1150–1159.
188. Tredici G, Miloso M, Nicolini G, Galbiati S, Cavaletti G, Bertelli A. Resveratrol, map kinases and neuronal cells: might wine be a neuroprotectant? *Drugs Exp Clin Res*. 1999;25:99–103.
189. Chen LW, Wang YQ, Wei LC, Shi M, Chan YS. Chinese herbs and herbal extracts for neuroprotection of dopaminergic neurons and potential therapeutic treatment of Parkinson's disease. *CNS Neurol Disord Drug Targets*. 2007;6:273–281.
190. Lofrumento DD, Nicolardi G, Cianciulli A, et al. Neuroprotective effects of resveratrol in an MPTP mouse model of Parkinson's-like disease: possible role of SOCS-1 in reducing pro-inflammatory responses. *Innate Immun*. 2014;20(3):249–260.
191. Fu W, Zhuang W, Zhou S, Wang X. Plant-derived neuroprotective agents in Parkinson's disease. *Am J Transl Res*. 2015;7(7):1189–1202.
192. Tellone E, Galtieri A, Russo A, Giardina B, Ficarra S. Resveratrol: a focus on several neurodegenerative diseases. *Oxid Med Cell Longev*. 2015;2015:392169.
193. Chimento A, De Amicis F, Sirianni R, et al. Progress to improve oral bioavailability and beneficial effects of resveratrol. *Int J Mol Sci*. 2019;20(6):pii: E1381.
194. Bungau S, Abdel-Daim MM, Tit DM, et al. Health benefits of polyphenols and carotenoids in age-related eye diseases. *Oxid Med Cell Longev*. 2019;2019: 9783429.
195. Breuss JM, Atanasov AG, Uhrin P. Resveratrol and its effects on the vascular system. *Int J Mol Sci*. 2019;20(7):pii: E1523.
196. Abdelgawad IY, Grant MKO, Zordoky BN. Leveraging the cardio-protective and anticancer properties of resveratrol in cardio-oncology. *Nutrients*. 2019;11(3):pii: E627.
197. van Andel MM, Groenink M, Zwinderman AH, Mulder BJM, de Waard V. The potential beneficial effects of resveratrol on cardiovascular complications in Marfan syndrome patients insights from rodent-based animal studies. *Int J Mol Sci*. 2019;20(5):pii: E1122.
198. Joshi T, Singh AK, Haratipour P, et al. Targeting AMPK signaling pathway by natural products for treatment of diabetes mellitus and its complications. *J Cell Physiol*. 2019; https://doi.org/10.1002/jcp.28528. [Epub ahead of print].
199. Carlos-Reyes Á, López-González JS, Meneses-Flores M, et al. Dietary compounds as epigenetic modulating agents in cancer. *Front Genet*. 2019;10:79. https://doi.org/10.3389/fgene.2019.00079. eCollection 2019. Review.
200. Mishra S, Verma SS, Rai V, et al. Long non-coding RNAs are emerging targets of phytochemicals for cancer and other chronic diseases. *Cell Mol Life Sci*. 2019;76: 1947–1966.

201. De Amicis F, Chimento A, Montalto FI, Casaburi I, Sirianni R, Pezzi V. Steroid receptor signallings as targets for resveratrol actions in breast and prostate cancer. *Int J Mol Sci*. 2019;20(5):pii: E1087.
202. Murgia D, Mauceri R, Campisi G, et al. *Biomolecules*. 2019;9(3):pii: E94.
203. Zainabadi K. Drugs targeting SIRT1, a new generation of therapeutics for osteoporosis and other bone related disorders? *Pharmacol Res*. 2019;143:97–105.
204. Harrison NL, Simmonds MA. Quantitative studies on some antagonists of N-methyl D-aspartate in slices of rat cerebral cortex. *Br J Pharmacol*. 1985;84:381–391.
205. Berman RM, Cappiello A, Anand A, et al. Antidepressant effects of ketamine in depressed patients. *Biol Psychiatry*. 2000;47:351–354.
206. Zarate Jr CA, Singh JB, Quiroz JA, et al. A double-blind, placebo-controlled study of memantine in the treatment of major depression. *Am J Psychiatry*. 2006;163: 153–155.
207. Iadarola ND, Niciu MJ, Richards EM, et al. Ketamine and other N-methyl-D-aspartate receptor antagonists in the treatment of depression: a perspective review. *Ther Adv Chronic Dis*. 2015;6:97–114.
208. Kavalali ET, Monteggia LM. How does ketamine elicit a rapid antidepressant response? *Curr Opin Pharmacol*. 2015;20:35–39.
209. Maeng S, Zarate Jr CA, Du J, et al. Cellular mechanisms underlying the antidepressant effects of ketamine: role of alpha-amino-3-hydroxy-5-methylisoxazole-4-propionic acid receptors. *Biol Psychiatry*. 2008;63:349–352.
210. Koike H, Iijima M, Chaki S. Involvement of AMPA receptor in both the rapid and sustained antidepressant-like effects of ketamine in animal models of depression. *Behav Brain Res*. 2011;224:107–111.
211. Freudenberg F, Celikel T, Reif A. The role of α-amino-3-hydroxy-5-methyl-4-isoxazolepropionic acid (AMPA) receptors in depression: central mediators of pathophysiology and antidepressant activity? *Neurosci Biobehav Rev*. 2015;52:193–206.
212. Akinfiresoye L, Tizabi Y. Antidepressant effects of AMPA and ketamine combination: role of hippocampal BDNF, synapsin, and mTOR. *Psychopharmacology (Berl)*. 2013;230(2):291–298.
213. Cavalleri L, Merlo Pich E, Millan MJ, et al. Ketamine enhances structural plasticity in mouse mesencephalic and human iPSC-derived dopaminergic neurons via AMPAR-driven BDNF and mTOR signaling. *Mol Psychiatry*. 2018;23(4):812–823.
214. Lepack AE, Fuchikami M, Dwyer JM, Banasr M, Duman RS. BDNF release is required for the behavioral actions of ketamine. *Int J Neuropsychopharmacol*. 2014;18(1). pii: pyu033.
215. Björkholm C, Monteggia LM. BDNF—a key transducer of antidepressant effects. *Neuropharmacology*. 2016;102:72–79.
216. Getachew B, Tizabi Y. Ketamine and NBQX both normalize alcohol-withdrawal induced depressive-like characteristics in rats. *J Drug Alc Res*. 2019;8:1–7.
217. Moda-Sava RN, Murdock MH, Parekh PK, et al. Sustained rescue of prefrontal circuit dysfunction by antidepressant-induced spine formation. *Science*. 2019;364: eaat8078.
218. Taniguchi T, Shibata K, Yamamoto K. Ketamine inhibits endotoxin-induced shock in rats. *Anesthesiology*. 2001;95:928–932.
219. Taniguchi T, Kanakura H, Takemoto Y, Yamamoto K. The antiinflammatory effects of ketamine in endotoxemic rats during moderate and mild hypothermia. *Anesth Analg*. 2004;98:1114–1120.
220. Simma N, Bose T, Kahlfuß S, et al. NMDA-receptor antagonists block B-cell function but foster IL-10 production in BCR/CD40-activated B cells. *Cell Commun Signal*. 2014;12:75.
221. De Kock M, Loix S, Lavand'homme P. Ketamine and peripheral inflammation. *CNS Neurosci Ther*. 2013;19:403–410.

222. Haroon E, Miller AH. Inflammation effects on brain glutamate in depression: mechanistic considerations and treatment implications. *Curr Top Behav Neurosci.* 2017;31:173–198.
223. Cui W, Ning Y, Hong W, Wang J, Liu Z, Li MD. Crosstalk between inflammation and glutamate system in depression: signaling pathway and molecular biomarkers for ketamine's antidepressant effect. *Mol Neurobiol.* 2018;56:3484–3500.
224. Emnett CM, Eisenman LN, Taylor AM, Izumi Y, Zorumski CF, Mennerick S. Indistinguishable synaptic pharmacodynamics of the N-methyl-D-aspartate receptor channel blockers memantine and ketamine. *Mol Pharmacol.* 2013;84(6):935–947.
225. Hudetz JA, Pagel PS. Neuroprotection by ketamine: a review of the experimental and clinical evidence. *J Cardiothorac Vasc Anesth.* 2010;24(1):131–142.
226. Kakinohana M. Protective effects of anesthetics on the spinal cord. *Curr Pharm Des.* 2014;20(36):5744–5750.
227. Bell JD. In vogue: ketamine for neuroprotection in acute neurologic injury. *Anesth Analg.* 2017;124(4):1237–1243.
228. Dhote F, Carpentier P, Barbier L, et al. Combinations of ketamine and atropine are neuroprotective and reduce neuroinflammation after a toxic status epilepticus in mice. *Toxicol Appl Pharmacol.* 2012;259(2):195–209.
229. Dorandeu F, Dhote F, Barbier L, Baccus B, Testylier G. Treatment of status epilepticus with ketamine, are we there yet? *CNS Neurosci Ther.* 2013;19(6):411–427.
230. Du E, McAllister P, Venna VR, Xiao L. Clinically relevant concentrations of ketamine inhibit osteoclast formation in vitro in mouse bone marrow cultures. *J Cell Biochem.* 2017;118(4):914–923.
231. Torres G, Hoehmann CL, Cuoco JA, et al. Ketamine intervention limits pathogen expansion in vitro. *Pathog Dis.* 2018;76(2):fty006.
232. Getachew B, Aubee JI, Schottenfeld RS, Csoka AB, Thompson KM, Tizabi Y. Ketamine interactions with gut-microbiota in rats: relevance to its antidepressant and anti-inflammatory properties. *BMC Microbiol.* 2018;18(1):222.
233. Kuehner JN, Bruggeman EC, Wen Z, Yao B. Epigenetic regulation in neuropsychiatric disorders. *Front Genet.* 2019;10:268.
234. Hyeon DY, Kim JH, Ahn TJ, Cho Y, Hwang D, Kim S. Evolution of the multi-tRNA synthetase complex and its role in cancer. *J Biol Chem.* 2019;294(14):5340–5351.
235. Gill H, Kang S, Lee Y, et al. The long-term effect of bariatric surgery on depression and anxiety. *J Affect Disord.* 2019;246:886–894.
236. Seitz J, Trinh S, Herpertz-Dahlmann B. The microbiome and eating disorders. *Psychiatr Clin North Am.* 2019;42(1):93–103.
237. Ojha R, Nandani R, Chatterjee N, Prajapati VK. Emerging role of circular RNAs as potential biomarkers for the diagnosis of human diseases. *Adv Exp Med Biol.* 2018;1087:141–157.
238. Valdés P, Schneider BL. Gene therapy: a promising approach for neuroprotection in Parkinson's disease? *Front Neuroanat* 2019;10:123.

CHAPTER TWO

Effects of cannabidiol (CBD) in neuropsychiatric disorders: A review of pre-clinical and clinical findings

Sonja Elsaid[a,b], Stefan Kloiber[b,c,d,e], Bernard Le Foll[a,b,e,f,g,h,*]

[a]Translational Addiction Research Laboratory, Centre for Addiction and Mental Health, Toronto, ON, Canada
[b]Institute of Medical Science, Faculty of Medicine, University of Toronto, Toronto, ON, Canada
[c]General Adult Psychiatry and Health Systems Division, Centre for Addiction and Mental Health, Toronto, ON, Canada
[d]Campbell Family Mental Health Research Institute, Centre for Addiction and Mental Health, Toronto, ON, Canada
[e]Department of Psychiatry, Faculty of Medicine, University of Toronto, Toronto, ON,Canada
[f]Department of Family and Community Medicine, University of Toronto, Toronto, ON, Canada
[g]Department of Pharmacology and Toxicology, University of Toronto, Toronto, ON, Canada
[h]Acute Care Program, Centre for Addiction and Mental Health, Toronto, ON, Canada
*Corresponding author: bernard.lefoll@camh.ca

Contents

Abstract

Cannabis sativa (cannabis) is one of the oldest plants cultivated by men. Cannabidiol (CBD) is the major non-psychomimetic compound derived from cannabis. It has been proposed to have a therapeutic potential over a wide range of neuropsychiatric

Progress in Molecular Biology and Translational Science, Volume 167
ISSN 1877-1173
https://doi.org/10.1016/bs.pmbts.2019.06.005

disorders. In this narrative review, we have summarized a selected number of pre-clinical and clinical studies, examining the effects of CBD in neuropsychiatric disorders. In some pre-clinical studies, CBD was demonstrated to potentially exhibit anti-epileptic, anti-oxidant, anti-inflammatory anti-psychotic, anxiolytic and anti-depressant properties. Moreover, CBD was shown to reduce addictive effects of some drugs of abuse. In clinical studies, CBD was shown to be safe, well-tolerated and efficacious in mitigating the symptoms associated with several types of seizure disorders and childhood epilepsies. Given that treatment with CBD alone was insufficient at managing choreic movements in patients with Huntington's disease, other cannabis-derived treatments are currently being investigated. Patients with Parkinson's disease (PD) have reported improvements in sleep and better quality of life with CBD; however, to fully elucidate the therapeutic potential of CBD on the symptoms of PD-associated movement disorders, larger scale, randomized, placebo-controlled studies still need to be conducted in the future. Currently, there are no human studies that investigated the effects of CBD in either Alzheimer's disease or unipolar depression, warranting further investigation in this area, considering that CBD was shown to have effects in pre-clinical studies. Although, anxiolytic properties of CBD were reported in the Social Anxiety Disorder, antipsychotic effects in schizophrenia and anti-addictive qualities in alcohol and drug addictions, here too, larger, randomized, placebo-controlled trials are needed to evaluate the therapeutic potential of CBD.

1. Introduction

Cannabis sativa or indica (cannabis) is one of the oldest plants known to men. The first evidence of its cultivation was found in China in 4000 years BC, which indicated that cannabis was used by Chinese manufacturers to make ropes, textiles and paper.[1,2] In one of the world's oldest pharmacopeias compiled in the first century AD, records were found of cannabis being used for medicinal purposes, such as to treat malaria, rheumatic pain, intestinal constipation, menstrual and surgical pain.[3,4] Since 1000 BC, cannabis was used as a sacred plant in religious rituals in South Asia and it was referred to as a source of happiness and joy.[3] The medicinal and recreational use of this plant has spread from India to the Middle East and Africa.[4] Earliest Arabic medical books mentioned its curative properties for digestive, urinary and respiratory ailments.[5] Cannabis was first brought to America to Brazil by African slaves, who used it in their popular religious rituals.[6] To the Western medicine, cannabis was first introduced in the 19th century by Dr. William O'Shaughnessy, who traveled to India with the British.[5] Dr. O'Shaughnessy was the first scientist to test the toxicological and therapeutic properties of cannabis in animals and human studies.[5,7]

Cannabis was widely used in the New World as a sedative/hypnotic, analgesic and appetite stimulant up until the late 19th century, when many medications became available, substituting its use.[8] Because of its psychoactive properties, it was removed from the American Pharmacopeia in 1941, limiting its use.[7,9] However, because of these properties, the recreational use of cannabis increased from approximately 5% in the 1960s to up to 70% in the 1980s in the young adult United States (US) population, prompting more stringent restriction laws for cannabis distribution and use.[10,11]

According to the US federal Controlled Substance Act (CSA), cannabis is classified as Schedule I substance. Schedule I indication implies that (1) cannabis has no currently accepted medical uses in the USA, (2) it has high a potential for abuse and (3) it has a lack of sufficient safety for use under the medical supervision.[12] However, the US federal and state laws are conflicting, since although prohibited at the federal level, cannabis is currently legal for the medical use in the 33 states in the USA.[12,13]

The major compound in *Cannabis* responsible for its psychoactive properties is δ-9-tetrahydrocannabinol (THC). The chemical structure of THC was discovered in the 1960s. THC is a partial agonist of both endocannabinoid receptors, CB_1 and CB_2 reviewed by Russo.[14] Previous studies with THC have demonstrated its potential for inducing feelings of "high," anxiety, perceptual alterations, cognitive deficits, and paranoia. The psychoactive and addictive properties of THC have been attributed primarily to the THC-induced dopamine system, via agonism of the CB_1 receptor.[15] Because of the THC's beneficial antiemetic and the appetite inducing properties, currently two synthetic THC formulations are marketed for the therapeutic use in the USA. Nabilone (Cesamet™) is indicated for the treatment of nausea and vomiting in cancer patients undergoing chemotherapy. Second, THC-containing-compound, dronabinol (Marinol®) is approved by the Food and Drug Administration (FDA) for treatment of anorexia, associated with weight loss in patients with Acquired Immune Deficiency Syndrome (AIDS) and, like nabilone, for alleviating treatment-resistant nausea and vomiting in cancer patients.

The second major compound found in *Cannabis sativa* is cannabidiol (CBD), which unlike THC, is a non-psychomimetic compound. CBD was first isolated from the cannabis plant in the late 1930s; however, it is not until 1963 that its chemical structure was characterized reviewed by Mechoulam et al.[16] Given that CBD exhibits multiple mechanisms of action, it was investigated for its therapeutic potential in neuropsychiatric disorders, especially in epilepsy, Alzheimer's disease (AD), Parkinson's disease (PD), Huntington's disease (HD), schizophrenia, anxiety, mood disorders and drug addictions reviewed by Pisanti et al.[17]

According to the US federal law, CBD was not listed as a separate substance, but rather as a derivative of cannabis up until December 2018.[12,13] As such, CBD retained its Schedule I substance designation, although up until then, the abuse potential of CBD had not been demonstrated.[12,13,18] However, with the introduction of the 2018 Farm Bill by the US federal government, hemp-derived CBD was legalized for purchase in all 50 states, if it contained $<0.3\%$ of THC.[13] Recently, Epidiolex®, a brand version of CBD, has been approved by the FDA for the treatment of seizures associated with Lennox-Gastaut syndrome (LGS), Dravet syndrome (DS) and tuberous sclerosis complex (TSC) in patients 2 years of age or older. CBD in combination with THC is also marketed as Sativex® and currently used as the adjunctive treatment for the symptomatic relief of neuropathic pain in multiple sclerosis in Canada and several European countries, including the United Kingdom (UK), Germany and Spain. At this point, Sativex® is not approved for the medicinal use in the USA.

2. Pharmacology of cannabidiol

The pharmacological effects of CBD were first tested in epilepsy in the 1970s[16] and the process of elucidating its full therapeutic potential continues even today, considering CBD's quite complex pharmacology.

2.1 Pharmacokinetics

Cannabidiol is a highly lipid-soluble drug with $K_{o/w}$ of 6–7[19] and it is relatively well absorbed in the small intestine. Once absorbed in the small intestine, CBD undergoes extensive first-pass metabolism before entering the systemic circulation. Due to the high first pass effect, the bioavailability of orally administered CBD is low, with estimates ranging from 6% to up to 13–19%.[20] When administered sublingually, the bioavailability of CBD appears to be slightly superior.[20,21] When administering an oral dose of 10 mg of CBD, maximal plasma concentration (C_{max}) of 3 ± 3.1 μg/L is observed, and this peak concentration is achieved at T_{max} of 2.8 ± 1.3 h.[22–24] Maximal concentration of CBD is achieved faster following intravenous administration, or when CBD is inhaled or administered orally in the fed state.[20,21] In fact, the systemic availability of CBD increases threefold, when CBD is co-administered with lipids.

In the systemic circulation, CBD is rapidly distributed in the brain, adipose tissue and other organs due to its highly lipid soluble properties. The volume of distribution of CBD was reported to be 32 L/kg[23–25] after oral administration. CBD is highly distributed to fat tissues and although its plasma levels do not show accumulation after repeated dosing, CBD accumulation in fat tissues is expected.[20,21] CBD is highly protein bound and approximately 10% of it is bound to red blood cells.[20,23]

Harvey et al. have identified 53 CBD metabolites of which most are the products of oxidation, β-oxidation, hydroxylation, glucuronide conjugation and epoxidation.[20,26,27] During the Phase I metabolism, CBD is oxidized by CYP 1A (1/2), CYP 2C (9/19), CYP 2D6 and CYP 3A (4/5) enzymes.[20,24,28] The major product of cytochrome P450 metabolism is 7-hydroxy CBD, which is primarily catalyzed by CYP 2C19 and CYP3A4.[28] The other seven metabolites include (1/2/3/4/5)-hydroxy CBD, 6α CBD, and 6β CBD.[28] Other, more prevalent metabolites are acids. These acids are mostly the products of CBD-7-oic acid, β-oxidation reactions and related compounds with degraded side-chain in which carboxylic acid is located at the C-7.[26,27]

Half-life ($t_{1/2}$) of CBD largely depends on the dose and the route of administration. In a few studies in which CBD was used as an oro-mucosal spray, $t_{1/2}$ of 1.4 to 10.9 h was observed, whereas oral chronic administration of CBD reported $t_{1/2}$ of 2–5 h. When administered intravenously and when smoked, CBD has respective $t_{1/2}$ of 24 and 31 h.[20,21] CBD is primarily eliminated in the bile; however, it is also eliminated in the urine in its free state and as a glucuronide derivative.[20,27] A plasma clearance rate of CBD was reported to be 960–1500 mL/min.[20,23]

Studies investigating the safety and tolerability of CBD have demonstrated no significant side effects with either acute or the chronic administration of CBD oral doses up to 1500 mg/day, or intravenous (i.v.) doses of 30 mg.[18] In the most recent study in which up to 20 mg/day or CBD was administered concomitantly with the regular anti-epileptic treatment to pediatric patients with LGS, the most commonly observed side effects were somnolence, decreased appetite and diarrhea. However, whether these side effects occurred due to the action of CBD or its additive or synergistic effects with other anti-epileptic treatments is not clearly understood.[29]

Given that CBD is a potent inhibitor of CYP 2C and CYP 3A enzymes, possible drug-drug interactions with CBD need to be carefully considered, especially when administered in conjunction with other drugs metabolized by these enzymes.[20,27,28] For instance, in a clinical trial with Lennox-Gastaut patients, CBD co-administered with clobazam was demonstrated to increase levels of clobazam's metabolite N-desmethyl clobazam by threefold, which may have potentiated the therapeutic effect of clobazam, but also may have contributed to the greater number of side effects observed in the trial. The suspected mechanism of this drug-drug interaction is the inhibition of CYP2C19 by CBD. CYP2C19 metabolizes N-Desmethylclobazam, thus the inhibition of this enzyme by CBD may have led to the drastic increases in the concentrations of the metabolite. Furthermore, three times more than normal concentrations of aminotransferase occurred more often in patients receiving concomitant valproate. However, since CBD has no effect on the plasma concentrations of valproate, suggestions were made that these drug-drug interactions may have been at the site of the drug action.[29]

2.2 Pharmacodynamics

CBD has a complex pharmacodynamic profile, to say the least, as it is known to interact with a wide variety of molecular targets. All these targets have not been fully investigated; however, it is believed that most of them (49%) are enzymatic, 20% are membrane and cellular transporters, 15% are receptors, while another 15% are ion channels.[30] Although the exact mechanisms of action of CBD have not been elucidated in experimental animal models and human studies, CBD has been shown to have analgesic, anti-psychotic, anxiolytic, anticonvulsant, anti-inflammatory, anti-apoptotic, anti-oxidant and neuroprotective properties reviewed by Campos et al.[31] In this section, we have only reviewed the mechanism of action of CBD on several main molecular targets.

Although classified as phyto-cannabinoid, CBD has a very low affinity for CB_1 and CB_2 receptors. The concentration of CBD needed to inhibit the CB_1 receptor activity by half (Ki) is 4359 nM, when measured in the rat brain tissue. For the CB_2 receptors found in rat spleen membranes, Ki of CBD is 2860 nM.[20,32] Furthermore, CBD has been suggested to reverse effects of cannabinoid receptor aginists such as THC.[30,33] However, it should be noted that the evidence of CBD attenuating psychoactive properties of THC is not conclusive.[34,35]

It has been proposed that CBD may inhibit the fatty acid amide hydrolase (FAAH), and by inhibiting FAAH, CBD may prevent the catabolism of anandamide; an endocannabinoid that influences physiological systems, such as pain, appetite regulation, pleasure and reward.[30,36] CBD was reported to act as the allosteric modulator of the $5HT_{1A}$ receptor in the dorsal periaqueductal gray (dPAG), basal ganglia, in the prelimbic system and at the dorsal raphe nucleus where it promotes the agonist-related stimulation of the GTPγS binding.[37] CBD interaction with the $5HT_{1A}$ receptor is believed to mediate its anxiolytic, anti-psychotic and antidepressant effects.[17] There is evidence that CBD activates and desensitizes transient receptor potential vanilloid (TRPV) 1–2 receptors, possibly exhibiting anti-epileptic, anxiolytic and anti-hyperalgesic effects.[38,39] CBD has been shown to inhibit ventricular tachycardia, exerting an antiarrhythmic effect by agonizing the adenosine A_1 receptor. By its mechanism of action on adenosine A_1 and A_2 receptors, CBD has been described to moderate anti-inflammatory effects.[30,40] Because of CBD's antagonistic properties on the μ and δ opioid receptors, its therapeutic effects have been investigated in animal and human studies of opioid addiction.[41,42] Furthermore, CBD has been revealed to modulate glycine-mediated synaptic transmission by the inhibition of pentameric glycine ionotropic receptors, known to mediate neuropathic pain and inflammation.[30,43] CBD also potentially inhibits synaptosomal uptake of noradrenaline, dopamine, serotonin, and γ-aminobutyric acid (GABA).[44] CBD was shown to activate and increase the protein expression of the peroxisome proliferator-activator receptor-γ (PPAR-γ) receptor, an important target studied in the experimental models of Alzheimer's disease.[45,46] Activation of PPAR-γ has anti-inflammatory and anti-oxidant properties, warranting further investigation of potential therapeutic effects of CBD in neurodegenerative disorders.[45] CBD has been demonstrated to be an antagonist of the G protein receptor 55 (GPR55). When activated, GPR55 increases the release of calcium from the intracellular stores, further upregulating the release of excitatory neurotransmitters, such as glutamate. Furthermore, activation of GPR55 stimulates the calcineurin-nuclear factor of activated T cells (NFAT) pathway, implicated in boosting the immune response and the neurodegenerative processes.[47] Although, not fully elucidated, blocking the GPR55 receptor could have beneficial properties by which CBD could downregulate the inflammatory and the overactive neuronal pathways.[30,47] CBD may block the activity of the calcium and sodium gated ion channels (VGCCs and VGSCs) and the α-7-nicotinic acetylcholine receptor.[48–50] However, although it is thought that inhibitory action by CBD could have the anti-convulsant activity on these channels, this still needs to be demonstrated in experimental models of epilepsy.[48–50]

Recently, in *in vitro* studies, CBD was shown to occupy the dopamine 2 High receptor (D2High) in the rat brain striatum at the doses of 11 nM. At a dose of 2800 nM, CBD occupies the dopamine 2 Low receptor. The biphasic effects of CBD on the two states of dopamine D2 receptors were compared to the effects of the known anti-psychotic aripiprazole.[51] However, here too, the anti-psychotic properties of CBD need to be investigated further and confirmed in future experiments.

3. Neurological disorders

3.1 Effects of cannabidiol in epilepsy

3.1.1 Evidence from in vitro *studies*

CBD has been shown to have anticonvulsive properties in epileptic neuron cell cultures.

Using the patch clamp technique, Iannotti et al. have demonstrated the occurrence of changes in voltage and the reversal of the emergence of action potentials when HEK293 rat neuronal cells over-expressing $TRPV_1$ were incubated in CBD.[52] This effect was abolished with the pre-treatment with the $TRPV_1$ antagonist, capsazepine (CAPZ), indicating CBD's mechanism of action by inhibition of the $TRPV_1$ receptor. Similar effects by CBD were observed on the $TRPV_2$ and $TRPA_1$ receptors, in which CBD effects were also reversed using an antagonist for each receptor.[52] The importance of this model was to show that overly sensitized TRP ion channels readily bind CBD, becoming desensitized with the repeated exposure.

In addition, CBD's effect on general convulsions was measured in Mg^+ free and 4-aminopyridine (4-AP) treated rat hippocampal brain slices, used as models of generalized epilepsy. In both cell cultures, CBD attenuated epileptiform, local field potential (LFP) amplitudes and the LFP duration without affecting the rate of the signal spread across the cell membranes. Moreover, the greatest effect of CBD was observed in the CA_1 cellular region, intimately involved in the propagation or epileptic activity.[53]

The model also tested effects of CBD on the inhibitory interneurons, such as fast spiking (FS) parvalbumin (PV) expressing and adapting chlecystokinin (CCK)-expressing interneurons known to be dysfunctional in epilepsy. In the CBD incubated pyramidal cells, CBD reduced the amplitude of the excitatory post-synaptic potentials (EPSPs) in the cell synapse and the amplitude of depolarized pyramidal cells. Moreover, CBD exhibited a greater magnitude of voltage reduction at more depolarized levels of excitation. However, in contrast to these findings, the inhibitory postsynaptic potentials (IPSPs) elicited by FS and CCK cells were, rather enhanced than inhibited by CBD. Because CBD reduced EPSPs of pyramidal cell cultures in a voltage-dependent manner, as earlier seen, the authors speculated that CBD may have produced the observed effects by activating Kv 7.2 M-type K^+ channels, like those seen with the THC administration.[54] However, this is yet to be confirmed in future studies.

3.1.2 Evidence from animal models of epilepsy

The anti-convulsant effects of CBD were further elucidated in the animal models of epilepsy. CBD anti-epileptic properties were observed in rats exposed to either the maximal electroshock seizures (MES) or the audiogenic seizures (AS) by administering respective doses of CBD of 12 and 17 mg/kg. Furthermore, anti-epileptic effects of CBD were compared to other pharmacotherapies used for treating epilepsy. In the experiments using MES, CBD was equipotent to phenytoin, slightly less potent than phenobarbital, but more potent than either trimethadione or ethosuximide. In the AS tests, CBD exhibited greater potency than phenytoin, but was less potent than phenobarbital. CBD also reduced the anti-convulsant properties of chlordiazepoxide, trimethadione, clonazepam and ethosuximide, most likely by inducing the metabolism of these drugs.[55]

In another animal seizure model elucidating the possible mechanisms of action of anti-convulsant properties of CBD, it was demonstrated that 60 mg/kg of intraperitoneal (i.p.) CBD reduced tonic seizures induced by $GABA_A$ receptor inhibitors, such as pentylenetetrazol (PTZ) or picrotoxin, when exposing mice to the transcorneal electroshock.[56] The results were not surprising, since it was previously determined that CBD disinhibited $GABA_A$-receptors, exerting the anti-convulsant effect.[57] Furthermore, the possible mechanism of action on big potassium (BK) channels by CBD was demonstrated in the PTZ-induced seizure models in mice. In these models, CBD seizure-preventing properties were diminished by the co-administration of paxilline, which is also a known antagonist of the BK channels.[58]

CBD was shown to be more effective at treating the partial and the generalized seizures, rather than seizures localized to the temporal lobe. Consequently, in the generalized seizure model it was demonstrated that injecting 100 mg/kg of CBD i.p. to the to PTZ-seizure induced rats decreased the number of tonic-clonic seizures and seizure-induced rat mortality rates. Similarly, when using penicillin (a selective antagonist of $GABA_A$ receptors) to induce partial seizures, CBD also decreased the number of seizures and the mortality rates. However, in the same experiment when pilocarpine, a muscarinic acetylcholine receptor (mAChR) agonist, was used it had no effect on the animal mortality rates.[57] The finding, therefore, supports the CBD's disinhibition of $GABA_A$ receptors, while CBD's reducing actions on the mAChR-mediated seizures are suspected to be limited.

When administering 20 and 50 mg/kg of CBD for 28 consecutive days to rats with PTZ-induced seizures, a reduction in the seizure activity was observed. Compared to the vehicle-treated rats, lower neuronal death in the CA_1 and CA_3 hippocampal regions and decreases of the hippocampal astrocyte hyperplasia were noted in the CBD treated animals. Moreover, the downregulating effect of CBD was observed on the expression of the NMDA subunit 1 in the hippocampus, indicating the inhibiting effects of CBD on NMDA receptors.[59]

Similar to the *in vitro* studies, the neuroprotective effects were observed by the intravenous injection of 0.1 mg/kg of CBD to newborn pigs exposed to the hypoxic-ischemic (HI) damage 4 h prior to the CBD administration. Seizures appeared only in 50% of the pigs, which was significantly lower than in the vehicle-treated pigs. Moreover, upon the histological examination, the CBD-treated pigs exhibited reduced HI tissue damage, accompanied by the 50% decrease in the amounts of brain tissue degeneration and in the HI-induced loss of cells in the cerebral cortex and the hippocampus.[60]

In a recent study using an epileptic model as seen in the DS, effects of CBD on the NMDA-mediated seizures via its action on the σ-1 receptors were tested in the epileptic mouse model and in an assay, using brain membranes and cells with the forced σ-1 receptor expression. NMDA-induced clonic-tonic seizures led to the death of 15–20% of the mouse colony. However, CBD protected the mice from the clonic convulsions by >50%, while tonic seizures were observed only in 20% of mice. No CBD pre-exposed mice died. To confirm the suspected antagonistic effect of CBD on the σ-1 receptor, *in vitro* studies were conducted showing that CBD's anticonvulsant properties occur by diminishing the σ-1 receptor coupling to the NR_1 subunit of the NMDA receptor; the mechanism of action, which most likely prevented cell excitation.[61]

As numerous evidence of anti-epileptic effects of CBD was observed in *in vitro* and *in vivo* animal experimental models, CBD was subsequently investigated as a treatment for epilepsy in the clinical trials.

3.1.3 Evidence from clinical studies

CBD was tested on patients with epilepsy, as early as in the 1970s. In one of the first placebo-controlled studies, nine patients with uncontrolled epilepsy were randomized to receive either 200 mg/day of CBD or the placebo for 3 months, together with their regular anti-epileptic drug (AED). In the CBD group, 2/4 patients achieved a full recovery from the symptoms of epilepsy. The third patient showed partial improvement, while the fourth did not respond to the treatment. The epileptic symptoms of patients in the placebo group were unchanged. None of the patients in the CBD group reported having any significant side-effects.[62] In a study, conducted between 1976 and 1978, 15 epileptic patients were enrolled, who were previously diagnosed with the temporal lobe epilepsy, including the secondary, generalized seizures. Considerable improvements were observed with the daily treatment of 200–300 mg of CBD for 16 weeks in 4/7 patients, with no major side effects reported other than somnolence.[63] In a 4-week, placebo-controlled study with 12 patients diagnosed with mental retardation and epilepsy in which 6 patients received up to 300 mg daily dose of CBD, no significant differences were noted in the seizure frequencies between the two groups.[64] The promising results observed in these studies prompted scientists to further test CBD in studies with more severe forms of epilepsy.

Consequently, an open-labeled study with patients with severe intractable, childhood-onset, treatment-resistant epilepsy was conducted for 12 weeks, in which 25–50 mg/kg of CBD was used in addition to co-administering their standard AED treatment or the vagus nerve stimulation. After completing the trial, CBD was reported to reduce monthly seizures by 36.5%. In this trial, more than 10% of patients experienced somnolence and 13% felt fatigued. Somnolence and fatigue occurred mainly in patients concomitantly taking clobazam. However, it was predicted that the side-effects resulted due to clobazam co-administration, since 60% elevations in plasma clobazam levels were seen only in patients experiencing these side-effects. As suspected, the side-effects were resolved after adjusting doses of clobazam.[65] In the subset of the patient population enrolled in the abovementioned study, a quality of life (QoL) survey was administered to patient caregivers. The result of the survey indicated improvements in the overall QoL, as well as in the social and health-related domains of QoL. However, the correlation between the frequency of seizures and the QoL survey results was not observed in this trial, indicating that improvements in social and health-related QoL may have occurred independently from the potential CBD effects on reducing seizures.[66]

The most recent clinical trials with CBD in epilepsy were conducted in patients with three different complex forms of epilepsy. In the first trial, children and young adults with DS were randomized in the double-blind, placebo-controlled fashion to receive either 20 mg/kg of CBD or placebo with their standard AED regimen for 14 weeks. After study completion, 43% of patients taking CBD had at least 50% reduction in convulsive seizures, while 5% became seizure free.[67] In the second double-blind, placebo-controlled trial, patients with LGS were enrolled to receive either 10 or 20 mg/kg of CBD or placebo, plus their stable antiepileptic treatment for 14 weeks. The primary efficacy outcome measure was the percentage change in drop seizures, and it was reported to be reduced by 41.9% in the 20 mg/kg of CBD group, 37.2% in the 10 mg/kg group, compared to 17.2% in the placebo group. However, the authors cautioned against over interpreting these results, since approximately 50% of the patients in all study groups were also taking clobazam during the drug trial.[29] Currently, the same group of investigators are examining effects of CBD on serum concentration of clobazam. In the third study, seven children with Febrile Infection-related Epilepsy (FIRES) were titrated up to 25 mg/kg/day of CBD, while taking their regular AED for 48 weeks. The assessments of the frequency and the duration of epileptic seizures were made at 4 and 48 weeks after initiating the CBD treatment. At both assessment time points, 5/7 patients showed improvements in symptoms of epilepsy as early as after 4 weeks of treatment. Moreover, on average, four concomitant AEDs were discontinued.[68] Overall, CBD was shown to have promising anti-epileptic properties, as seen in these clinical trials. Table 1 provides a brief overview of the clinical studies mentioned above.

3.2 Effects of cannabidiol in Alzheimer's disease

3.2.1 Evidence from in vitro *studies*

In vitro AD models were specifically focused on examining the effects of CBD on tau and amyloid-β (Aβ) pathology, amyloid-β-induced toxicity and on microglial function known to be impaired in AD.

The direct effect of CBD on Aβ pathology was assessed in human neuroblastoma SHSY5Y^{APP+} cells that normally produce high levels of Aβ. When incubating SHSY5Y^{APP+} neuroblastoma cells with 10^{-9} to 10^{-6} M of CBD, significant decreases in C83, C99 and Aβ were demonstrated. Moreover, a greater appearance of the full length, non-cleaved, ubiquitinated amyloid precursor proteins (APPs) was observed with CBD co-incubation. This finding was not surprising since it was previously described that attaching ubiquinone residue to the APP proteins protects them from undergoing cleavage by secretases. Furthermore, CBD enhanced a greater survival rate of the SHSY5YAPP + cells and decreased the cellular apoptosis rates cells. CBD seemed to have exerted its neuroprotective effect via its direct action on PPAR-γ, given that the protective effects of CBD were reversed by administering the antagonist of PPAR-γ, GW9662.

Table 1 A summary of clinical studies examining effects of CBD in neurological disorders.

References	Study sample (patient population, sample size, gender and age)	Treatment schedule	Study design	Primary findings	Adverse events
Epilepsy/seizures					
Mechoulam and Carlini[62]	Subjects with temporal lobe epilepsy and uncontrolled secondary generalized epilepsy refractory to treatment N=9 Gender and age not reported	200 mg of CBD or placebo, administered orally for 3 months	Randomized double-blind, placebo-controlled study	In the CBD group, 2 patients exhibited full improvement (had no epileptic seizures), 1 patient showed partial improvement, while another patient showed no improvement	None reported
Carlini and Cunha[63]	Subjects with temporal lobe epilepsy with secondary generalized epilepsy N=15 (11F) Mean age 24 years	Up to 300 mg of CBD or placebo, administered orally with the concomitant AED for 18 weeks	Randomized, double-blind, placebo-controlled study	In the CBD group, 4 patients had complete absence of epileptic seizures during the trial	Most commonly observed was somnolence
Ames et al.[64]	Subjects with epileptic seizures and mental retardation N=12 Age and gender not reported	300 mg of CBD or placebo for the first week, followed by 200 mg of CBD for the next 3 weeks, administered orally	Randomized, double-blind, placebo-controlled study	No effects on seizure frequency in the CBD group	Most commonly observed was mild drowsiness
Devinsky et al.[65] and Rosenberg et al.[66]	Subjects with severe intractable, childhood-onset treatment-resistant epilepsy N=214 (51% F) Mean age 10.5 years Age range 1–30 years	Up to 50 mg/kg/day of CBD with regular AED, administered orally for 12 weeks	Open label study	The mean reduction in epileptic seizures observed after completion of the trial Improvements in quality of life in a subset of 48 patients No correlation between improvement in quality of life and reduction in seizure frequency	Most commonly observed somnolence, decreased appetite, diarrhea and fatigue

Continued

Table 1 A summary of clinical studies examining effects of CBD in neurological disorders.—cont'd

References	Study sample (patient population, sample size, gender and age)	Treatment schedule	Study design	Primary findings	Adverse events
Devinsky et al.[67]	Subjects with Dravet syndrome and medication resistant seizures N = 120 (58% F) Mean age 9.8 ± 4.8 years Age range 2–18 years	Up to 20 mg/kg/day or placebo with regular AED, administered orally for 14 weeks	Double-blind, placebo-controlled study	Reduction in convulsive-seizure frequency in the CBD group	Most commonly observed diarrhea, vomiting, fatigue, pyrexia, somnolence and abnormal liver function
Gofshteyn et al.[68]	Subjects with Febrile Infection-related epilepsy syndrome (FIRES) with treatment refractory epilepsy N = 7 (2F)	Up to 25 mg/kg/day of CBD with regular AED, administered orally for up to 48 weeks	Open label case study	In 6 patients, seizures improved in frequency and duration On average, four AEDs were discontinued	Dizziness in 2/7, decreased appetite and weight loss in 1/7 and nausea and vomiting in 1/7 patients
Devinsky et al.[29]	Subjects with the Lennox-Gestaut syndrome N = 225 (32F in the placebo; 33F in the 10 mg/kg of CBD and 31F in the 20 mg/kg of CBD group) Age range 2–55 years	10 mg/kg, 20 mg/kg of daily CBD or placebo with standard AED, administered orally for 14 weeks	Randomized, double-blind, placebo-controlled study	Reduction in the frequency of drop seizure in both, 10 and 20 mg/kg CBD groups	Most commonly observed somnolence, decreased appetite and diarrhea
Huntington's disease					
Consroe et al.[69]	Subjects with HD N = 15 (7F) Mean age: 47.8 ± 15.3 years	10 mg/kg/day of CBD or placebo, administered orally for 6 weeks	Randomized, double-blind, placebo-controlled, cross-over study	No effects on symptoms of HD were observed in the CBD group	No major side effects

Parkinson's disease					
Zuardi et al.[70]	Subjects with psychosis and PD N = 6 (2F) Mean age: 58.8 ± 14.9 years	Mean oral dose 400 mg of CBD with 1050 mg of L-dopa administered orally for 4 weeks	Open label pilot study	Significant improvements on positive and negative symptoms, and on symptoms of sleep disturbance No changes in cognitive function	None reported
Chagas et al.[71]	Subjects with PD N = 4, (4M) Subject 1: 61 year old Subject 2: 59 year old Subject 3: 63 year old Subject 4: 71 year old	Subjects 1–3 received 75 mg of CBD, while subject 4 received 300 mg of CBD, administered orally for 6 weeks	Case series	Symptoms of associated with rapid eye movement sleep behavior disorder completely improved in three subjects receiving 75 mg of CBD, while these symptoms partially improved in subject receiving 300 mg of CBD	None reported
Chagas et al.[72]	Subjects with Parkinson's disease N = 21 (6F) Age range 51–82 years	75 or 300 mg of CBD or placebo, administered orally for 6 weeks	Randomized, double-blind, placebo-controlled study	No improvements in motor or general symptoms of PD in both CBD groups Subjects receiving 300 mg of CBD showed improvements in quality of life	None observed

N, number of subjects; CBD, cannabidiol; F, female; AED, anti-epileptic drug; HD, Huntington's disease; PD, Parkinson's disease; M, male.

PPAR-γ is virtually inactive during normal physiological conditions, but it becomes over-expressed during the AD pathogenesis with a function to enhance the intra- and extra-neuronal cell clearance of the Aβ deposits. It is believed that post-transcriptional activation of PPAR-γ also mediates APP ubiquitination, preventing further buildup of the Aβ deposits.[45] The protective effects of CBD mediated by PPAR-γ were also demonstrated in mouse CA_1 hippocampal slides in experiments assessing the long-term potentiation (LTP), which is activity-dependent on the increases in the synaptic efficacy. This model is used for studying the Aβ mediated deficits on the LTP and memory retention. Exposure to Aβ triggered the decrease in the LTP levels, indicating reduced levels of the neurotransmitter release. However, when incubating hippocampal brain slides pre-exposed to Aβ with 10 μM of CBD, the dampening effects of Aβ were reversed. Like in the previous experiment, the pre-treatment of brain slides with the PPAR-γ-antagonist, GW9662, blocked CBD protective effects on the LTP, suggesting that activation of the PPAR-γ is essential for retaining normal physiological levels of LTP.[73]

The effects of CBD on tau pathology were examined by exposing PC_{12} neuronal cells to Aβ. CBD was observed to inhibit the formation of the activated form of glycogen synthase kinase 3 beta (GSK 3β) and phosphorylated β-catenin, otherwise known to induce apoptosis in neuronal cells. Moreover, the levels of hyperphosphorylated tau proteins responsible for formation of neurofibrillary tangles (NFTs) were decreased in the CBD pre-treated PC_{12}, after being exposed to Aβ.[74]

Effects of CBD on the Aβ-induced toxicity were examined in the Aβ peptide treated PC12 rat neuroblastoma cells, exhibiting low levels of the survival rate several hours after the exposure. The effect on cell death was associated with the increased levels of DNA fragmentation and cellular lipid peroxidation. Also, higher than normal accumulation of reactive oxygen species (ROS) and intracellular calcium were seen. Activation of caspases 3 was observed, indicating the initiation of pro-apoptotic pathways that would eventually lead to programmed cells death. CBD incubated PC_{12} rat neuroblastoma cells appeared to be protected from ROS-induced cellular damage, since lower levels of DNA fragmentation and lipid peroxidation were observed. Moreover, lower levels of caspases 3 and accumulated calcium were measured in the CBD treated cells, alluding to both, anti-apoptotic and anti-oxidant properties of CBD.[75]

Similarly, in the experiment with the Aβ transfected SHSY5Y human neuroblastoma cells exhibiting high levels of ROS and lipid peroxidation, CBD pre-treatment attenuated lipid peroxidation. The anti-oxidant effects of CBD were demonstrated by the mechanism of action of CBD's upregulation of cellular enzymes, such as superoxide dismutase, catalase, and glutathione reductase, involved in the detoxification of ROS.[76]

The anti-apoptotic effects of CBD were also tested in the N13 neonatal rat cortex microglial cells. After being exposed to 400 μM of ATP, dying microglial cells were observed to release and accumulate higher than normal levels of intracellular calcium. However, cell incubation with CBD prevented calcium release, reversing the programmed cells death.[77] In the subsequent experiments, co-incubation with A2A receptor antagonist reversed the protective effects of CBD, suggesting that CBD blockade of ATP-induced calcium influx was A2A-receptor mediated.[77]

3.2.2 Evidence from animal models

In mice inoculated with Aβ in the right dorsal hippocampus, CBD administered at the doses of 2.5 and 10 mg/kg for 7 days was demonstrated to dose-dependently impair expression of inducible Nitric Oxide Synthase (iNOS), which generates ROS and reactive nitrogen species (RNS). CBD also suppressed the expression of interleukin-1β (IL-1β), which is associated with neurodegenerative processes, and it attenuated the levels of glial fibrillary acidic proteins (GFAP). In reactive gliosis, GFAP serves as an indicator of activated astrocytes. The two main features of activated astrocytes are hypertrophy and the constant proliferation. Furthermore, higher than normal levels of nitric oxide (NO) responsible for the formation of neurofibrillary tangles (NFT) were also ameliorated by CBD, possibly as a result of the downregulation of iNOS.[78]

The protective effects of CBD on cognitive impairment and spatial learning and memory were tested in another AD murine model, using Aβ inoculated mice. In this test, CBD-treated mice demonstrated improved cognitive performance compared to those treated with the vehicle. Furthermore, lower levels of pro-inflammatory cytokines, tumor necrosis factor α (TNF-α) and interleukin 6 (IL-6) were measured in the homogenized brains of the CBD treated mice.[77]

CBD effects were also tested in transgenic mice with the mutated APP and PS1 genes, which is one of the animal models of AD. APP × PS1 transgenic mice exhibit the accelerated amyloid pathophysiology due to the mutations of the two AD genes. Physiologically, these mice have elevated NO and TNF-α implicated to be responsible for the cognitive deficits observed in these mice. Mice treated with 20 mg/kg of CBD for 3 weeks showed significant improvements in social recognition and object recognition memory. Furthermore, CBD treated group exhibited reduced levels of NO and TNF-α.[79]

Long-term, preventive properties of CBD were also tested in the same transgenic mice. In this experiment, APP × PS1 mice were treated with 20 mg/kg of CBD for 8 months. The prevention of the social recognition deficits were observed, post-treatment; however, these findings were not associated with the amyloid load or the oxidative damage, as the social recognition deficits remained unchanged 8 months after initiating the CBD treatment. Instead, decreased levels of the neuroinflammatory factors were observed indicating that neuroinflammatory, more so than the oxidative changes in the transgenic mice might have led to the observed deficits.[80]

Up until now, there have been no clinical studies evaluating the effects of CBD in patients with AD; however, given the evidence of neuroprotective effects of CBD were repeatedly observed in a number of *in vivo* and *in vitro* models of AD, conducting clinical trials might be warranted.

3.3 Effects of cannabidiol in Huntington's disease

3.3.1 Evidence from experimental models and clinical studies

The pre-clinical studies investigating the effects of CBD in the models of HD have produced mixed results. In the cell cultures involving HD, mutant $STHd^{Q111/Q111}$ medium, spiny neurons, normally observed to have 50% of depleted ATP and the excessive GABA release, CBD was demonstrated to further deplete ATP while enhancing the GABA release.[81] The depletion of cell energy storage is quite detrimental to cell survival, leading to the eventual cell death.

Conversely, in $STHd^{Q111/Q111}$ striatal-derived cell lines, CBD increased cell survival rates by 40%, while reducing the membrane permeability and the rates of cell death by 20%. In $STHdh^{Q7/Q7}$, which is another HD *in vitro* model, CBD promoted cAMP response element-binding protein (CREB) phosphorylation, upregulating the gene expression of the proteins implicated in the cell survival.[81] Similarly, 5 mg/kg of CBD co-administered with a neuronal toxin, 3-nitropropionic acid (3-NP), completely reversed reduced mRNA levels of the anti-oxidant enzyme, superoxide dismutase (SOD-2) and the neuronal specific enolase (NSE), which is a marker of the neuronal integrity. The subsequent tests to explicate the possible mechanism of CBD action in the HD implied that CBD's anti-oxidant properties may have been responsible for the observed results, rather than CBD receptor-mediated action on the signaling pathways via CB_1, TRPV or A2A receptors.[82]

Considering the mixed evidence of the benefits of CBD in the HD experimental models, it was not surprising that CBD was ineffective in ameliorating the choreic movements in the HD patients, as demonstrated by the double-blind, placebo-controlled clinical trial (Table 1).[69] However, better results were seen with using nabiximols (a 1:1 combination of CBD and THC), which is a compound, today marketed as Sativex® anti-oxidant. In these subsequent experiments, Sativex® was shown to exhibit both, anti-oxidant and the anti-inflammatory effects in the HD animal models.[83,84] However, larger-scale randomized clinical trials with higher Sativex® doses are necessary to demonstrate the therapeutic potential of CBD in patients with the HD.

3.4 Effects of cannabidiol in Parkinson's disease

3.4.1 Evidence from in vitro *studies and experimental animal models*

CBD has been shown to be both effective and to have no effect on ameliorating several features of the PD, depending on the specific PD model used.

In the experiments designed to assess the effects of CBD on the progression of PD, rats were injected with in the forebrain bundle 6-hydroxyldopamine (6-OHDA). A 2-week injection treatment with 3 mg/kg of CBD led to the increases in the dopamine content in the caudate—putamen.[85] The neuroprotective effects of CBD were not reversed by co-administering CB_1 antagonists, suggesting the anti-oxidant, rather than the CB_1-receptor-mediated mechanism of CBD. In the successive study using a similar methodology, CBD recovered the 6-OHDA-induced dopamine depletion, in addition to upregulating previously reduced levels of the anti-oxidant enzyme superoxide dismutase. However, these effects were only observed after the immediate, post-6-OHDA treatment with CBD.[86] The results suggested two specific characteristics of CBD protection in models of PD; one being its anti-oxidant properties, and the other suggesting CBD's greater propensity for slowing the disease progression in earlier, rather than later stages.

The neuroprotective effects of CBD were observed in two more experimental models of PD, in which mitochondrial complex I inhibitors were used, *in vitro*. In murine mesencephalic cell cultures, a 20% rotenone-induced degeneration of dopaminergic neurons was reversed by 10 μM of CBD 48 h after the co-incubation period. The tests also found that glutathione levels were restored with the CBD co-treatment to the cell cultures further confirming the anti-oxidant properties of CBD. However, in the same experiment using N18TG2 murine neuroblastoma cells, CBD had no effect on reversing the rotenone-induced rise of the superoxide radicals.[87] The results indicated that anti-oxidant properties of CBD in the brain were cell- or tissue-specific. In the second model of mitochondrial-induced neuron toxicity, CBD co-incubation with 1-methyl-4-phenyl pyridium (MPP+) for 2 days also protected dopaminergic neuron degeneration and increased neuron count by 117% in the murine mesencephalic cell cultures.[88]

Contrary to this evidence, a 5-week co-treatment of mice with 1-methyl-4-phenyl-1,2,3,6-tetrahydropyridine (MPTPp) and 5 mg/kg of CBD had no effects on the degeneration of the dopaminergic nerve terminals, nor it had any effects on preserving the dopamine levels in the striatum. Also, no effects of CBD on the symptoms relief of PD-related movement disorders were noted, as mice co-injected with MPTPp and CBD had comparable performance to those exposed to the neurotoxin alone in the rotarod experiment.[89] Conversely, in the same study, protective effects on the symptom relief and the progression of symptoms were observed with the co-application of MPTPp and CBD. These protective effects were reversed by using the GPR55 antagonists, suggesting the GPR55-mediated mechanism of action. GPR55 is found in the striatum, globus pallidus, subthalamic nucleus, substantia nigra and the cortex; the areas of the brain all affected by PD where GPR55 is believed to exhibit its anti-inflammatory properties. Given that CBD is an antagonist, rather than an agonist for GPR55, it was not surprising that CBD had no effects on attenuating MPTPp-induced toxicities. Moreover, unlike 6OHDA, which generates free radicals in the PD animal models, MPTPp acts primarily as an inhibitor of the mitochondrial complex I necessitating a different neuroprotective pathway. However, to further explain their findings, the scientists who conducted the experiment also argued that CBD may have had different effects on the different tissues, hence CBD attenuated the morphological changes in microglial cells, but not in astrocytes.[89]

The positive effects of CBD on restoring the mitochondrial damage caused by iron loading were reported in rats injected with 10 mg/kg of CBD. Accumulation of iron is often observed in the basal ganglia of those affected with PD, where iron over-load may lead to oxidative stress and cellular damage. Here, CBD completely reversed the iron-induced effects on the two proteins necessary for maintaining the homeostasis in the mitochondria. Also, CBD co-administration to iron-loaded rats reduced caspase 3 levels inhibiting apoptosis in the hippocampal tissue.[90] The results further strengthened the evidence supporting the anti-oxidant and the anti-apoptotic properties of CBD.

CBD was also demonstrated to attenuate the effects of catalepsy produced by three different pro-cataleptic compounds; haloperidol, L-nitro-N-arginine, which is a non selective inhibitor of NOS, and the CB_1 receptor agonist (WIN55,212-2), and these effects were dose-dependent. Subsequently, it was shown that 30 mg/kg of CBD mitigated the cataleptic effects of haloperidol by its mechanism of action on the $5HT_{1A}$ receptor, given that co-administration of haloperidol and CBD with the $5HT_{1A}$ receptor antagonist reversed the protective effects of CBD.[91]

3.4.2 Evidence from clinical studies

To date, three clinical studies with patients with PD were conducted in which therapeutic properties of CBD were tested (Table 1). In one study, 6 PD patients exhibiting psychosis were enrolled to receive a concomitant treatment of 400 mg of CBD and 1050 mg of L-dopa for 4 weeks. In the CBD treated group, significant reductions in the negative and the positive symptoms of psychosis were observed. However, no effects were seen on the cognitive function.[70]

The effects of CBD on the REM sleep behavioral disorder (RBD) were tested for 6 weeks. In this study, patients reported reduced nightmares and decreased levels of agitation and aggressive behavior 6 weeks after being treated with CBD.[71] The third study was a randomized, double-blind placebo-controlled trial, using two different doses (75 and 300 mg/day) of CBD and the placebo. While significant improvements on the measures of patient well-being were observed, no effects on symptoms of PD were found.[72] However, larger randomized-controlled trials with higher doses of CBD are necessary to fully investigate a true potential of CBD to control the movement disorders associated with PD.

4. Psychiatric disorders

4.1 Effects of cannabidiol in schizophrenia

4.1.1 Evidence from experimental models

Most of the experimental models of schizophrenia were designed to simulate the occurrence of the specific behavioral symptoms, thus for the purposes of the review, they are described below, accordingly.

One of the symptoms observed in patients with schizophrenia is psychomotor agitation. In animal studies, these symptoms are modeled by the hyper-locomotor activity, manifested by the increases in the distance of movement in a particular area. Hyper-locomotion can be induced by the administration of dopamine receptor agonists, such as amphetamine or ketamine while the application of anti-psychotics can reverse this effect. Consequently, in the animal experiments of schizophrenia treatment with 60 mg/kg of CBD normalized the hyper-locomotion in mice previously induced by dexamphetamine or ketamine.[92] Similar findings were observed with the chronic administration of 50 mg/kg/day of CBD, which was observed to decrease the hyper-locomotor activity in mice injected with dexamphetamine. However, contrary to the previous findings, the acute treatment of CBD had no effect on the hyper-locomotive behavior in C57Bl/6 mice.[93]

Another symptom normally observed in patients with schizophrenia is the inability to filter out the unnecessary information. This symptom has been linked to the abnormalities of the sensorimotor gating. In animal studies, the sensorimotor gating is measured by the pre-pulse inhibition (PPI) of the startle response characterized by the reduction of the startle reflex to the acoustic stimulus when immediately preceded by the lower-intensity sounds.[94,95] In one study, CBD attenuated the amphetamine-induced PPI impairment when the drug was systemically injected in the nucleus accumbens in Swiss mice.[95]

Similar findings were reported in another study using Sprague-Dawley rats in which CBD infusion normalized the amphetamine-induced hyper-locomotion and the PPI. In the subsequent biochemical experiments, it was established that protective effects of CBD may have been induced by the upregulation of the mammalian target of rapamycin (mTOR) signaling without increasing the signaling in the Wingless/Integrated (Wnt) pathway, as contrarily seen with typical anti-psychotics. According to the researchers of this study, the observed difference may have accounted for the absence of the extrapyramidal side-effects when administering CBD.[94]

In several subsequent studies, CBD did not seem to produce catalepsy, which is a side effect observed with the haloperidol administration.[92,95,96] When administered in combination with haloperidol, CBD attenuated haloperidol-induced catalepsy.[96] Moreover, increases in the prolactin levels resulting from the haloperidol-mediated D2 receptor blockade were only observed at extremely high doses of CBD that had noxious, rather than the therapeutic effects.[95]

A few studies have been conducted using MK-801, which is a non-competitive, NMDA-receptor antagonist determined to induce hyper-locomotion, impairments in PPI and social withdrawal.[96] As previously mentioned, the ionotropic NMDA-receptor hypofunction in the prefrontal cortex ultimately up-regulates the cortical dopamine neurotransmission[97] making CBD a potentially suitable compound for alleviating the noxious symptoms of the dopamine up-regulation.

In a study using male Swiss mice, the i.p., injection of 5 mg/kg of CBD successfully reversed the MK-801-induced PPI impairment. Interestingly, the effect of CBD on the PPI seems to have been mediated by the $TRPV_1$ receptor, since capsazepine, an antagonist of the TRPV1 receptor, reversed the protective effects of CBD.[98]

While generally positive effects of CBD were observed on the MK-801-induced impairments in the mouse-animal model of schizophrenia, mixed results were seen in the experiments using rats, where CBD administered at doses of 3–30 mg/kg was ineffective or disruptive on the PPI.[99,100] However, at lower doses of 1 and 3 mg/kg, CBD was demonstrated to normalize hyper-locomotion and social interaction in male rats.[101]

All abovementioned animal models used chemical compounds to induce symptoms while simulating the effects of the dopamine or the NMDA-receptor mediated pathogenesis. However, very little research has been conducted using genetic animal models of schizophrenia. In one study, neuregulin 1 (NRG1) mutant mice (phenotypically exhibiting the hyper-locomotion, the PPI deficits and the reduced $5HT_{2A}$ receptor density) were treated with either 1, 50 and 100 mg/kg/day of CBD for 21 days. The abovementioned phenotypes, however, were not reversed by CBD, although CBD was shown to be the enhancer of the social interaction in the NRG1 mutant mice. The lack of effects of CBD in this study may have indicated the NRG1 modulation of the neurobehavioral effects of CBD.[102]

Recently, two studies, using the pre-natal infection models examined the effects of the chronic CBD treatments. In these tests, pregnant rodents were infected with the polyinosinic: polycytidylic acid (poly I:C) virus known to cause the impairment in social interactions, cognitive symptoms, and the PPI in the offspring. In the first study, male offspring born to infected poly I:C pregnant rats were treated with 10 mg/kg of CBD for 3 weeks. CBD significantly improved social interaction, recognition, and working memory.[103] In the second study, pregnant C57Bl/6 mice were also infected with the poly I:C virus and their offspring also were treated with 1 mg/kg of CBD 1–2 months, postnatally. Following the treatment period, the tested male mice displayed the reduced hyper-locomotor activity.[104]

Although many experiments have been conducted in animal models of schizophrenia, more research is needed to elucidate the effects of CBD.

4.1.2 Evidence from clinical studies

CBD was subsequently tested in human studies (Table 2). The first time CBD was given in a clinical setting was to a patient with schizophrenia, who received his experimental treatments in the following order: Placebo

Table 2 A summary of clinical studies examining effects of CBD in psychiatric disorders.

References	Study sample (patient population, sample size, gender and age)	Treatment schedule	Study design	Primary findings	Adverse events
Schizophrenia/psychosis					
Zuardi et al.[105]	A subject with schizophrenia (M) Age: 19-year old	For 4 days, placebo For 4 weeks, up to 1500 mg of CBD For 4 days, placebo For 4 weeks, up to 12.5 mg of haloperidol, all administered orally All daily doses	Case report	Symptoms of thought disturbance and hostility and suspiciousness subsided with CBD treatment	None observed
Zuardi et al.[106]	Subjects with treatment resistant schizophrenia N=3 (1F) Age range, 22–23 years	For 5 days, the placebo For 30 days, up to 1280 mg of CBD For 5 days, the placebo For 15 days, 20 mg of olanzapine All daily doses	Three case reports	The first patient improved on CBD, deteriorated on placebo and partially improved on olanzapine The second patient did not improve on CBD and partially improved on olanzapine The third patient minimally improved on CBD and had no improvement on olanzapine	None observed
Hallak et al.[107]	Subjects with schizophrenia N=28 (10F) No mean age reported	300 mg, 600 mg of CBD or placebo daily administered, orally 60 min before the assessment	Double-blind, placebo, controlled study	No improvements on selective attention were observed with either doses of CBD	None reported
Leweke et al.[108]	Subjects with schizophrenia and schizophreniform psychosis N=42 (5F in CBD group; 2F in amisulpride group) Age range: 18–50 years	Up to 800 mg of CBD or 800 mg of amisulpride daily, administered orally for 28 days	Double-blind, randomized, parallel-group, controlled study	CBD was as effective as the amisulpride in treating the symptoms of psychosis CBD had no effect on negative symptoms	Fewer extrapyramidal symptoms, less weight gain and lower prolactin increase in the CBD group

Boggs et al.[109]	Subjects with chronic schizophrenia N = 41 (33.3% F in CBD group and 27.8% F in the placebo group) Mean age in CBD group 48.4 ± 9.3 years Mean age in the placebo group 46.4 ± 9.5 years	600 mg of CBD or placebo daily in addition to regular anti-psychotic treatment, administered orally for 6 weeks	Randomized, double-blind, placebo-controlled parallel group, fixed-dose study	Patient augmented with CBD showed no improvement in positive, negative and cognitive symptoms of schizophrenia	No major side effect
McGuire et al.[110]	Patients with schizophrenia N = 88 (37F) Mean age in CBD group 40.9 ± 12.49 years Mean age in placebo group 40.8 ± 11.69 years	1000 mg of CBD or placebo daily, in addition to a regular anti-psychotic treatment, administered orally for 6 weeks	Randomized, double-blind, placebo-controlled parallel group study	Patients augmented with CBD showed improvement in positive and no improvements in negative and cognitive symptoms of schizophrenia	Gastrointestinal events
Anxiety and mood disorders					
Crippa et al.[111]	Treatment naïve subjects with SAD N = 10, (10M) Mean age, 24.2 ± 3.7 Age range: 20–30 years	400 mg of CBD or placebo administered orally 110 min before SPECT analysis	Randomized, double-blind, placebo controlled, cross-over study	Decreases in state anxiety in the CBD group rCBF was attenuated in the left parahippocampal gyrus and hippocampus and increased in right posterior cingulate gyrus in the CBD group	None observed
Bergamaschi et al.[112]	Treatment naïve subjects with SAD N = 24 (12 with SAD and 6 healthy controls) 6F in each group Mean age: 22.9 ± 2.4 years, placebo group 24.6 ± 3.6 years, CBD group 23.3 ± 1.7 years, healthy controls	600 mg of CBD or placebo administered to SAD patients, orally Healthy controls received no treatment Treatments administered 150 min before the public speaking test	Randomized, double-blind, placebo-controlled setting for SAD patients	Reduction in symptoms of anxiety, cognitive impairment and discomfort caused by public speech and decreased levels of arousal during anticipatory speech in the CBD-treated SAD subjects	None observed
Shannon et al.[113]	Subjects with anxiety (diagnosis unspecified) and disturbed sleep, except for PTSD	25–75 mg of daily CBD with other concomitant unspecified psychiatric	Retrospective chart review	Symptoms of anxiety decreased over the 3-month period	Mild sedation and fatigue

Continued

Table 2 A summary of clinical studies examining effects of CBD in psychiatric disorders.—cont'd

References	Study sample (patient population, sample size, gender and age)	Treatment schedule	Study design	Primary findings	Adverse events
	N=47 (19F) Mean age 34 years Age range 18–70 years	medications, administered orally for 3 months			
Zuardi et al.[114]	Subjects with BAD I, with mania and psychotic features N=2 (2F) Case 1: 35 years old Case 2: 36 years old	*Case 1*: First 5 days placebo For 7 days, 600 mg of CBD and 10 mg of olanzapine For 7 days, 600 mg of CBD and 15 mg of olanzapine For 7 days, 900 mg of CBD For 7 days, 1200 mg of olanzapine For 5 days, placebo *Case 2*: First 5 days, placebo For 7 days, 600 mg of CBD For 7 days, 900 mg of CBD For 14 days, 1200 mg of CBD For 5 days, placebo All daily doses	Two case reports	*Case 1*: Reduction in symptoms of mania, anxiety and psychosis during the olanzapine and CBD phase *Case 2*: No symptom improvement during CBD treatment	None observed
Substance use disorders/addiction					
Haney et al.[34] and Babalonis et al.[115]	Non-treatment seeking, healthy cannabis smokers N=31 (14F) Mean age: 29.1 ± 9.1 years	200, 400, 800 mg of CBD or the placebo administered orally 90 min prior to smoking active or inactive cannabis cigarette	Randomized, placebo-controlled, cross-over study	No reduction in "liking," having a "good effect" or the desire to smoke cannabis again at any dose of CBD	No abuse potential with CBD

Solowij et al.[116]	Frequent cannabis users N = 20 (4F) Mean age: 25.1 years Age range: 20.6–46.8 years	200 mg of daily CBD administered orally for 10 weeks, while subjects continued using cannabis	Open label study	Significant reduction in psychosis, depressive and cognitive symptoms No changes in the quantity of cannabis use	No major side effects
Crippa et al.[117]	A female patient diagnosed with cannabis dependence 19 years of age	300 mg of CBD for 1 day, 600 mg of CBD for 9 days, and 300 mg for 1 day, administered orally	Single case report	Reduction in symptoms of anxiety and cannabis withdrawal	None reported
Manini et al.[118]	Subjects, previously exposed to opioids N = 17 (47% F) Mean age 38.5 ± 2.2 years	400, 800 mg of daily CBD or placebo for 1 week, administered orally, with either 0.5 or 1.0 μg/kg of intravenous fentanyl	Randomized, double-blind, placebo-controlled, cross-over study	CBD did not exacerbate adverse effects associated with the fentanyl administration	None reported
Hurd et al.[119]	Subjects with opioid dependence, abstinent for at least 7 days Number of subjects, gender and age not reported	400, 800 mg of daily CBD or placebo for 3 consecutive days, administered orally	Randomized, double-blind, placebo-controlled pilot study	Both doses of CBD reduced the cue-induced heroin craving 1 and 24 h after administering the first dose and 7 days later after receiving the final dose of CBD	None reported
Consroe et al.[120]	Social drinkers N = 10 (4F) Ager range: 21–33 years	Either 1 g/kg of alcohol, 200 mg of CBD, 1 g/kg of alcohol + 200 mg of CBD or placebo, administered orally prior to testing	Randomized, double-blind, placebo-controlled, cross-over study	CBD did not improve alcohol-induced reduction in psychomotor performance and subjective perceptions Blood alcohol levels reduced with CBD in the alcohol + CBD group	None reported
Morgan et al.[121]	Regular smokers (>10 cigarettes/day)	400 μg of daily CBD or placebo inhaled for 7 days CBD/placebo inhalant was	Randomized, double-blind, placebo-	Cigarette consumption reduced in the CBD group No effects observed on nicotine craving	None reported

Continued

Table 2 A summary of clinical studies examining effects of CBD in psychiatric disorders.—cont'd

References	Study sample (patient population, sample size, gender and age)	Treatment schedule	Study design	Primary findings	Adverse events
	N=24 (12F) Age range: 18–35 years	administered when subjects felt urge to smoke	controlled study		
Hindocha et al.[122]	Nicotine-dependent smokers N=30 (14F) Mean age: 28.07±8.66 years	800 mg of CBD or placebo, administered orally, followed by an overnight abstinent session	Randomized, double-blind, placebo-controlled, cross-over study	Reduction in salience and pleasantness of cigarette cues in the CBD group No effects on tobacco craving or symptoms of withdrawal	None reported
Hindocha et al.[123]	Nicotine-dependent smokers N=30 (15F) Mean age: 28.07±8.66 years	800 mg of CBD or placebo, administered orally, followed by overnight abstinent sessions	Randomized, double-blind, placebo-controlled, cross-over study	CBD had no effect on impulsivity to smoke during the tobacco abstinence phase No improvements in verbal or spatial working memory in nicotine dependent subjects	None reported

CBD, cannabidiol; N, number of subjects; F, female; SAD, social anxiety disorder; M, male; SPECT, single-photon emission computed tomography; rCBF, regional cerebral blood flow; PTSD, Post-traumatic stress disorder; BAD, bipolar affective disorder.

for 4 weeks, followed by 1500 mg/day of CBD for 4 weeks, then a 4-day placebo treatment, and the subsequent treatment with haloperidol for 4 weeks. Reduction in symptoms of suspiciousness, thought disturbance and hostility were observed after treating the patient for 4 weeks with CBD. Furthermore, the improvements observed with the CBD treatment were not increased with the haloperidol treatment.[105] Moreover, even at high doses of 1500 mg/day, no major side effects were reported with CBD administration.

In a case report study, CBD of up to 1280 mg/day was administered to three patients diagnosed with the treatment-resistant schizophrenia. First, placebo was given for 5 days, followed by CBD for 30 days, then placebo administered for another for 5 days, followed by olanzapine of 20 mg/day for 15 days. The first patient improved on CBD; however, deteriorated during the second placebo phase after which his symptoms only partially improved with the olanzapine therapy. The second patient showed no symptom improvements on CBD and a partial improvement while receiving olanzapine. The third patient demonstrated minimal improvements in the positive and the negative symptoms of schizophrenia with CBD and no improvements while taking olanzapine.[106]

Considering the bi-phasic effect of CBD demonstrated in the animal studies[124] with the lower doses exhibiting the therapeutic potential, while the higher doses demonstrated to be inefficient, the same group of researchers tested lower doses of CBD in the subsequent two studies.

A lack of effect of CBD was observed on the improvement of cognitive symptoms of schizophrenia in the double-blind, placebo-controlled study, in which 300 mg, 600 mg of CBD and placebo were given to three separate groups of patients with schizophrenia.[107]

Efficacy and safety of CBD were further compared to the atypical anti-psychotic, amisulpride in a double-blind, randomized controlled trial with patients diagnosed with schizophrenia or schizophreniform psychosis. In this trial, CBD was shown to be as effective as amisulpride in treating the symptoms of psychosis. Moreover, patients in the CBD-treated group had fewer extrapyramidal symptoms, less weight gain and lower increases in prolactin levels compared to the amisulpride-treated group.[108]

More recently, two additional trials were published on the topic in which an anti-psychotic treatment insufficient in controlling the symptoms of schizophrenia was augmented with CBD. In one study, 600 mg of CBD or the placebo was co-administered in a randomized, double-blind, placebo-controlled trial for 6 weeks to investigate if the CBD add-on was effective on improving positive, negative and cognitive symptoms of schizophrenia. However, no observed effects of CBD were reported, although no significant adverse effects were seen in the CBD group.[109] In the second study with a similar study design, a maximal dose of 1000 mg/day of CBD was demonstrated to augment the anti-psychotic treatments. CBD was effective in improving positive symptoms of schizophrenia, while no effects were seen on managing negative and cognitive symptoms.[110]

Overall, the evidence from the clinical trials suggests that CBD demonstrated relatively good efficacy in treating the positive symptoms, moderate efficacy in alleviating the negative symptoms and relatively low efficacy on mitigating the cognitive symptoms of schizophrenia at a wide range of doses from 400 mg to 1000 mg/day. CBD was well tolerated in the clinical trials. However, to fully elucidate CBD's potential as an effective treatment for schizophrenia larger-scale studies with longer CBD treatment periods are needed.

4.2 Effects of cannabidiol in anxiety disorders

4.2.1 Evidence from experimental animal models

Cannabidiol has been studied in several animal models of anxiety; each model simulating the specific characteristics of a separate disorder. Considering many experiments were conducted in order to assess the anxiolytic properties of CBD, only a few studies describing each disorder have been described below for the purpose of this review.

Effects of anxiolytic drugs on the symptoms of Generalized Anxiety Disorder (GAD) are normally studied in the animal models using elevated plus maze (EPM), elevated T maze (ETM) and Vogel conflict (VTC) tests. In the ETM and EPM tests, animals normally avoid the open arm entries to the maze. Previous studies examining several anxiolytic agents have demonstrated increases in the open arm entries to the maze. In one of the first experiments, doses of 2.5, 5, 10 or 20 mg/kg of CBD were injected in rats prior to their exposure to the EPM. CBD exhibited anxiolytic properties, as the rats entered the open-arms of the maze more frequently following the pre-treatment with CBD. The anxiolytic effects of CBD did not seem to be associated with impaired locomotion or sedation, since the number of total entries into the arms were comparable to the control. However, CBD exhibited a bell-shaped, dose response curve with the anxiolytic effects observed when injecting the first three doses, but not upon administering the high dose of 20 mg/kg of CBD.[124] This information implied the possibility of neutralizing the anxiolytic effects of CBD by another receptor-mediated mechanism that exclusively responded to higher doses of CBD. This receptor was later suspected to be the vanilloid, $TRPV_1$ receptor, which, once activated by high CBD doses generated an increased glutamate release facilitating anxiety-promoting neurotransmission. Thereupon, in similar, subsequent experiments, the bell-shaped dose-response curve of CBD was replicated since the direct injection of 30 nmol, but not 60 nmol of CBD in the periaqueductal gray (PAG) (the brain region activated by anxiety) increased the percentage of rodent entries into the open arm of the ETM. However, when repeating the experiment in which CBD was co-administered with the $TRPV_1$ antagonist, capsazepine, the rodent exploration in the open arm of the maze was increased, even at the higher doses of CBD implicating the $TRPV_1$ receptor-mediated anxiogenic properties at higher doses of CBD.[39]

The VCT exposes water-deprived rodents to the conflicting situation in which they are forced to make a choice either to drink the water, while receiving a painful foot shock or to remain thirsty in order to avoid the shock. In these experiments, the application of the anxiolytic drugs produces an anticonflicting effect by increasing the number of punished, water-drinking attempts. Consequently, 10 mg/kg of CBD injected to male Wistar rats increased the punished water licking. However, anxiolytic effects of CBD were not inhibited by the co-injection with flumazenil, which is a benzodiazepine receptor antagonist, indicating another mechanism of CBD's anxiolytic action independent of $GABA_A$ receptors.[125]

A chronic treatment with CBD for 21 days attenuated the inhibitory acquisition,[126] while the microinjection with CBD to the dorsal PAG in the EPM, VCT and the ETM, and to the bed nucleus of the stria terminalis (BNST; the region coordinating the fear response) in the VCT and EPM was demonstrated to be anxiolytic by the mechanism of activation of the $5HT_{1A}$ receptor, but not the CB_1 receptor.[127–129]

In the stress-induced anxiety models simulating traumatic stress exposure as a model of PTSD, the animals are experimentally restrained. The restraining related stress subsequently reduces the exploratory activity, observed in the open-fields of the EPM or the ETM. Administration of 30 nmol of CBD to the cisterna magna reduced the increases in mean arterial pressure and heart rate, previously induced by the restrained stress. Similarly, pre-treatment with CBD dampened the symptoms of anxiety in these mice, as they were spending more time in the open arm of the EPM.[130] Similar effects were observed with the intraperitoneal injections of 1–20 mg/kg of CBD to Wistar rats. In this experiment, CBD dose-dependently attenuated the consequences of the restrained stress. These anxiolytic CBD effects were, moreover, inhibited upon the co-administration of CBD with $5HT_{1A}$ antagonists, also indicating $5HT_{1A}$-receptor mediated mechanism of anxiolytic action of CBD in this model.[131]

In the animal models of induced contextual anxiety, anxiolytic effects of CBD were assessed on the symptoms of PTSD and specific phobias. Namely, a conditioned fear to the specific context is produced by re-exposing an animal to the environment previously paired with the aversive or the unpleasant stimulus. Subsequent animal re-exposures to the conditioned stimulus lead to the freezing behavior, associated with the autonomic response, such as increased mean arterial pressure (MAP) and the heart rate. Like in the other models of anxiety, the pre-treatment with anxiolytic drugs alleviates symptoms induced by the contextual fear. In an experiment in which CBD was microinjected into the prefrontal cortex, the freezing response to the electric shock was reduced. In this case, the electric shock chamber served as the conditioned stimulus, as mice were conditioned to fear the chamber by receiving the electric shock. More specifically, the pre-treatment with CBD decreased the freezing response to the electric shock chamber while decreases in the MAP and the heart rate were also measured.[132]

In another experiment, CBD facilitated extinction of the contextual fear memory in rats previously exhibiting the conditioned freezing behavior in the EPM. Interestingly, in this case, the effects of CBD were reversed by the co-administration of CBD with the CB_1 receptor antagonist, but not the antagonist for the $TRPV_1$ receptor.[128] A similar mechanism of action of CBD was implicated in the experiment assessing the effects of CBD on the fear memory reconsolidation blockade in which intraperitoneal 10 mg/kg injection of CBD was able to disrupt 7-days-old memories after the immediate memory retrieval. The fear memories disrupted by the CBD did not show the reinstatement or the spontaneous recovery up to 3 weeks after the CBD treatment indicating a relatively lasting effect of CBD on the fear memory reconsolidation blockade.[133]

Panicolytic effects of CBD were studied in the predator-prey model in which panic response was elicited by placing mice in the physical proximity to a wild snake. When injected with 3 mg/kg of CBD, the fear-related behaviors manifested by the freezing behavior and the explosive escape were reduced compared to the mice treated with the vehicle. The panicolytic properties of CBD seem to have been $5HT_{1A}$-receptor mediated, as the pre-treatment with the $5HT_{1A}$ inhibitor blocked the panicolytic properties of CBD.[134] In another experiment, CBD administration to the dPAG inhibited the escape response generated by the electrical stimulation to the midbrain and these effects were also blocked by the $5HT_{1A}$ receptor inhibition.[129] Moreover, panic-like responses were attenuated by the CBD administration to the substantia nigra in rats receiving a $GABA_A$ receptor inhibitor eliciting panic. However, in this case, the effects of CBD were partially inhibited by the CB_1 receptor antagonist while the effects of the $5HT_{1A}$ antagonists were not specifically studied in this experiment.[135]

Marble-burying experiments normally are used for studying rodent repetitive behaviors and this model mimics the compulsions observed in patients with Obsessive Compulsive Disorder (OCD). In these experiments, rodents placed in a cage filled with marbles would automatically start burying them. When 30 mg/kg of CBD was administered to Swiss mice, the marble burying was attenuated indicating anti-compulsive effects of CBD. Furthermore, when CBD was co-administered with meta-chlorophenyl-piperazine (mCPP); an agonist of the $5HT_{1A}$ receptor, which at high doses induces repetitive burying behavior in mice, CBD reversed the marble-burying effects of mCPP. The blockade of the mCPP-induced repetitive effects by CBD suggested the anti-compulsive properties of CBD were mediated by the $5TH_{1A}$ receptor.[136]

In summary, the evidence from the animal studies strongly supports the therapeutic potential of CBD in various anxiety disorders. Activation of $5HT_{1A}$ receptors appears to mediate anxiolytic, panicolytic, anti-compulsive properties, while CB_1 receptor appears to have a limited role, by potentially moderating the enhancement of fear extinction and reconsolidation blockade of CBD. However, more studies are needed to demonstrate the effects of chronic exposure of CBD.

4.2.2 Evidence from clinical studies

Despite the plentiful evidence of the effects of CBD on ameliorating the symptoms of anxiety in animal studies, only three human studies (Table 2) investigated the efficacy and safety of CBD. In a double-blind, placebo-controlled, cross-over study, treatment-naïve patients with Social Anxiety Disorder (SAD) and no other co-morbid psychiatric disorder were randomized to receive either placebo, or 400 mg of CBD, 110 min before the single-photon emission computed tomography (SPECT) analysis. SPECT is an imaging technique, was used in this experiment for the purpose of comparing the effects of CBD and the placebo on the resting regional cerebral blood flow (rCBF). Furthermore, the SPECT scan was considered as the anxiety-provoking event in patients with SAD. Subjective anxiety levels were measured 30 min before, at the time and 60, 75 and 140 min after the drug ingestion. The results showed a significant decrease in the levels of the subjective anxiety in the CBD group, 60, 75 and 140 min after the drug ingestion. Furthermore, reduced level of rCBF were noted in the left parahippocampal gyrus, the hippocampus, and in the inferior temporal gyrus. Increased rCBF in the posterior cingulate gyrus was also observed and this finding was consistent with the anxiolytic properties of CBD. No correlation was found between the subjective anxiety ratings and the rCBF measures.[111]

In the second study, patients with SAD either received 600 mg of CBD or placebo 150 min before being subjected to a public speaking test. The healthy control subjects were also recruited and subjected to the same test, without receiving any treatments. Subjects with SAD were assessed 80, 94, 105 and 150 min after receiving either CBD or the placebo. The healthy controls were also assessed at these time points. The pubic speaking test is known to be the anxiety-provoking event in patients with SAD. Compared to the placebo, CBD was demonstrated to reduce anxiety, cognitive impairment, discomfort, as well as the arousal during the anticipation of speech and the public speaking event. These ratings were comparable to the ratings observed in the healthy control group, validating the effectiveness of CBD in treating symptoms of SAD in the anxiety-induced experimental setting.[112]

Most recently, a retrospective chart review of 47 patients diagnosed with various, unspecified anxiety disorders was published, indicating that concomitant administration of 25 mg to 75 mg of CBD to the existing psychiatric medications was effective at attenuating the levels of anxiety. However, given that this was not a placebo-controlled study and the anti-anxiety effects may have resulted due to the concomitant psychiatric medication use, the study findings should be interpreted with caution.[113]

Efficacy, safety, and tolerability of CBD in anxiety disorders need to be more thoroughly examined in the randomized, placebo-controlled setting. These clinical studies are needed to properly elucidate the effects of CBD, not only in SAD but also in other anxiety disorders.

4.3 Effects of cannabidiol in mood disorders

4.3.1 Evidence from experimental animal models

The anti-depressant effects of CBD were tested in the genetically modified mice with olfactory bulbectomy (OBX) to induce the symptoms of depression. In one study, 50 mg/kg of CBD was demonstrated to reverse the

OBX-induced hyperactivity and anhedonia.[137] In another animal model of depression, the effects of CBD were investigated in genetically modified Wistar-Kyoto rats. These animals exhibit symptoms of depression, such as behavioral despair, anhedonia, increased immobility and decreased exploration of novel objects. The results showed the pro-hedonic effects of 30 mg/kg of CBD in the Saccharine Preference Test in which the drinking of the sweetened water solution mimicks the animal's ability to experience pleasure. Animals treated with 45 mg/kg of CBD were more inclined to explore novel objects. Increased locomotion was observed at CBD doses of 15 and 45 mg/kg indicating anti-depressant properties.[138] However, these properties are yet to be tested in human studies.

There is an only a limited number of studies investigating the effects of CBD in animal models of mania. Valvassori et al. tested the effects of CBD on D-amphetamine (D-AMPH)-treated rats at total daily doses of 30, 60 and 120 mg/kg.[139] In this experiment, D-AMPH was used for the purposes of inducing the symptoms reflecting mania. In the symptom-reversal experiment, the animals received either saline or the D-AMPH i.p. injection of 2 mg/kg/day for 14 days. During this time, the rats also received CBD between days 8 and 14. Here, although 30 mg/kg/day of CBD was able to reverse the previously, AMPH-induced neuronal damage in the hippocampus and increase the levels of brain derived neuronal factor (BDNF), the CBD treatment was ineffective in reversing the hyperactivity, which in rats is considered as a behavioral symptom of mania.[139]

Similar findings were observed in a second, preventive experiment. In this experiment, the same daily doses of CBD as above were administered for 14 days while neurotoxicity was induced by the injections of D-AMPH during the last 7 days of the CBD treatment. Here too, no effects of CBD were observed on the behavioral symptoms.[139] Given the evidence from the pre-clinical studies on the effects of CBD in models of mania is scarce, further investigations are needed to elucidate a potential therapeutic potential of CBD.

4.3.2 Evidence from clinical studies

Two case reports examining female patients with Bipolar Affective Disorder (BAD) I investigated the efficacy of CBD to treat the symptoms of mania (Table 2). In the first case, the patient received the placebo version of CBD for the first 6 days followed by CBD treatment with up to 1200 mg/day for the next 3.5 weeks. To ensure patients received the adequate treatments, 10–15 mg of olanzapine was also added during the first

2 weeks of the CBD treatment. Subsequent to the CBD treatment, the placebo was given for additional 5 days until the end of the study. While the symptoms of mania and psychosis were reduced during the combined treatment, no further improvements were seen with the CBD monotherapy. In the second case, CBD monotherapy was administered following the same dosing schedule; however, no improvements were observed in this case.[114] Overall, the data showed no effect of CBD on BAD I and, specifically in the manic phase. However, given that only two case studies were reported, double-blind, placebo-controlled trials would be required to elucidate the effects of CBD.

4.4 Effects of cannabidiol in addictions

4.4.1 Evidence from experimental animal models

CBD was investigated for its anti-addictive properties in several animal models of cannabis, opioid, alcohol, and nicotine addictions.

In experiments of opioid use, the effects of CBD were examined on heroin intoxication, morphine abstinence syndrome, and morphine withdrawal. Using the drug self-administration model, the impact of CBD was evaluated in rats trained to press a lever to self-administer heroin. Although no effects of CBD were evident on heroin intake at doses of 5–20 mg/kg, CBD was effective in ameliorating the heroin-seeking behavior, which was stimulated by the exposure to the conditioned stimulus cues.[140] This response to CBD was maintained even 2 weeks after the administration.

Effects of CBD on the morphine reward in the conditioned place preferences paradigm were tested on mice receiving either saline or morphine in combination with CBD. The morphine treated mice exhibited drug-place conditioning, which was attenuated by the pre-treatment with 10 mg/kg of CBD suggesting the ability of CBD to block the opioid reward.[141]

CBD effect on the THC-induced attenuation of the morphine abstinence syndrome was investigated, after inducing morphine dependence in rats. Animals were pre-treated with either the vehicle or 10 mg/kg of CBD, followed by either 2 mg/kg of THC or the vehicle injection, before receiving naloxone, which precipitated abstinence. Subsequently, animal abstinence was evaluated by examining the signs of animal wet shakes, escapes, number of fecal boluses, presence of teeth chattering, chewing, diarrhea, vocalization, abnormal posture, and ear blanching. The results showed that CBD alone did not affect the abstinence scores; however, in synergy with THC, it reduced the abstinence scores more so than the THC alone.[142]

Mixed evidence was demonstrated with respect to the CBD attenuating effect on the morphine withdrawal. In one study, the responses to CBD at doses of 5, 10 and 20 mg/kg were tested on morphine-dependent mice and on their withdrawal symptoms, previously induced by naloxone. The dose of naloxone needed to induce the withdrawal symptoms in 50% of animals was determined for each dose of CBD. The results showed that CBD inhibited the withdrawal-induced jumping, reduced defecation and the rearing behavior in the morphine dependent mice.[143] In another study, the effects of 5, 20 or 80 mg/kg of CBD were tested on rats given morphine and, subsequently naloxone to induce the quasi-morphine withdrawal syndrome. However, CBD at all doses was ineffective at ameliorating the behavioral signs of the withdrawal.[144]

The effects of CBD on the outcomes of intoxication, withdrawal and relapse, were further tested in the animal models of cannabis addictions.

In one study, the effects of CBD were examined on the THC drug discrimination and the conditioned place preference (CPP) in mice and rats, after several combinations of CBD and THC were administered at different doses. The results showed that when administered alone, CBD did not produce the THC discrimination stimulus. However, at low doses of 1 and 10 mg/kg of CBD and in combination with THC, CBD reversed the conditioned place aversion, induced by the 10 mg/kg injection of THC.[145]

More recently, the CBD effect on the behavioral alterations induced by cannabis withdrawal was studied in mice pre-treated with either 5, 10 or 20 mg/kg of CBD. The results showed that CBD significantly blocked the symptoms of withdrawal, indicated by the increases in the number of motor activities, rearings, rubbings, and jumping, and the decreases in the number of grooming sessions.[146]

Initial studies investigating the effects of CBD on the alcohol consumption in animals, focused on treating alcohol-induced neurodegeneration, because CBD's previously demonstrated anti-oxidant properties.[147,148]

However, more recently, CBD was also evaluated for its effects on the ethanol reinforcement, motivation, and relapse in C57Bl mice. CBD was demonstrated to reduce the ethanol consumption and the conditioned preference. Furthermore, decreases in the number of self-administration ethanol responses and in the ethanol-induced relapse were observed upon the CBD administration.[149]

In yet another study, rats were trained to self-administer alcohol at 24-h intervals for 7 days and were administered CBD. The results showed that CBD attenuated the stress-induced, drug seeking behavior, the sedative effects and interfered with the motivation to administer alcohol. These effects of CBD were maintained for up to 5 months, although the plasma levels of CBD were detectable for only 3 days. Furthermore, CBD reduced the rates of impulsivity and anxiety in the alcohol dependent rats.[150]

The positive effects of CBD observed in the animal models of alcohol addictions warranted further investigating therapeutic effects of CBD in patients seeking treatment from alcohol use disorder.[151,152]

Several studies investigated effects of CBD on psychostimulant addictive behaviors. Katsidoni et al. examined the effects on CBD on the brain reward function and the reward-facilitating effects of cocaine in rats trained to respond to the electrical intracranial self-stimulation (ICSS).[153] Cocaine lowered ICSS thresholds, indicating its reward facilitating effects. However, the pre-treatment with CBD was ineffective at blocking the reward-facilitating effects of cocaine. Recently, the CBD effect on cocaine self-administration and cue-induced seeking was tested in rats trained to self-administer cocaine under fixed and the progressive ratio schedules of reinforcement. However, the results showed that either 5 or 10 mg/kg of CBD did not attenuate the reinforcing effects of the cocaine self-administration nor it did reduce the cue-induced cocaine seeking after the 14-day withdrawal period.[154]

Parker et al. tested 5 mg/kg of CBD on the cocaine and the amphetamine induced CPP in rats in the extinction trial. Contrary to the previous negative findings, the administration of CBD potentiated the extinction of both, the cocaine and the amphetamine induced CPP learning without having any effects of the stimulant-establishing CPP effects.[155]

Furthermore, 20-, 40- and 80 mg/kg-doses of CBD were investigated on rats trained to self-administer methamphetamine under the fixed ratio schedule. The highest dose of 80 mg/kg of CBD was demonstrated to be effective at reducing the motivation to administer methamphetamine. Also, the same dose of CBD was shown to be effective at reducing the methamphetamine-primed relapse to the drug-seeking behavior after the extinction.[156]

4.4.2 Evidence from clinical studies

Haney et al. has demonstrated that when CBD was administered at the doses of 200, 400 and 800 mg or the placebo 90 min prior to smoking cannabis, no alteration in the subjective, reinforcing or cardiovascular effects of smoked cannabis were seen relative to the placebo.[34] In this study, 31 healthy, regular cannabis users were asked to complete 8 sessions in the cross-over design in which they received either the placebo or three different doses of

CBD prior to smoking either the inactive or the active cannabis cigarette. Subjects were asked to complete a scale in which they had to indicate if they felt "high" when smoking the active cannabis cigarette after being pre-treated with any administered dose of CBD or the placebo. Similar effects were found from the Marijuana Rating Form, which indicated that the active cannabis smokers rated "liking" the effect of cannabis, having a "good effect" and having the "desire to take again" the drug.[34] The CBD pre-administration at any dose did not change the ratings of smoking the active cannabis. Furthermore, when administered at all three doses, CBD did not display any signs of the abuse liability further substantiating the safety of orally ingested CBD.[115]

The first study testing chronically administered CBD was an open label, 10-week, clinical trial in which 20 frequent cannabis users received 200 mg of daily CBD. After completing the treatment, participants reported having much fewer psychotic-like and depressive symptoms compared to the pretreatment period (the baseline). Furthermore, during the cognitive testing, subjects improved in attentional switching, as well as in verbal learning, and memory. The study also demonstrated that CBD plasma levels were positively associated with the improvements in the attentional control. Greater effects of CBD were observed in those cannabis users who met the criteria for cannabis dependence.[116]

To our knowledge, there are only two studies to date, investigating the effects of CBD on the cannabis withdrawal syndrome. In a single case report study, Crippa et al. administered CBD to a 19-year-old female with a history of cannabis dependence. Prior to the enrollment, the patient smoked 4–8 cannabis cigarettes/day, and while attempting was unable to quit the cannabis use.[117] The patient was hospitalized, during which she was administered 300 mg of CBD on the first day, 600 mg of CBD for the next 9 days and tapered down to 300 mg of CBD on day 11. While in the study, the patient was assessed daily for her symptoms of cannabis withdrawal and anxiety, which all decreased through the treatment period.[117]

CBD was also tested in the human translational studies for its effects on the opioid craving. In combination with fentanyl, safety of CBD was examined, when it was administered to abstinent heroin abusers. Two doses of fentanyl (0.5 and 1.0 μg/kg) on two different occasions were given to three groups of patients, each group receiving either 400 mg or 800 mg of CBD or the placebo. The study demonstrated that CBD was well tolerated at all doses and that co-administration with fentanyl did not produce the symptoms of respiratory depression or any cardiovascular complications, suggesting the potential of CBD to be potentially used as a treatment for the opioid abuse.[118]

A small, double-blind study was conducted investigating the effects of CBD on the opioid dependent individuals, who abstained from heroin use for at least 7 days. The subjects were randomized to receive either 400 mg, or 800 mg of CBD, or the placebo for three consecutive test days before participating in the laboratory sessions. Both doses of CBD reduced the cue-induced heroin craving 1 and 24 h after the first administered dose, as well as 7 days after the final dose.[119] Given that the intensity of heroin craving may predict the relapse, the data suggests the potential of CBD to prevent opioid relapse.

Impact of CBD on alcohol intoxication was studied in 10 healthy volunteers administered either the placebo, 1 g/kg of alcohol, 200 mg of CBD or both, alcohol and CBD in the cross-over double-blind design. Compared to the placebo, the participants receiving alcohol and CBD and those on alcohol alone experienced significant impairments of the motor and the psychomotor performance and reported significant overestimations of time. The study results suggested either negligible or minimally enhancing effects of the alcohol consumption in combination with CBD. Interestingly, although virtually no effects of CBD on objective or subjective observations of intoxication were observed, the combination of alcohol and CBD resulted in significant lowering of the blood alcohol levels compared to when alcohol was given alone.[120] Currently, large randomized, placebo-controlled studies are being conducted as reported on the clinicaltrial.gov website, thus the therapeutic potential of CBD to treat the alcohol addictions is yet to be determined.

Although CBD has never been studied in animal models of the nicotine dependence, it has been investigated in three clinical studies in which the effects of CBD were tested on managing the symptoms of nicotine addiction. In the first clinical study using CBD in patients with the desire to quit smoking, 24 smokers were randomized in the double-blind, placebo-controlled manner to either receive the inhaler with CBD or the placebo for 7 days. The subjects were told to use the inhaler every time they felt the urge to smoke. Compared to the placebo group, patients using the CBD inhaler reduced the number of cigarettes smoked by 40% during the treatment. However, although craving plays a pivotal role in the relapse, no effects due to CBD were found on the craving response to nicotine. Moreover, CBD was well tolerated throughout the trial, as no obvious signs of depression, sedation or anxiety were reported.[121]

In addition to the nicotine withdrawal, similar effects by CBD on craving were found in another, more recent, randomized, double-blind, cross-over trial in which either 800 mg of CBD or placebo were administered to participants, smoking >10 cigarettes/day, for over a year. In this study, the participants were asked to smoke as their usual at the baseline assessment, but abstained from smoking overnight, (12 h) before the test session. After abstaining from smoking, the subjects were tested before and 180 min after the study drug administration. The results showed that although, CBD had no effects on nicotine craving, it reversed the attention bias to the cigarette cues in these abstained smokers so that no difference was observed compared to the baseline when subjects were satiated. Furthermore, during the abstinence period, the subjects rated the smoking cigarette stimuli as less pleasant after receiving CBD, rather than the placebo. No significant side effects were observed in this trial.[122] The attention bias to the smoking cues is known to be the predictor of the relapse to smoking in those trying to quit, thus the fact that CBD attenuated the attention bias may be indicative of its potential to prevent the relapse to smoking cigarettes.

Furthermore, effects of CBD on two the other predictors of relapse were examined in the nicotine dependent individuals. In one study, 30, non-treatment seeking, nicotine- dependent individuals were randomized to receive either 800 mg of CBD or the placebo after abstaining from smoking for approximately 12 h. Like in the previous study, subjects were assessed at the baseline and pre- and post-drug administration. The results showed no improvements in either the symptoms of cognition or impulsivity during tobacco abstinence due to the CBD acute administration phase.[123] However, given that only the acute administration of CBD was tested in this study, studies investigating the effects of the chronic, repeated dosing of CBD on its potential to curb the impulsive and modulate the impaired, relapse-provoking cognition are yet to be conducted in the randomized, placebo-controlled setting. All the above mentioned human studies in this section are summarized in Table 2.

5. Conclusions

Because of the wide range of pharmacodynamic effects of CBD, it was extensively studied in both, the pre-clinical and clinical studies. These studies have demonstrated that CBD has a wide range of potential therapeutic applications in the neuropsychiatric disorders. The studies reviewed in this article consistently indicated the good safety and tolerability of CBD at oral doses of up to 1500 mg/day when administered to human subjects. Special attention should be given to potentially occurring drug-drug interactions with other drugs metabolized by CYP 2C and CYP 3A enzymes, since CBD has been shown to be a potent inhibitor of these enzymes. Considering that CBD is currently marketed as Epidolex® as the adjuvant treatment for specific childhood epilepsies, careful drug monitoring of plasma levels of other AEDs may be required.

The results of the several experimental and human studies conducted in neurological disorders, including epilepsy, AD, HD, and PD were reviewed in this chapter. The effects of CBD on ameliorating the symptoms of epilepsy have been studied most prevalently. Although, CBD has demonstrated antioxidant, anti-inflammatory and antiapoptotic properties in the animal models of AD, its potential therapeutic effects still need to be investigated in human studies of AD. Because mixed results were found with the CBD monotherapy in the experimental models of HD and no effects on minimizing the symptoms of HD were observed in the human studies, new cannabinoid compounds, are currently being investigated. The protective effects of CBD were also demonstrated in the experimental models of PD. However, although limited effects of CBD on motor symptoms of PD were observed in the human studies using rather small sample sizes, these results should be deemed insufficient to fully elucidate the therapeutic effect of CBD in the PD. A larger-scale, randomized, placebo-controlled studies using higher CBD doses are needed, especially those assessing the effects of CBD on the motor symptoms associated with PD.

When examining the effects of CBD in the psychiatric disorders, CBD was demonstrated to be more effective in treating the positive rather than negative and the cognitive symptoms of schizophrenia. Considering that the latest evidence indicates that CBD in combination with other anti-psychotics was effective at attenuating the symptoms of psychosis, CBD may have a future role as an adjuvant treatment. However, long-term, randomized, clinical trials with larger sample sizes are needed to examine the efficacy and safety of CBD in this patient population. The Anxiolytic properties of CBD were found after the acute administration of CBD to patients with SAD; however, given that animal studies indicated effects of CBD in animal models of OCD, GAD, specific phobias and PTSD, the therapeutic potential of CBD is yet to be studied in patients diagnosed with anxiety spectrum disorders. Furthermore, longer CBD administration in patients with anxiety is needed disorder, as well as randomized, placebo-controlled studies to substantiate the anxiolytic properties of CBD. So far, CBD was deemed to be ineffective in treating the symptoms of mania in both animal and human studies, and although it was tested in a few animal models of depression, there too, larger scale, randomized, placebo-controlled trials are needed to investigate potential treatment properties of CBD in mood disorders. Similarly, these types of trials would be also necessary to investigate the effects of CBD in opioid-, cannabis-, alcohol- and stimulant-induced addictions, focusing on the potential of CBD to mitigate the symptoms of alcohol and substance-induced withdrawal, abstinence and relapse.

References

1. Li HL, Lin H. An archaeological and historical account of cannabis in China. *Econ Bot.* 1974;28(4):437–447.
2. Li HL. Hallucinogenic plants in Chinese herbals. *J Psychodelic Drugs.* 1978;10(1):17–26.
3. Touwn M. The religious and medicinal uses of Cannabis in China, India and Tibet. *J Psychoactive Drugs.* 1981;13(1):23–34.
4. Zuardi AW. History of cannabis as a medicine: a review. *Rev Bras Psiquiatr.* 2006;28(2):153–157.
5. Fankhauser M. History of cannabis in Western medicine. In: Grotenhermen F, Russo E, eds. *Cannabis and Cannabinoids.* New York: The Haworth Integrative Healing Press; 2002:37–51. [chapter 4].
6. Pinho AR. Social and medical aspects of the use of cannabis in Brazil. In: Rubin V, ed. *Cannabis and Culture.* Paris: Mounton Publishers; 1975:293–302.
7. Mikuriya TH. Marijuana in medicine: past, present and future. *Calif Med.* 1969;110(1):34–40.
8. Aldrich M. History of therapeutic cannabis. In: Mathre ML, ed. *Cannabis in Medical Practice.* Jefferson, NC: Mc Farland; 1997:35–55.
9. Grinspoon L, Bakalar JB. Marihuana: the forbidden medicine. In: *History of Cannabis.* New Haven: Yale University Press; 1993:1–23.
10. Harris LS. Cannabis: a review of progress. In: Lipton MA, Dimascio A, Killam KF, eds. *Psychopharmacology: A Generation of Progress.* New York: Raven Press; 1978:1565–1574.
11. Kandel DB. Marihuana users in young adulthood. *Arch Gen Psychiatry.* 1984; 41(2):200–209.
12. Mead A. The legal status of cannabis (marijuana) and cannabidiol (CBD) under U.S. law. *Epilepsy Behav.* 2017;70(pt. B):288–291.
13. Sandler LN, Beckerman JL, Whitford F, Gibson KA. Cannabis as conundrum. *Crop Prot.* 2019;117:37–44.
14. Russo EB. History of cannabis and its preparations in saga, science, and sobriquet. *Chem Biodivers.* 2007;4(8):1614–1648.
15. D'Souza DC, Perry E, MacDougall L, et al. The psychotomimetic effects of intravenous delta-9-tetrahydrocannabinol in healthy individuals: implications for psychosis. *Neuropsychopharmacology.* 2004;29(8):1558–1572.
16. Mechoulam R, Parker LA, Gallily R. Cannabidiol: an overview of some pharmacological aspects. *J Clin Pharmacol.* 2002;42(S1):11S–19S.
17. Pisanti S, Malfitano AM, Ciaglia E, et al. Cannabidiol: state of the art and new challenges for therapeutic applications. *Pharmacol Ther.* 2017;175:133–150.
18. Bergamaschi MM, Queiroz RHC, Zuardi AW, Crippa JAS. Safety and side effects of cannabidiol, a Cannabis sativa constituent. *Curr Drug Saf.* 2011;6(4):237–249.
19. Huestis MA. Pharmacokinetics and metabolism of the plant cannabinoids, delta9-tetrahydrocannabinol, cannabidiol and cannabinol. *Handb Exp Pharmacol.* 2005;168: 657–690.

20. Hawksworth G, McArdle K. *Metabolism and Pharmacokinetics of Cannabinoids*. London, UK: Pharmaceutical Press; 2004.
21. Millar SA, Stone NL, Yates AS, O'Sullivan SE. A systematic review on the pharmacokinetics of cannabidiol in humans. *Front Pharmacol.* 2018;9:1365.
22. Guy G, Robson P. A phase I, open label, four-way crossover study to compare the pharmacokinetic profiles of a single dose of 20 mg of a cannabis based medicine extract (CBME) administered on 3 different areas of the buccal mucosa and to investigate the pharmacokinetics of CBME per oral in healthy male and female volunteers (GWPK0112). *J Cannabis Ther.* 2004;3:79–120.
23. Devinsky O, Cilio MR, Cross H, et al. Cannabidiol: pharmacology and potential therapeutic role in epilepsy and other neuropsychiatric disorders. *Epilepsia.* 2014;55(6):791–802.
24. Reddy DS. The utility of cannabidiol in the treatment of refractory epilepsy. *Clin Pharmacol Ther.* 2017;101(2):182–184.
25. Ohlsson A, Lindgren J, Andersson S, et al. Single dose kinetics of cannabidiol in man. In: Agurell S, Dewey WL, Willette RE, eds. *The Cannabinoids: Chemical, Pharmacologic, and Therapeutic Aspects*. Orlando: Academic Press; 1984:219–225.
26. Harvey DJ, Samara E, Mechoulam R. Comparative metabolism of cannabidiol in dog, rat and man. *Pharmacol Biochem Behav.* 1991;40(3):523–532.
27. Ujváry I, Hanuš L. Human metabolites of cannabidiol: a review on their formation, biological activity, and relevance in therapy. *Cannabis Cannabinoid Res.* 2016;1(1):90–101.
28. Jiang R, Yamaori S, Takeda S, Yamamoto I, Watanabe K. Identification of cytochrome P450 enzymes responsible for metabolism of cannabidiol by human liver microsomes. *Life Sci.* 2011;89(5–6):165–170.
29. Devinsky O, Patel AD, Cross JH, et al. Effect of cannabidiol on drop seizures in the Lennox-Gastaut syndrome. *N Engl J Med.* 2018;378(20):1888–1897.
30. Ibeas Bih C, Chen T, Nunn AVW, Bazelot M, Dallas M, Whalley BJ. Molecular targets of cannabidiol in neurological disorders. *Neurotherapeutics.* 2015;12(4):699–730.
31. Campos AC, Fogaça MV, Scarante FF, et al. Plastic and neuroprotective mechanisms involved in the therapeutic effects of cannabidiol in psychiatric disorders. *Front Pharmacol.* 2017;8:269.
32. Pertwee RG. The diverse CB1 and CB2 receptor pharmacology of three plant cannabinoids: delta9-tetrahydrocannabinol, cannabidiol and delta9-tetrahydrocannabivarin. *Br J Pharmacol.* 2008;153(2):199–215.
33. Thomas A, Baillie GL, Phillips AM, Razdan RK, Ross RA, Pertwee RG. Cannabidiol displays unexpectedly high potency as an antagonist of CB1 and CB2 receptor agonists in vitro. *Br J Pharmacol.* 2007;150:613–623.
34. Haney M, Malcolm RJ, Babalonis S, et al. Oral cannabidiol does not alter the subjective, reinforcing or cardiovascular effects of smoked cannabis. *Neuropsychopharmacology.* 2016;41:1974.
35. Boggs DL, Nguyen JD, Morgenson D, Taffe MA, Ranganathan M. Clinical and preclinical evidence for functional interactions of cannabidiol and delta(9)-tetrahydrocannabinol. *Neuropsychopharmacology.* 2018;43(1):142–154.
36. De Petrocellis L, Ligresti A, Moriello AS, et al. Effects ofcannabinoids and cannabinoid-enriched Cannabis extracts on TRP channels and endocannabinoid metabolic enzymes. *Br J Pharmacol.* 2011;163:1479–1494.
37. Russo EB, Burnett A, Hall B, Parker KK. Agonistic properties of cannabidiol at 5-HT_{1A} receptors. *Neurochem Res.* 2005;30(8):1037–1043.
38. Bisogno T, Hanus L, De Petrocellis L, et al. Molecular targets for cannabidiol and its synthetic analogues: effect on vanilloid VR1 receptors and on the cellular uptake and enzymatic hydrolysis of anandamide. *Br J Pharmacol.* 2001;134:845–852.

39. Campos AC, Guimarães FS. Evidence for a potential role for TRPV1 receptors in the dorsolateral periaqueductal gray in the attenuation of the anxiolytic effects of cannabinoids. *Prog Neuropsychopharmacol Biol Psychiatry*. 2009;33(8):1517–1521.
40. Gonca E, Darıcı F. The effect of cannabidiol on ischemia/reperfusion-induced ventricular arrhythmias: the role ofadenosine A1 receptors. *J Cardiovasc Pharmacol Ther*. 2015;20:76–83.
41. Kathmann M, Flau K, Redmer A, Tränkle C, Schlicker E. Cannabidiol is an allosteric modulator at mu- and delta-opioid receptors. *Naunyn Schmiedebergs Arch Pharmacol*. 2006;372:354–361.
42. Wiese B, Wilson-Poe AR. Emerging evidence for cannabis' role in opioid use disorder. *Cannabis Cannabinoid Res*. 2018;3(1):179–189.
43. Lynch JW. Molecular structure and function of the glycine receptor chloride channel. *Physiol Rev*. 2004;84:1051–1095.
44. Zhornitsky S, Potvin S. Cannabidiol in humans-the quest for therapeutic targets. *Pharmaceuticals (Basel, Switzerland)*. 2012;5(5):529–552.
45. Scuderi C, Steardo L, Esposito G. Cannabidiol promotes amyloid precursor protein ubiquitination and reduction of beta amyloid expression in SHSY5YAPP+ cells through PPARgamma involvement. *Phytother Res*. 2014;28(7):1007–1013.
46. Karl T, Garner B, Cheng D. The therapeutic potential of the phytocannabinoid cannabidiol for Alzheimer's disease. *Behav Pharmacol*. 2017;28(2–3):142–160.
47. Abdul HM, Sama MA, Furman JL, et al. Cognitive decline in Alzheimer's disease is associated with selective changes in calcineurin/NFAT signaling. *J Neurosci*. 2009;29(41):12957.
48. Mahgoub M, Keun-Hang SY, Sydorenko V, et al. Effects of cannabidiol on the function of α7-nicotinic acetylcholine receptors. *Eur J Pharmacol*. 2013;720:310–319.
49. Ross HR, Napier I, Connor M. Inhibition of recombinant human T-type calcium channels by Delta9-tetrahydrocannabinol and cannabidiol. *J Biol Chem*. 2008;283:16124–16134.
50. Hill AJ, Jones NA, Smith I, et al. Voltage-gated sodium (NaV) channel blockade by plant cannabinoids does not confer anticonvulsant effects per se. *Neurosci Lett*. 2014;566:269–274.
51. Seeman P. Cannabidiol is a partial agonist at dopamine D2High receptors, predicting its antipsychotic clinical dose. *Transl Psychiatry*. 2016;6(10):e920.
52. Iannotti FA, Hill CL, Leo A, et al. Nonpsychotropic plant cannabinoids, cannabidivarin (CBDV) and cannabidiol (CBD), activate and desensitize transient receptor potential vanilloid 1 (TRPV1) channels in vitro: potential for the treatment of neuronal hyperexcitability. *ACS Chem Nerosci*. 2014;5(11):1131–1141.
53. Jones NA, Hill AJ, Smith I, et al. Cannabidiol displays antiepileptiform and antiseizure properties in vitro and in vivo. *J Pharmacol Exp Ther*. 2010;332(2):569.
54. Khan AA, Shekh-Ahmad T, Khalil A, Walker MC, Ali AB. Cannabidiol exerts antiepileptic effects by restoring hippocampal interneuron functions in a temporal lobe epilepsy model. *Br J Pharmacol*. 2018;175(11):2097–2115.
55. Consroe P, Wolkin A. Cannabidiol—antiepileptic drug comparisons and interactions in experimentally induced seizures in rats. *J Pharmacol Exp Ther*. 1977;201(1):26.
56. Consroe P, Benedito MAC, Leite JR, Carlini EA, Mechoulam R. Effects of cannabidiol on behavioral seizures caused by convulsant drugs or current in mice. *Eur J Pharmacol*. 1982;83(3):293–298.
57. Jones NA, Hill AJ, Weston SE, et al. Cannabidiol exerts anti-convulsant effects in animal models of temporal lobe and partial seizures. *Seizure*. 2012;21(5):344–352.
58. Shirazi-zand Z, Ahmad-Molaei L, Motamedi F, Naderi N. The role of potassium BK channels in anticonvulsant effect of cannabidiol in pentylenetetrazole and maximal electroshock models of seizure in mice. *Epilepsy Behav*. 2013;28(1):1–7.

59. Mao K, You C, Lei D, Zhang H. High dosage of cannabidiol (CBD) alleviates pentylenetetrazole-induced epilepsy in rats by exerting an anticonvulsive effect. *Int J Clin Exp Med*. 2015;8(6):8820–8827.
60. Alvarez FJ, Lafuente H, Carmen Rey-Santano M, et al. Neuroprotective effects of the nonpsychoactive cannabinoid cannabidiol in hypoxic-ischemic newborn piglets. *Pediatr Res*. 2008;64:653.
61. Rodriguez-Munoz M, Onetti Y, Cortes-Montero E, Garzon J, Sanchez-Blazquez P. Cannabidiol enhances morphine antinociception, diminishes NMDA-mediated seizures and reduces stroke damage via the sigma 1 receptor. *Mol Brain*. 2018;11(1):51.
62. Mechoulam R, Carlini EA. Toward drugs derived from cannabis. *Naturwissenschaften*. 1978;65(4):174–179.
63. Carlini EA, Cunha JM. Hypnotic and antiepileptic effects of cannabidiol. *J Clin Pharmacol*. 1981;21(S1):417S–427S.
64. Ames FR, Cridland S. Anticonvulsant effect of cannabidiol. *S Afr Med J*. 1986; 69(1):14.
65. Devinsky O, Marsh E, Friedman D, et al. Cannabidiol in patients with treatment-resistant epilepsy: an open-label interventional trial. *Lancet Neurol*. 2016;15(3):270–278.
66. Rosenberg EC, Louik J, Conway E, Devinsky O, Friedman D. Quality of life in childhood epilepsy in pediatric patients enrolled in a prospective, open-label clinical study with cannabidiol. *Epilepsia*. 2017;58(8):e96–e100.
67. Devinsky O, Cross JH, Laux L, et al. Trial of cannabidiol for drug-resistant seizures in the Dravet syndrome. *N Engl J Med*. 2017;376(21):2011–2020.
68. Gofshteyn JS, Wilfong A, Devinsky O, et al. Cannabidiol as a potential treatment for febrile infection-related epilepsy syndrome (FIRES) in the acute and chronic phases. *J Child Neurol*. 2017;32(1):35–40.
69. Consroe P, Laguna J, Allender J, et al. Controlled clinical trial of cannabidiol in Huntington's disease. Special Issue: Pharmacological, chemical, biochemical and behavioral research on cannabis and the cannabinoids. *Pharmacol Biochem Behav*. 1991;40(3):701–708.
70. Zuardi AW, Crippa JAS, Hallak JEC, et al. Cannabidiol for the treatment of psychosis in Parkinsons disease. *J Psychopharmacol*. 2009;23(8):979–983.
71. Chagas MHN, Eckeli AL, Zuardi AW, et al. Cannabidiol can improve complex sleep-related behaviours associated with rapid eye movement sleep behaviour disorder in Parkinson's disease patients: a case series. *J Clin Pharm Ther*. 2014;39(5):564–566.
72. Chagas MHN, Zuardi AW, Tumas V, et al. Effects of cannabidiol in the treatment of patients with Parkinson's disease: an exploratory double-blind trial. *J Psychopharmacol*. 2014;28(11):1088–1098.
73. Hughes B, Herron CE. Cannabidiol reverses deficits in hippocampal LTP in a model of Alzheimer's disease. *Neurochem Res*. 2019;44(3):703–713.
74. Esposito G, De Filippis D, Carnuccio R, Izzo AA, Iuvone T. The marijuana component cannabidiol inhibits beta-amyloid-induced tau protein hyperphosphorylation through Wnt/beta-catenin pathway rescue in PC12 cells. *J Mol Med (Berl)*. 2006;84(3):253–258.
75. Iuvone T, Esposito G, Esposito R, Santamaria R, Di Rosa M, Izzo AA. Neuroprotective effect of cannabidiol, a non-psychoactive component from Cannabis sativa, on beta-amyloid-induced toxicity in PC12 cells. *J Neurochem*. 2004;89(1): 134–141.
76. Briggs GH. Cannabidiol, a non-psychoactive marijuana component, exhibits antioxidant and neuroprotective effects in neuronal Alzheimer's cell model SH-5Y. *FASEB J*. 2018;32(1 suppl 1):Abstract no. 552.6.
77. Martin-Moreno AM, Reigada D, Ramirez BG, et al. Cannabidiol and other cannabinoids reduce microglial activation in vitro and in vivo: relevance to Alzheimer's disease. *Mol Pharmacol*. 2011;79(6):964–973.

78. Esposito G, Scuderi C, Savani C, et al. Cannabidiol in vivo blunts beta-amyloid induced neuroinflammation by suppressing IL-1beta and iNOS expression. *Br J Pharmacol.* 2007;151(8):1272–1279.
79. Cheng D, Low JK, Logge W, Garner B, Karl T. Chronic cannabidiol treatment improves social and object recognition in double transgenic APPswe/PS1E9 mice. *Psychopharmacology (Berl).* 2014;231(15):3009–3017.
80. Cheng D, Spiro AS, Jenner AM, Garner B, Karl T. Long-term cannabidiol treatment prevents the development of social recognition memory deficits in Alzheimer's disease transgenic mice. *J Alzheimers Dis.* 2014;42(4):1383–1396.
81. Laprairie RB, Bagher AM, Kelly MEM, Denovan-Wright EM. Biased type 1 cannabinoid receptor signaling influences neuronal viability in a cell culture model of Huntington diseases. *Mol Pharmacol.* 2016;89(3):364–375.
82. Sagredo O, Ramos JA, Decio A, Mechoulam R, Fernandez-Ruiz J. Cannabidiol reduced the striatal atrophy caused 3-nitropropionic acid in vivo by mechanisms independent of the activation of cannabinoid, vanilloid TRPV1 and adenosine A2A receptors. *Eur J Neurosci.* 2007;26(4):843–851.
83. Sagredo O, Pazos MR, Satta V, Ramos JA, Pertwee RG, Fernandez-Ruiz J. Neuroprotective effects of phytocannabinoid-based medicines in experimental models of Huntington's disease. *J Neurosci Res.* 2011;89(9):1509–1518.
84. Valdeolivas S, Satta V, Pertwee RG, Fernández-Ruiz J, Sagredo O. Sativex-like combination of phytocannabinoids is neuroprotective in malonate-lesioned rats, an inflammatory model of Huntington's disease: role of CB1 and CB2 receptors. *ACS Chem Nerosci.* 2012;3(5):400–406.
85. Lastres-Becker I, Molina-Holgado F, Ramos JA, Mechoulam R, Fernandez-Ruiz J. Cannabinoids provide neuroprotection against 6-hydroxydopamine toxicity in vivo and in vitro: relevance to Parkinson's disease. *Neurobiol Dis.* 2005;19(1–2):96–107.
86. García-Arencibia M, González S, de Lago E, Ramos JA, Mechoulam R, Fernández-Ruiz J. Evaluation of the neuroprotective effect of cannabinoids in a rat model of Parkinson's disease: importance of antioxidant and cannabinoid receptor-independent properties. *Brain Res.* 2007;1134:162–170.
87. Moldzio R, Pohn B, Krewenka C, Kranner B, Duvigneau JC. Phytocannabinoids tetrahydrocannabinol and cannabidiol ACT against rotenone induced damages in murine cell cultures. *Parkinsonism Relat Disord.* 2012;18(suppl 2):S205–S206.
88. Moldzio R, Pacher T, Krewenka C, et al. Effects of cannabinoids DELTA(9)-tetrahydrocannabinol, DELTA(9)-tetrahydrocannabinolic acid and cannabidiol in MPP+ affected murine mesencephalic cultures. *Phytomedicine.* 2012;19(8–9):819–824.
89. Celorrio M, Rojo-Bustamante E, Fernandez-Suarez D, et al. GPR55: a therapeutic target for Parkinson's disease? *Neuropharmacology.* 2017;125:319–332.
90. da Silva VK, de Freitas BS, da Silva DA, et al. Cannabidiol normalizes caspase 3, synaptophysin, and mitochondrial fission protein DNM1L expression levels in rats with brain iron overload: implications for neuroprotection. *Mol Neurobiol.* 2014;49(1):222–233.
91. Gomes FV, Del Bel EA, Guimaraes FS. Cannabidiol attenuates catalepsy induced by distinct pharmacological mechanisms via 5-HT_{1A} receptor activation in mice. *Prog Neuropsychopharmacol Biol Psychiatry.* 2013;46:43–47.
92. Moreira FA, Guimarães FS. Cannabidiol inhibits the hyperlocomotion induced by psychotomimetic drugs in mice. *Eur J Pharmacol.* 2005;512(2):199–205.
93. Long LE, Chesworth R, Huang X-F, McGregor IS, Arnold JC, Karl T. A behavioural comparison of acute and chronic Δ9-tetrahydrocannabinol and cannabidiol in C57BL/6JArc mice. *Int J Neuropsychopharmacol.* 2010;13(7):861–876.
94. Renard J, Loureiro M, Rosen LG, et al. Cannabidiol counteracts amphetamine-induced neuronal and behavioral sensitization of the mesolimbic dopamine pathway through a novel mTOR/p70S6 kinase signaling pathway. *J Neurosci.* 2016;36(18):5160–5169.

95. Pedrazzi JFC, Issy AC, Gomes FV, Guimaraes FS, Del-Bel EA. Cannabidiol effects in the prepulse inhibition disruption induced by amphetamine. *Psychopharmacology (Berl).* 2015;232(16):3057–3065.
96. Zuardi AW, Antunes Rodrigues J, Cunha JM. Effects of cannabidiol in animal models predictive of antipsychotic activity. *Psychopharmacology (Berl).* 1991;104(2):260–264.
97. Sonego AB, Gomes FV, Del Bel EA, Guimaraes FS. Cannabidiol attenuates haloperidol-induced catalepsy and c-Fos protein expression in the dorsolateral striatum via 5-HT_{1A} receptors in mice. *Behav Brain Res.* 2016;309:22–28.
98. Long LE, Malone DT, Taylor DA. Cannabidiol reverses MK-801-induced disruption of prepulse inhibition in mice. *Neuropsychopharmacology.* 2005;31:795.
99. Gururajan A, Taylor DA, Malone DT. Effect of cannabidiol in a MK-801-rodent model of aspects of schizophrenia. *Behav Brain Res.* 2011;222(2):299–308.
100. Deiana S, Watanabe A, Yamasaki Y, et al. MK-801-induced deficits in social recognition in rats: reversal by aripiprazole, but not olanzapine, risperidone, or cannabidiol. *Behav Pharmacol.* 2015;26(8 Specl pt. 3):748–765.
101. Gururajan A, Taylor DA, Malone DT. Cannabidiol and clozapine reverse MK-801-induced deficits in social interaction and hyperactivity in Sprague-Dawley rats. *J Psychopharmacol.* 2012;26(10):1317–1332.
102. Long LE, Chesworth R, Huang X-F, et al. Distinct neurobehavioural effects of cannabidiol in transmembrane domain neuregulin 1 mutant mice. *PLoS One.* 2012;7(4): e34129.
103. Osborne AL, Solowij N, Babic I, Huang X-F, Weston-Green K. Improved social interaction, recognition and working memory with cannabidiol treatment in a prenatal infection (poly I:C) rat model. *Neuropsychopharmacology.* 2017;42(7):1447–1457.
104. Peres FF, Diana MC, Suiama MA, et al. Peripubertal treatment with cannabidiol prevents the emergence of psychosis in an animal model of schizophrenia. *Schizophr Res.* 2016;172(1–3):220–221.
105. Zuardi AW, Morais SL, Guimaraes FS, Mechoulam R. Antipsychotic effect of cannabidiol. *J Clin Psychiatry.* 1995;56(10):485–486.
106. Zuardi AW, Hallak JEC, Dursun SM, et al. Cannabidiol monotherapy for treatment-resistant schizophrenia. *J Psychopharmacol.* 2006;20(5):683–686.
107. Hallak JEC, Machado-de-Sousa JP, Crippa JAS, et al. Performance of schizophrenic patients in the stroop color word test and electrodermal responsiveness after acute administration of cannabidiol (CBD). *Rev Bras Psiquiatr.* 2010;32(1):56–61.
108. Leweke FM, Piomelli D, Pahlisch F, et al. Cannabidiol enhances anandamide signaling and alleviates psychotic symptoms of schizophrenia. *Transl Psychiatry.* 2012;2:e94.
109. Boggs DL, Surti T, Gupta A, et al. The effects of cannabidiol (CBD) on cognition and symptoms in outpatients with chronic schizophrenia a randomized placebo controlled trial. *Psychopharmacology (Berl).* 2018;235(7):1923–1932.
110. McGuire P, Robson P, Cubala WJ, et al. Cannabidiol (CBD) as an adjunctive therapy in schizophrenia: a multicenter randomized controlled trial. *Am J Psychiatry.* 2018;175(3):225–231.
111. Crippa JAS, Derenusson GN, Ferrari TB, et al. Neural basis of anxiolytic effects of cannabidiol (CBD) in generalized social anxiety disorder: a preliminary report. *J Psychopharmacol.* 2011;25(1):121–130.
112. Bergamaschi MM, Queiroz RHC, Chagas MHN, et al. Cannabidiol reduces the anxiety induced by simulated public speaking in treatment-naive social phobia patients. *Neuropsychopharmacology.* 2011;36(6):1219–1226.
113. Shannon S, Lewis N, Lee H, Hughes S. Cannabidiol in anxiety and sleep: a large case series. *Perm J.* 2019;23:18–041.
114. Zuardi A, Crippa J, Dursun S, et al. Cannabidiol was ineffective for manic episode of bipolar affective disorder. *J Psychopharmacol.* 2010;24(1):135–137.

115. Babalonis S, Haney M, Malcolm RJ, et al. Oral cannabidiol does not produce a signal for abuse liability in frequent marijuana smokers. *Drug Alcohol Depend.* 2017;172:9–13.
116. Solowij N, Broyd SJ, Beale C, et al. Therapeutic effects of prolonged cannabidiol treatment on psychological symptoms and cognitive function in regular Cannabis users: a pragmatic open-label clinical trial. *Cannabis Cannabinoid Res.* 2018;3(1):21–34.
117. Crippa JAS, Hallak JEC, Machado-de-Sousa JP, et al. Cannabidiol for the treatment of cannabis withdrawal syndrome: a case report. *J Clin Pharm Ther.* 2013;38(2):162–164.
118. Manini AF, Yiannoulos G, Bergamaschi MM, et al. Safety and pharmacokinetics of oral cannabidiol when administered concomitantly with intravenous fentanyl in humans. *J Addict Med.* 2015;9(3):204–210.
119. Hurd YL, Yoon M, Manini AF, et al. Early phase in the development of cannabidiol as a treatment for addiction: opioid relapse takes initial center stage. *Neurotherapeutics.* 2015;12(4):807–815.
120. Consroe P, Carlini E, Zwicker A, Lacerda L. Interaction of cannabidiol and alcohol in humans. *Psychopharmacology (Berl).* 1979;66(1):45–50.
121. Morgan CJA, Das RK, Joye A, Curran HV, Kamboj SK. Cannabidiol reduces cigarette consumption in tobacco smokers: preliminary findings. *Addict Behav.* 2013; 38(9):2433–2436.
122. Hindocha C, Freeman TP, Grabski M, et al. Cannabidiol reverses attentional bias to cigarette cues in a human experimental model of tobacco withdrawal. *Addiction.* 2018;113(9):1696–1705.
123. Hindocha C, Freeman TP, Grabski M, et al. The effects of cannabidiol on impulsivity and memory during abstinence in cigarette dependent smokers. *Sci Rep.* 2018;8(1):7568.
124. Guimarães F, Chiaretti T, Graeff F, Zuardi A. Antianxiety effect of cannabidiol in the elevated plus-maze. *Psychopharmacology (Berl).* 1990;100(4):558–559.
125. Moreira FA, Aguiar DC, Guimaraes FS. Anxiolytic-like effect of cannabidiol in the rat Vogel conflict test. *Prog Neuropsychopharmacol Biol Psychiatry.* 2006;30(8):1466–1471.
126. Campos AC, de Paula SV, Carvalho MC, et al. Involvement of serotonin-mediated neurotransmission in the dorsal periaqueductal gray matter on cannabidiol chronic effects in panic-like responses in rats. *Psychopharmacology (Berl).* 2013;226(1):13–24.
127. Gomes F, Reis D, Alves F, Corrêa F, Guimarães F, Resstel L. Cannabidiol injected into the bed nucleus of the stria terminalis reduces the expression of contextual fear conditioning via 5-HT_{1A} receptors. *J Psychopharmacol.* 2012;26(1):104–113.
128. Bitencourt RM, Pamplona FA, Takahashi RN. Facilitation of contextual fear memory extinction and anti-anxiogenic effects of AM404 and cannabidiol in conditioned rats. *Eur Neuropsychopharmacol.* 2008;18(12):849–859.
129. Soares VP, Campos AC, VCd B, Zangrossi H, Guimarães FS, Zuardi AW. Intra-dorsal periaqueductal gray administration of cannabidiol blocks panic-like response by activating 5-HT_{1A} receptors. *Behav Brain Res.* 2010;213(2):225–229.
130. Granjeiro ÉM, Gomes FV, Guimarães FS, Corrêa FMA, Resstel LBM. Effects of intracisternal administration of cannabidiol on the cardiovascular and behavioral responses to acute restraint stress. *Pharmacol Biochem Behav.* 2011;99(4):743–748.
131. Resstel LBM, Tavares RF, Lisboa SFS, Joca SRL, Corrêa FMA, Guimarães FS. 5-HT_{1A} receptors are involved in the cannabidiol-induced attenuation of behavioural and cardiovascular responses to acute restraint stress in rats. *Br J Pharmacol.* 2009;156(1):181–188.
132. Fogaça MV, Reis FMCV, Campos AC, Guimarães FS. Effects of intra-prelimbic prefrontal cortex injection of cannabidiol on anxiety-like behavior: involvement of 5HT_{1A} receptors and previous stressful experience. *Eur Neuropsychopharmacol.* 2014;24(3):410–419.
133. Stern CAJ, Gazarini L, Takahashi RN, Guimarães FS, Bertoglio LJ. On disruption of fear memory by reconsolidation blockade: evidence from cannabidiol treatment. *Neuropsychopharmacology.* 2012;37:2132.

134. Twardowschy A, Castiblanco-Urbina MA, Uribe-Marino A, et al. The role of 5-HT1A receptors in the anti-aversive effects of cannabidiol on panic attack-like behaviors evoked in the presence of the wild snake *Epicrates cenchria crassus* (Reptilia, Boidae). *J Psychopharmacol.* 2013;27(12):1149–1159.
135. da Silva JA, Biagioni AF, Almada RC, et al. Dissociation between the panicolytic effect of cannabidiol microinjected into the substantia nigra, pars reticulata, and fear-induced antinociception elicited by bicuculline administration in deep layers of the superior colliculus: the role of CB1-cannabinoid receptor in the ventral mesencephalon. *Eur J Pharmacol.* 2015;758:153–163. Complete.
136. Nardo M, Casarotto PC, Gomes FV, Guimarães FS. Cannabidiol reverses the mCPP-induced increase in marble-burying behavior. *Fundam Clin Pharmacol.* 2014; 28(5):544–550.
137. Linge R, Jiménez-Sánchez L, Campa L, et al. Cannabidiol induces rapid-acting antidepressant-like effects and enhances cortical 5-HT/glutamate neurotransmission: role of 5-HT_{1A} receptors. *Neuropharmacology.* 2016;103:16–26. Complete.
138. Shoval G, Shbiro L, Hershkovitz L, et al. Prohedonic effect of cannabidiol in a rat model of depression. *Neuropsychobiology.* 2016;73(2):123–129.
139. Valvassori SS, Elias G, de Souza B, et al. Effects of cannabidiol on amphetamine-induced oxidative stress generation in an animal model of mania. *J Psychopharmacol.* 2009;25(2):274–280.
140. Ren Y, Whittard J, Higuera-Matas A, Morris CV, Hurd YL. Cannabidiol, a nonpsychotropic component of cannabis, inhibits cue-induced heroin seeking and normalizes discrete mesolimbic neuronal disturbances. *J Neurosci.* 2009;29(47): 14764–14769.
141. Markos JR, Harris HM, Gul W, Elsohly MA, Sufka KJ. Effects of cannabidiol on morphine conditioned place preference in mice. *Planta Med.* 2018;84(4):221–224.
142. Hine B, Torrelio M, Gershon S. Differential effect of cannabinol and cannabidiol on THC-induced responses during abstinence in morphine-dependent rats. *Res Commun Chem Pathol Pharmacol.* 1975;12(1):185–188.
143. Bhargava HN. Effect of some cannabinoids on naloxone-precipitated abstinence in morphine-dependent mice. *Psychopharmacology (Berl).* 1976;49(3):267–270.
144. Chesher GB, Jackson DM. The quasi-morphine withdrawal syndrome: effect of cannabinol, cannabidiol and tetrahydrocannabinol. *Pharmacol Biochem Behav.* 1985;23(1):13–15.
145. Vann RE, Gamage TF, Warner JA, et al. Divergent effects of cannabidiol on the discriminative stimulus and place conditioning effects of Δ'9-tetrahydrocannabinol. *Drug Alcohol Depend.* 2008;94(1–3):191–198.
146. Navarrete F, Aracil-Fernandez A, Manzanares J. Cannabidiol regulates behavioural alterations and gene expression changes induced by spontaneous cannabinoid withdrawal. *Br J Pharmacol.* 2018;175(13):2676–2688.
147. Liput DJ, Hammell DC, Stinchcomb AL, Nixon K. Transdermal delivery of cannabidiol attenuates binge alcohol-induced neurodegeneration in a rodent model of an alcohol use disorder. *Pharmacol Biochem Behav.* 2013;111:120–127.
148. Hamelink C, Hampson A, Wink DA, Eiden LE, Eskay RL. Comparison of cannabidiol, antioxidants, and diuretics in reversing binge ethanol-induced neurotoxicity. *J Pharmacol Exp Ther.* 2005;314(2):780–788.
149. Viudez-Martinez A, Garcia-Gutierrez MS, Navarron CM, et al. Cannabidiol reduces ethanol consumption, motivation and relapse in mice. *Addict Biol.* 2018;23(1):154–164.
150. Gonzalez-Cuevas G, Martin-Fardon R, Kerr TM, et al. Unique treatment potential of cannabidiol for the prevention of relapse to drug use: preclinical proof of principle. *Neuropsychopharmacology.* 2018;43(10):2036–2045.

151. Nona CN, Hendershot CS, Le Foll B. Effects of cannabidiol on alcohol-related outcomes: a review of preclinical and human research. *Exp Clin Psychopharmacol.* 23 May 2019;1–12.
152. Turna J, Syan SK, Frey BN, et al. Cannabidiol as a novel candidate alcohol use disorder pharmacotherapy: a systematic review. *Alcohol Clin Exp Res.* 2019;43(4):550–563.
153. Katsidoni V, Anagnostou I, Panagis G. Cannabidiol inhibits the reward-facilitating effect of morphine: involvement of 5-HT_{1A} receptors in the dorsal raphe nucleus. *Addict Biol.* 2013;18(2):286–296.
154. Mahmud A, Gallant S, Sedki F, D'Cunha T, Shalev U. Effects of an acute cannabidiol treatment on cocaine self-administration and cue-induced cocaine seeking in male rats. *J Psychopharmacol.* 2017;31(1):96–104.
155. Parker LA, Burton P, Sorge RE, Yakiwchuk C, Mechoulam R. Effect of low doses of Δ9-tetrahydrocannabinol and cannabidiol on the extinction of cocaine-induced and amphetamine-induced conditioned place preference learning in rats. *Psychopharmacology (Berl).* 2004;175(3):360–366.
156. Hay GL, Baracz SJ, Everett NA, et al. Cannabidiol treatment reduces the motivation to self-administer methamphetamine and methamphetamine-primed relapse in rats. *J Psychopharmacol.* 2018;32(12):1369–1378.

CHAPTER THREE

Stress, psychiatric disorders, molecular targets, and more

Fatin Atrooz[a], Hesong Liu[b], Samina Salim[a,*]
[a]Department of Pharmacological and Pharmaceutical Sciences, College of Pharmacy, University of Houston, Houston, TX, United States
[b]Baylor College of Medicine, Houston, TX, United States
*Corresponding author: e-mail address: ssalim@uh.edu

Contents

Abstract

Mental health is central to normal health outcomes. A widely accepted theory is that chronic persistent stress during adulthood as well as during early life triggers onset of neuropsychiatric ailments. However, questions related to how that occurs, and why are some individuals resistant to stress while others are not, remain unanswered. An integrated, multisystemic stress response involving neuroinflammatory, neuroendocrine, epigenetic and metabolic cascades have been suggested to have causative links. Several theories have been proposed over the years to conceptualize this link including the cytokine hypothesis, the endocrine hypothesis, the oxidative stress hypothesis and the oxido-neuroinflammation hypothesis. The data discussed in this review describes potential biochemical basis of the link between stress, and stress-induced neuronal, behavioral and emotional deficits, providing insights into potentially novel drug targets.

Progress in Molecular Biology and Translational Science, Volume 167
ISSN 1877-1173
https://doi.org/10.1016/bs.pmbts.2019.06.006

1. Introduction

Maintenance of body homeostasis is critical for overall wellbeing of man. Any external or internal stressful stimuli, which disrupts body homeostasis, might lead to a state of allostasis. Allostasis includes all stress response mediators that work together to reestablish the steady state. However, prolonged activation or insufficient activation of stress response mediators lead to allostatic load or overload, a state of wear and tear, which can elicit adverse physiological, behavioral and emotional outcomes. Stressors may be physical or psychological. Physical injury, hunger, or thirst, are examples of physical stress while psychological stress includes fear, anger, disappointment, traumatic encounters, etc.[1–3] The physical and/or psychological stressors can have severe health outcomes especially when the stress is inflicted during developmental stages of early life, as the key components of the adaptive stress response system, including the autonomic nervous system (ANA) and the hypothalamic-pituitary-adrenal (HPA) axis, critical for stress-coping mechanisms, are not fully developed during early life.[4,5] In fact, the period between prenatal to the late adolescence phase is considered as critical period for formation and development of stress response system.[6] Furthermore, postnatal brain development including cortical gray matter expansion/regression and white matter development are important for the optimization of neuronal connectivity and maturation of neurobehavioral functions.[7] Stress induces structural changes in dendritic branching and synapse density and stability leading to disruption in neural connectivity in critical brain regions. Disruption of functional connectivity during early life, in brain regions that regulate stress response mediators and emotional and behavioral functions including prefrontal cortex, hippocampus and amygdala increases the sensitivity to future stressful events.[8–10] Additionally, early life stress may induce alteration in neuroendocrine and autonomic stress response matrix, rendering the individual vulnerable to stress-related disorders.[11,12] Moreover, elevation of stress hormones following stress exposure may induce inflammation, oxidative stress, and modify gene expression profile.[13–17] However, not everyone who experiences early life adverse events develops stress-related disorders later in life, in fact, successive adaptation following early life stress might promote resiliency. Interaction between genetic factors and environment, epigenetic programming of stress response mediators, and the presence of a window of sensitive period during development are suggested as potential factors that determine the vulnerability or resiliency to early life stress-related disorders.[18,19] Recent research indicates that both early life environment and genetic predispositions act as co-determinants which predict the response to stress, and also can be indicative of negative behavioral outcomes.[18] Here, we review the existing literature on the behavioral and biochemical consequences of early life stress. A discussion of potential biochemical basis of the link between stress, and stress-induced neuronal, behavioral and emotional deficits, also is provided along with current hypotheses and some insights into potential novel drug targets.

2. Early life stress

Early life stress has been linked to later life physical and mental ailments, including diabetes and cardiovascular disease, neurodegenerative disease, anxiety and mood disorders, and also early death.[20] Children subjected to stressful events are reported to exhibit behavioral and cognitive deficits later in life, often leading to post-traumatic stress disorder (PTSD) and major depression.[21] Many studies have suggested an association between the risk of developing psychopathologies during adult life with a history of early life abuse or neglect.[22,23] According to the Adverse Childhood Experiences (ACE) study conducted by the Center of Disease Control and Prevention, adverse childhood experiences including living in a stressful environment of violence, crime, abuse, poverty, and neglect, are strongly associated with the occurrence of physiological and psychological diseases later in life.

Exposure to violence is the most common source of stress for children. Approximately, one-quarter of children experience some forms of early life trauma.[24] Different forms of early life stress and their association with later life adverse health outcomes have been comprehensively studied. According to a population-based study conducted in developing countries, 5–35% of children were reported to experience physical abuse, 5–30% experienced sexual abuse, and 10–20% witnessed domestic violence.[25] A United States nationwide study indicated that 60.6% of children and youth experienced physical trauma or witnessed some types of traumatic events. 46.3% of these young victims were physically abused, 10.2% of them experienced child maltreatment, 6.1% were sexually abused, and 25.3% experienced indirect trauma.[26] Similarly, a study from the United Kingdom reported that approximately 20% of children experienced severe maltreatment. One in three children experienced sexual abuse. Additionally, a strong correlation between early life traumatic experiences and poor emotional wellbeing in later life was also observed.[27] A retrospective study conducted in 21 countries by The World Health Organization reported that 5–11% respondents reported a history of adverse experiences early in their life, 1–2% respondents recalled childhood experience of sexual abuse, and 4–8% reported violent household environment.[28] Patients with symptoms of mental disorders such as schizophrenia and autism reported a history of early life adversities including abuse, neglect, among others.[29,30] Specifically, loss of a family member during childhood increases the risk for major depression in adulthood.[29] Victims of suicide had higher rates of childhood physical and sexual abuse. Additionally, there is an established association between borderline personality disorders and early life sexual abuse.[31,32] In summary, the prevalence of early life traumatic events are worldwide and the strong associations between different forms of trauma and occurrence of later life negative health conditions is well described. Interestingly, not all individuals exposed to early life stress develop adverse consequences. In addition, some individuals remain resilience when exposed to additional stressors in adulthood.[33,34] The variation in outcomes in the effect of early life stress might due to presence of critical period throughout brain development with increased brain plasticity during which trauma induce strong persistent effects, in addition to predisposed genetic factors and epigenetic modulation as discussed below.[18]

3. Early life stress and brain development

Brain development starts prenatally and continues postnatally in all mammals. Postnatal brain development is characterized by cortical gray matter expansion/regression and white matter development. Gray matter maturation is indicated by a massive increase of synapse (synaptogenesis), which reaches approximately 50% greater than the level seen in adults by age of 2 years. This massive increase in synapse is followed by rapid elimination and refinement of excess synapses (synaptic pruning) which continues until adolescent stage. However, synaptogenesis and synaptic pruning are region dependent which reflects the brain maturation trajectories. For instance, synaptic density in prefrontal cortex peaks at childhood (2–4 years old), compared to 6–8 months of age in the primary visual cortex.[35] White matter development is indicated by neuronal myelination which starts prenatally and continues during early life stages (childhood and adolescence). Interestingly, myelination of axons in some regions such as temporal lobe and arcuate fasciculus within the cortex continues until adult stage.[35] These processes are important for optimization of neuronal connectivity and maturation of neurobehavioral functions.[7] Notably, the increase in brain maturation is associated with the increase in functional and connectivity between brain regions. The variation in the rate of brain region maturation and the extent of plasticity throughout postnatal brain development includes brain regions that regulate stress and emotion, namely, the prefrontal cortex, hippocampus, and amygdala.[18] Interestingly, these brain regions are also implicated in neuroendocrine and autonomic stress responses, which explains the variation in emotional and behavior responses to stress across development. The processes of postnatal brain maturation increase brain plasticity in temporal and regional dependent manner. During such periods, experiences are necessary for normal brain development and function; however, adverse experiences during this period might induce adverse persistent effects. The concept of existence of sensitive periods throuhout brain development which is characterized by increased neuroplasticity mechanisms, renders the brain amenable to environmental changes.[36–38] Therfore, encountering adverse experiences during critical periods of early life brain development has two probable courses. Exposure to adverse experiences might negatively influence the developmental trajectory and predispose the individual to later psychiatric diseases; or, exposure to favorable conditions or enriching environment might help the brain to adaptively cope with stress, thus developing resistance to future stressful events.[18,19]

It is reported that adverse experiences during brain developmental period disrupts the maturation of neuronal circuit, leading to the formation of pathological neuronal circuit and emergence of psychiatric disorders later in life.[39] This hypothesis was further supported by several preclinical and clinical studies. In animal models of early life stress, maternal separation or low maternal care at early life caused structural and functional changes in brain regions that are implicated in regulation of neuroendocrine, autonomic, and immune systems. These brain regions that are mostly affected by early life stress also are implicated in vigilance, emotional response, and fear conditioning. Changes in neural connectivity and function in these brain regions increased physiological and behavioral responses to subsequent stressors.[18] Moreover, rats raised in low maternal care environment during first 10 days of their life showed enhanced stress response and impaired behavioral and psychological response to restraint stress at adulthood when compared with rats raised by highly nurturing mothers at early life.[40] In monkeys, social isolation during the first 3 months after birth were reported to have long-lasting deficits in their social behavior.[41] Clearly, early environment has a critical role in neurobiological development and regulation of behavioral and cognitive functions.

In humans, adults with adverse early life experiences exhibit increased neuroendocrine and autonomic response to laboratory psychosocial stress, increased levels of inflammation, and glucocorticoid resistance.[17,42,43] Exposure to early life stress also impacts the key brain regions implicated in stress and depression, adults with early life adversity showed reduction in volume of medial prefrontal cortex, orbital-frontal cortical, or hippocampus. On the other hand, early life stress was associated with an increase of amygdala volume and enhancement of amygdala activity in response to emotional cue.[44–50] Thus, neurobiological changes induced by early life stress may reflect the risk for development of depression and other stress-related disorders later in life.

4. Early life stress and the hypothalamic-pituitary-adrenal (HPA) axis

Stress response system is one of the most widely studied and well-understood systems. Briefly, during stressful stimuli the brain triggers adaptive responses that are proportional to the nature of stressor. For example, physical stressors trigger the brainstem catecholaminergic neurons, which activates the autonomic nervous system leading to rapid increase in heart rate and blood pressure owing to the increase of the sympathetic tone. Shortly, activation of sympathetic system is weakened by the parasympathetic tone. Physical stressors also activate the HPA axis indicated by the elevation of systemic cortisol levels. On the other hand, psychological stressors trigger the limbic circuits of forebrain which coordinates, monitors, and calibrates the behavioral and the physiological responses to cope with stressor's based on prior experience or innate programs.[51–53] The locus coeruleus/norepinephrine (LC/NE)-sympathetic nervous system, the corticotropin-releasing hormone (CRH), and arginine-vasopressin (AVP) are the central components of the stress response system. The LC catecholaminergic cell groups are located in the medulla and pons of the brainstem. Upon activation, NE is released throughout the brain leading to enhanced arousal and vigilance, in addition to cardiovascular system excitation. The CRH and AVP are released by parvocellular neurons of the paraventricular nuclei (PVN) of the hypothalamus. The CRH releasing neurons are also found in the paragigantocellular and parabranchial nuclei of the medulla. Importantly, CRH and NE stimulate the secretion of each other through CRH receptors and $\alpha 1$-adrenergic receptors, respectively.[6,54]

Activation of HPA axis as a response to stress is triggered by hypothalamic secretion of CRH and AVP neuropeptides into the hypophyseal portal system which access the anterior pituitary gland, upon binding to CRH receptors (CRH-R1) of the pituitary corticotropes. CRH with synergetic effect of AVP stimulates the synthesis of adrenocorticotropic hormone (ACTH). The adrenal cortex is the principal target of ACTH, the latter regulates synthesis of glucocorticoids by zona fasciculata region of adrenal cortex.[54,55] Glucocorticoids (cortisol in humans and corticosterone in rodents), through their ubiquitously distributed receptors, regulate variable physiological and behavioral responses. Glucocorticoids play a principal role in regulation of duration and magnitude of HPA axis activation, glucocorticoids negative feedback of HPA axis is mediated by direct reduction of PVN neuronal activity via glucocorticoids receptors (GRs) that are highly expressed on hypophysiotropic neurons or indirectly via the limbic structures.[55–57] In fact, the limbic structures of forebrain including the prefrontal cortex, hippocampus, and amygdala have essential role in stress regulation. In addition to receiving sensory and memory associated information from higher order brain regions, they also receive inputs from areas in the brainstem that are involved in arousal and attention such as locus coeruleus and raphe nuclei. The limbic regions work in parallel and the outputs they generate converge on subcortical regions allowing top-down regulation of HPA axis and autonomic system responses to threatening stressors.[52,58,59]

Importantly, the LC/NE catecholaminergic neurons and the PVN CRH neurons of HPA axis send their projections to the mesocortical and mesolimbic dopaminergic system, hence, these systems are activated by catecholamines, and glucocorticoids during stress. The mesocortical system includes dopaminergic neurons that project from the ventral tegmentum to the prefrontal cortex area and is involved in anticipatory and cognitive functions. Activation of these neurons most likely suppresses the stress system response.[60] The ventral tegmentum dopaminergic neurons also send their projections to nucleus accumbens forming the mesolimbic system, the mesolimbic neurons are considered the central dopaminergic reward system, they are implicated in motivation and reinforcement behavior; therefore, it has been suggested that mesocorticolimbic system is implicated in euphoria and dysphoria.[61]

Notably, in addition to activation by inner emotional stressors, the limbic structures of amygdala and hippocampus are also activated by LC/NE ascending neurons during stress. Activation of amygdaloid nuclei is central for emotional analysis of any given stressor through retrieval of all relevant stored information. It is well known that amygdala has stimulatory effect on central stress system as well as the mesocorticolimbic dopaminergic system. Through the CRH neurons of the central amygdaloid nucleus, which send their projections to the PVN of hypothalamus and to the parabrachial nucleus of brain stem, the amygdala induces neuroendocrine, autonomic and behavioral responses to emotional stressors. On the other, the hippocampus has inhibitory effect on amygdala activity and the neuroendocrine stress response system.[54,61,62]

Interestingly, the LC/NE-catecholaminergic neurons of brain stem and the CRH/AVP neurons of hypothalamus have reciprocal connection with the proopiomelanocortin (POMC) producing neurons of the arcuate nucleus of hypothalamus. The neuroendocrine stress system stimulates POMC-neuron's secretion of opioid peptides including α-melanocyte-stimulating hormone and β-endorphin, which mediate feedback inhibition of both components of neuroendocrine stress response system. Additionally, the POMC-neurons promote analgesic effect by inhibition of ascending pain pathways through their projections to hindbrain and spinal cord.[54,63]

Stress response mediators have protective effect in cases of acute stress, they are adaptive responses that are rapidly activated and terminated. Indeed, exposure to stressful experience can be beneficial if the individual meets the demands of the stressors. Successful coping with threatening stressors can induce growth, adaptation, learning, and good health, promoting resiliency to future stressful events. However, overactivation or insufficient activity of stress response system, for example, in cases of chronic or repeated stress, can lead to immune system related health issues, cardiovascular related diseases, metabolic related diseases such as diabetes and obesity.[64–67] Moreover, in appropriate activation or termination of stress system as a response to stressful life events such as loss of love ones or traumatic incidents can lead to development of mental health issues such as mood and anxiety disorders, in addition to PTSD.[68,69]

Prenatal, infancy, childhood, and adolescence are sensitive periods to environmental changes, and exposure to stress during these critical periods can induce prolonged detrimental effect on the adaptive stress response system that might last for a lifetime. In humans, many studies addressed the psychological sequelae of exposure to stress during early life. Exposure to chronic or intense stress during childhood or adolescence increased the risk of psychological deficits such as, anxiety and mood disorders, PTSD, antisocial behavior, general distress, and learned helplessness.[70–73] There is evidence suggesting that early life stress can lead to modifications in the

HPA axis. Since the HPA axis also undergoes early development, the changes in the function of HPA axis persist into adulthood and increase the susceptibility of an individual to depression.[74] The effects are mostly mediated by stress hormones such as CRH and glucocorticoids which have substantial effects on the brain response to stress.[75] For instance, it has been suggested that exposure to stress during the stress hyporesponsive period in human infants (6 months to 12 months old) leads to elevation of basal cortisol level turning the HPA axis responsive to future stressors.[4] Relevant to this, adults with depressive disorders who have a history of childhood adverse experience show alterations in the stress response system.[17,22,76,77] Under stressful circumstances, children show prolonged elevation of cortisol. Both clinical and preclinical studies showed that long-term elevation in cortisol levels can affect the function and architecture of neural systems that are essential for learning and memory function.[78] For example, the brain-derived neurotrophic factor (BDNF), regulates neural plasticity, synaptic integrity, and neurogenesis, it has critical role in brain development.[79,80] Early life stress in humans can reduce BDNF levels at adulthood, in association with memory impairment.[81,82] In parallel, preclinical studies of early life stress in rodents reported that maternal separation decreases the expression levels of BDNF and glucocorticoid receptors in the hippocampus, these phenotypes together alter stress regulation mechanism and the synaptic integrity in the developing brain increasing vulnerability to stress-related disorders later in life.[83] Meaney et al.[84] found that rats raised by nurturing mothers have increased levels of central serotonin activity and hence increased expression of glucocorticoid receptor genes in the brain, when compared with rats that were raised with less nurturing mothers. Increased expression of glucocorticoid receptors in the limbic brain regions improved glucocorticoid feedback following stress exposure, and improved regulation of stress response system throughout the lifespan of laboratory rats. Clearly, early environment has a critical role in neurobiological development and regulation of behavioral and cognitive functions.

Prenatal stress is also linked to affective and emotional disorders later in life.[85] Furthermore, children born to mothers with high anxiety symptoms during pregnancy showed high rates of emotional problems at childhood, and others reported depressive symptoms at adolescence.[86–89] The underling mechanism might involve the stress hormone that cross the placenta and affect brain development, development of HPA axis and limbic brain regions are mostly affected by maternal anxiety and stress during pregnancy.[90]

5. Early life stress, epigenetic and genetic factors

Interestingly, recent studies have shown that genetic background also plays an important role in mediating the effect of early life stress. The Gene × Environment theory of early life stress demonstrated that during development, the predisposing genetic factors interact with environmental variables to promote vulnerability or resiliency to stress-related disorders such as depression and anxiety.[91–94] Brain imaging in humans showed that carriers for the *Bdnf* variant (Val66Met) have smaller hippocampus gray matter and poor cognitive performance compared to non-carriers.[95–98] Moreover, *Bdnf* Met carriers who reported early life adverse experiences showed reduced gray matter in hippocampal and associated lateral prefrontal cortex, structural remodeling was associated with elevation in body arousal and temperamental trait, predicting higher anxiety and depression symptoms.[99]

Exposure to early life trauma is associated with increased activity of CRH system during development, which further increases the risk of later life affective disorders.[100,101] Single-nucleotide polymorphisms within *CRHR1* gene is suggested as one of the candidate genes that promotes adult vulnerability to depression due to childhood exposure to stressful life event.[102,103] Remarkably, other *CRHR1* gene single-nucleotide polymorphisms (SNPs) were found to have protective effect against development of depression in adults following childhood adverse experience.[104] In a Chinese cohort study, SNPs at three loci in *CRHR1* gene are proposed to promote the response to antidepressant treatment in patients with major depression.[105] Another candidate gene that modifies vulnerability to stress-related psychiatric disorders is the glucocorticoid receptor co-chaperone (FKBP5). FKBP5 plays a key role in the regulation of glucocorticoid receptor activity by participating in the negative feedback mechanism. Binding of FKBP5 to the GR-hsp90 heterocomplex, lowers the affinity of cortisol to GR and reduce the efficiency of GR nuclear translocation, hence, lowering GR sensitivity.[106] Indeed, increased HPA axis activity and elevated GR sensitivity are consistently reported in patients with PTSD, bipolar disorder, and major depression.[24,106–108] Polymorphisms in *FKBP5* gene differentially modify GR sensitivity. In healthy individuals, alleles that enhance expression of FKBP5 following GR activation increased GR resistance and decreased negative feedback of glucocorticoids leading to prolonged activation of stress hormone system following stress exposure.[109,110] Remarkably, while some *FKBP5* variants are associated with increased recurrence of depressive episodes, they are also related to the faster response to antidepressant treatment in depressed individuals,[111] suggesting FKBP5 as a potential therapeutic target for treatment of depression disorders. Moreover, a cross-sectional study examining genetic and psychological risk factors, reported that four *FKBP5* SNPs interacted with childhood abuse increasing the risk of development of PTSD at adult life.[109] The role of gene-environment interaction in modifying vulnerability to depression has been reported in individuals with polymorphism in the promoter region of the serotonin transporter (*5-HTT*) gene. Individuals with one or two copies of the short (S) allele of the *5-HTT* showed increased susceptibility to depression and suicidality in relation to stressful life events than individuals homozygous for the long allele.[112]

Interestingly, it has been proposed that prenatal or early life stress exposure has a substantial influence on mental health later in life, most probably through alteration of epigenetic regulation of gene expression in the brain.[113] Epigenetic mechanisms including the level of non-coding RNA, DNA methylation, and histone modifications might mediate vulnerability or resilience to stress-related psychiatric disorders. DNA methylation at the CpG region of the promoters inhibits gene transcription by affecting the interaction of DNA with transcription factors and chromatin proteins. Indeed, many clinical and preclinical studies demonstrated that stress affects DNA methylation. For instance, acute psychological stress in humans induced DNA methylation of oxytocin receptor (*Oxtr*) gene which involved in endocrine regulation.[114] Additionally, patients with major depression showed distinguished DNA methylation profile of CpG region of *Bdnf* promoter when compared to healthy controls, indicating that DNA methylation profile of *Bdnf* can be a reliable diagnostic biomarker of major depression.[115,116] In rodent models of chronic stress, exposure to 10 days of social defeat stress induced long-lasting demethylation of *Crh* gene in the PVN of mice that exhibited social avoidance phenotype but not in resilient mice. Interestingly, site specific knockdown of *Crh* significantly attenuated the stress induced social avoidance behavior, these data suggested that functional DNA methylation of *Crh* potentially promotes resiliency to social stress in adult mice.[117] The interaction between environment and epigenetic modification of the expression of stress response genes was demonstrated by the comparative study between rats received high maternal care and rats received less maternal care during the first 10 days of their life and subsequent measuring their stress response system at adulthood. Offspring received less maternal care at early life showed increased levels of

CRHRs in amygdala in addition to impaired behavioral and psychological response to restraint stress at adulthood. In contrast, rats that received high maternal care at early life showed reduced HPA axis response to restraint stress at adulthood, increased density of GR in hippocampus, and decreased mRNA level of CRH in hypothalamus.[118–120] Interestingly, evidence of epigenetic reprograming of stress response genes were demonstrated by these studies; the increased levels of GR mRNA in hippocampus of rats that received high maternal care were related to low DNA methylation levels at exon 1_7 of the promoter region of *GR* gene, compared to high DNA methylation level at the same region of *GR* promoter in rats received less maternal care.[40] Variation in the DNA methylation of *GR* promoter region were also found in adopted offspring who received distinctive maternal care, providing evidence of non-genome transmission of maternal behavior on adopted offspring through epigenetic mechanisms.[40] Early life stress in rodents caused long-lasting increase in DNA methylation of *Bdnf* gene associated with decreased BDNF mRNA expression in the prefrontal cortex throughout the adulthood.[121] In another preclinical study, mice exposed to stress during the first 10 days of life showed high basal corticosterone level, and hypersecretion of corticosterone in response to acute stress at age of postnatal (PND) 42, PND90, and 1 year. Hypomethylation of the enhancer region of the *AVP* gene associated with high expression of the gene were observed in the hypothalamic PVN of mice exposed to early life stress.[122] Histone acetylation was also increased in the promoter region of *GR* gene in hippocampus of rats received high maternal care as compared to rats received less maternal care. Increased histone acetylation enhanced the binding of the transcription factor (NGFI-A) to the promoter region and hence, increased *GR* gene expression.[40] The above studies demonstrated the role of early life stress in modification of stress response system later in life potentially via epigenetic mechanisms.[113]

It is well known that small non-coding RNAs are involved in many regulatory mechanisms of gene expression. They also have a key role in regulation of synaptic plasticity.[123–125] A growing body of evidence suggests that several types of miRs (a type of non-coding RNA) are implicated in vulnerability to stress-related neuropsychiatric disorders such as depression and suicidal tendency.[126–128] Both acute and chronic stress paradigms in rats lead to upregulation of miR-34c in the amygdala, remarkably, one of miR-34c targets is *CRHR1* mRNA (the central component of HPA axis). In vitro studies demonstrated that miR-34c treatment reduces neuronal cell responsiveness to CRH, suggesting the significant role of miR-34c in regulation of stress response system.[129] In another study, acute stress in rats upregulated expression of miR-134 and miR-183 in amygdala, while chronic stress downregulated expression of miR-134 in both amygdala and hippocampus, interestingly, the changes in miRs expression were linked to the changes in alternative splicing of acetylcholinesterase. Indeed, miR-134 and miR-183 target the serine/arginine rich splicing factor 2(SC35), the later is upregulated in response to stress regulating the cholinergic neurotransmission in the brain.[130] Moreover, miRs are involved in regulation of expression of both types of glucocorticoid receptors; MRs and GRs in the brain. In chronically stressed rats, miR-18 regulates the expression of GRs in the PVN, and miR-124 and miR-136a might regulate the expression of MRs. Interestingly, glucocorticoids treatment induced the expression of miR-124, suggesting its involvement in feedback mechanisms that mediate neuronal plasticity.[131–134] Importantly, it has been proposed that exposure to early life stress leads to changes in miRs expression in the brain, which are associated with persistent alteration of expression of genes involved in stress response mechanisms increasing the risk for development of neuropsychiatric disorders later in life.[124,135] Early life maternal separation in rats induced long lasting upregulation of the neuronal plasticity regulators; miR-132, and miR-124 in the medial prefrontal cortex, suggesting persistent epigenetic regulation of neuronal plasticity in medial prefrontal cortex following stress exposure at early life.[135] The above data suggested microRNA as a therapeutic target in stress-related psychiatric disorders. Interestingly, circulating miRs profile are projected as a reliable biomarkers for identification and characterization of the broad phenotype of depression disorders, which might facilitate development of personalized approaches for depression treatment.[136]

6. Early life stress and the immune system

Similar to the central nervous system, the immune system undergoes postnatal developmental processes, which continue throughout childhood to adolescent period. The initial evidence that supports the association between early life stress and immune system function emerged from the observation that handling rats before weaning reduces the development of a transplanted tumor.[137] After this, emerging research in the developmental psychoneuroimmunology field started to pay attention to the association between early life stress and immune functions in later life. In both non-human primates and rodents, elevated production of pro-inflammatory cytokines in plasma and increased activity of macrophages were observed in animals that were exposed to maternal separation.[138–140] A longitudinal study with children in England and Wales showed that inflammatory biomarkers in blood were elevated in children who were maltreated and showed symptoms of depression by age 12 years when compared to controls.[141] Studies with patients with a history of early life childhood maltreatment reported that childhood maltreatment was associated with higher rate of major depression and increased levels of IL-6 and NF-κB DNA binding.[142,143]

As noted from above studies, early life stress is associated with elevated levels of inflammatory markers at adulthood, the critical question is how early experiences have such lasting consequences on inflammatory response? The interaction of stress system with immune system is complex and bidirectional. Activation of stress response system affects trafficking and function of leukocytes and suppress the production of proinflammatory cytokines, namely, TNF-α, IL-6, IL-1, IL-8, and IL-12. In fact, glucocorticoids are well known for their anti-inflammatory and immunosuppressive effects, they have been widely used for treatment of autoimmune diseases. On the other hand, proinflammatory cytokines mainly, TNF-α, IL-1, and IL-6 have stimulatory effect on the HPA axis leading to enhanced secretion of glucocorticoids, in return, glucocorticoids suppress the inflammatory reaction, providing an important negative feedback mechanism to protect the individual from inflammatory response overshoot.[75] Although glucocorticoids generally show inhibitory effect on inflammation, the persistence elevation of glucocorticoid levels can decrease the immune cells' response to glucocorticoids. When the immune cells become insensitive to glucocorticoids, production of proinflammatory cytokines increased as a response to environmental stimuli, enhancing inflammation. Epigenetic regulation of many genes throughout the genome, including genes of the immune system, has been proposed as a mechanism for long lasting effects of adverse experiences during early life. For example, adults who reported early life adversity showed genome-wide changes in DNA methylation, which further correlated with changes in inflammatory response of peripheral blood mononuclear cells to pathogen stimulation.[144] One of the hypothesized mechanisms by which inflammation may affect brain function could involve BDNF modulation. Many preclinical studies showed that proinflammatory cytokines cause significant reduction in mRNA and protein levels of BDNF in several brain regions including prefrontal cortex and hippocampus, among others.[145–147] Given the role of BDNF as an important mediator of neuroplasticity, neurogenesis, and synaptic integrity, neuroinflammation during critical period of brain development might be the underlying mechanisms for the later onset of stress-related mental disorders.

7. Oxidative stress and oxido-inflammation theory of stress

Oxidative stress is caused by an imbalance between the levels of reactive oxygen species (ROS) and the capacity of antioxidant systems to scavenge and detoxify ROS. The oxygen we breathe in plays an important role in generating energy in the form of ATP. Mitochondrial electron transport chain ($O_2 + 4e^- + 4H^+ \rightarrow 2H_2O$) drives the production of ATP, which makes the mitochondria a "powerhouse" for producing ATP from oxygen.[148] This process of ATP generation is accompanied with free radicals as side products; the oxygen can directly generate superoxide anion ($O_2^{\dot{-}}$), which can subsequently form hydrogen peroxide (H_2O_2) and hydroxyl radical ($HO^{\bullet}$). At physiologically low to moderate levels, ROS play significant roles in performing different cellular functions and are critical for optimum cellular health. For instance, ROS play an essential role in long-term potentiation (LTP) which is a form of synaptic plasticity considered essential for learning and memory.[149] ROS cause activation of mitogen-activated protein kinases (MAPKs), protein kinases that regulate cellular signal transduction.[150] Moreover, ROS are also a critical part of cellular immune response as they lead to activation of proinflammatory cytokines such as IL-1β and TNF-α.[151,152] Oxidants such as H_2O_2 oxidize proteins thereby modifying their structure and function which then regulate transcription of specific genes or activation of specific signaling cascades promoting normal cellular proliferation, maturation or controlled degradation.[153,154] However, elevation of ROS above the physiological level modifies critical components of the cells causing deleterious effects as discussed below.

To fight against the deleterious effects of ROS and to maintain cellular homeostasis, the cells have developed endogenous antioxidant mechanisms which maintain physiological level of ROS. Once the levels of ROS increase to the extent that the equilibrium between ROS and antioxidants is broken, a state of oxidative stress is produced. Oxidative stress leads to modifications in cellular macromolecules such as lipids, proteins, and DNA molecules. The lipids and lipoproteins are abundantly present in the cell membranes, making them vulnerable to free radical-mediated lipid peroxidation, this process generates toxic products such as malondialdehyde.[155] Free radical-mediated protein oxidation modifies protein structure and alters their activity, whereas DNA oxidation leads to breakage in DNA strands and increases mutation rate.[156] Homologous DNA repair after DNA damage has been reported to change DNA methylation patterns at the repaired sites. Therefore, oxidative stress induced by early life stress may induce DNA breakage, the subsequent DNA repair process modifies methylation pattern of the gene altering the expression of critical genes. Oxidative stress-mediated injury is one of the primary underlying mechanisms for several diseases involving; respiratory, renal, and cardiovascular diseases, as well as metabolic dysfunction related diseases such as diabetes, obesity, and liver diseases.[157–159]

The brain is considered highly vulnerable to oxidative stress. The reasons for this consideration are the following. The brain has high-energy demand, so there is high oxygen consumption in the brain, resulting in excessive ROS generation. Neuronal membranes are rich in polyunsaturated fatty acids, which are the preferred target for ROS-induced damage. The neurotransmitter, glutamate, can cause oxidative stress. Often generation of excessive intracellular Ca^{2+} may contribute to heightened oxidative stress. The antioxidant system in the CNS is not very robust with low levels of catalase, glutathione peroxidase, and vitamin E. Since the brain is comparatively more vulnerable to oxidative damage, oxidative stress might be a common underlying mechanism implicating many signaling cascades. In fact, many clinical studies showed a high correlation between oxidative stress and psychiatric disorders.[160–163] Increased oxidative stress markers such as NO metabolites, isoprostanes, and DNA damage were observed in patients with schizophrenia.[164–166] Depressed patients showed an increase in peripheral oxidative stress indicated by an increase in the lipid peroxidation marker, malondialdehyde (MDA), and a decrease in antioxidant enzymes in serum, as compared to age-matched control group.[167] Anxiety disorders in humans were also associated with elevation of lipid peroxidation products and alteration of antioxidant enzymes.[160] Additionally, children and adolescents with anxiety disorders showed plasma elevation of oxidative stress markers.[168] Moreover, decreased antioxidants such as Glutathione, SOD, and catalase were observed in patients with mental disorders.[169–171] Interestingly, clinical trials with antioxidant treatment for schizophrenia and other mental disorders showed that antioxidants improve negative symptoms and reverse the decrease in antioxidant.[172–174] Preclinical studies have suggested that oxidative stress plays a significant role in the development of several behavioral deficits including anxiety- and depression-like behaviors.[175] Several lines of studies prove this link. Pharmacologically-induced oxidative stress with L-Buthionine-(*S*, *R*)-sulfoximine (BSO) treatment increased anxiety-like behavior in rats.[15,176,177] Treatment with the antioxidant tempol reduced anxiety-like behavior caused by multiple anxiogenic drugs in rat.[178] Other studies also showed that psychological stress induced by chronic social defeat led to elevated oxidative stress as well as behavioral deficits in rats.[179,180]

Interestingly, many studies found that the increase in oxidative stress induces an inflammatory response suggesting the link between oxidative stress and inflammatory pathways. Oxidative stress induces the release of proinflammatory cytokines such as TNF-α and IL-6. Other mediators of inflammation such as nuclear factor-kappa B (NF-κB), vascular cell adhesion molecule-1 (VCAM-1) are also activated by ROS. This inflammatory response is an underlying causal mechanism for various diseases such as diabetes, chronic renal failure, hypertension, chronic obstructive pulmonary diseases (COPD) and rheumatoid arthritis, as well as neurodegenerative disorders like Parkinson's disease and multiple sclerosis.[181–185] A study conducted in epithelial cells suggested that pro-oxidants activate NF-κB pathway.[186] Activation of NF-κB pathway contributes to increased expression of proinflammatory cytokines.[187] And, antioxidant treatment suppressed NF-κB pathway and cytokine production.[188] Many redox-sensitive transcription factors such as nuclear factor erythroid 2 (NFE2)-related factor 2 (Nrf2) which induced antioxidants genes also have strong anti-inflammatory functions.[189] In addition, Nrf2 knockout mice have increased the risk of developing inflammatory lesions.[190] These data suggest that there is a crosstalk between Nrf2 regulated antioxidative pathway and NF-κB regulated proinflammatory pathway. The depressive phenotype is reported to be associated with a decrease in Nrf2 and an increase in NF-κB pathway. This is an interesting relationship. It seems that early life stress induces an initial rise in oxidative stress, which causes activation of both redox-sensitive transcription factors, Nrf2 and NF-κB; however, one dominates over the other and NF-κB with its pro-damage effects supersedes the pro-survival act of Nrf2. In summary, it clear that stressful stimuli at early life and also during adult stages of life cause increase in oxidative stress and inflammation,[191–199] and whether an individual will be resistant or susceptible to the stressful stimuli may be determined by the oxido-inflammation balance.

8. Potential drug targets

The regulation of cellular responses usually involves activation of transcription factors and their binding to the regulatory regions of target genes. Transcription factors that are in the cytoplasm are considered to reside in an inactive state. Once activated by signals such as phosphorylation, transcription factors translocate into the nucleus and promote transcription of target genes.[200] Redox-sensitive transcription factors such as nuclear factor erythroid 2 (NFE2)-related factor 2 (Nrf2) and nuclear factor kappa-light-chain-enhancer of activated B cells (NF-κB) have become the focus of research as they are being regarded as future therapeutic targets for various conditions. Nrf2 has a significant role in antioxidant defense. It is a transcription factor that binds to the promoter region of antioxidant response elements (ARE) genes, which regulate the expression of a group of antioxidants, such as superoxide dismutase (SOD), glutathione peroxidase (GPx), glutathione reductase, heme oxygenase-1 (OH-1), and other non-enzymatic antioxidants. Under normal conditions, Keap1 is constitutively active and interacts with Nrf2, promoting its Ubiquitination by Cullin 3 and subsequent proteasomal degradation. During oxidative stress, Nrf2/Keap1/Cullin3 interaction is disrupted, so Nrf2 can translocate into the nucleus and activate the expression of a large group of antioxidants through interaction with the promoter regions of ARE genes. If Keap1/Nrf2 pathway is disrupted, the antioxidative capacity of the cell will be dampened, leading to increased oxidative stress. Nrf2 is reported to play an important role in protecting against the harmful effects of oxidative stress in neurodegenerative diseases (Alzheimer's disease and Parkinson's disease),[201] airway diseases,[202] cardiovascular disease, and kidney diseases.[203] Importantly, Nrf2 is believed to be epigenetically regulated. The hypermethylation of Nrf2 promoter region reportedly leads to a decrease in the production of antioxidant enzymes and an increase in oxidative stress.[204]

NF-κB is another important transcription factor that responds directly to reactive oxygen species to regulate inflammatory-related genes such as tumor necrosis factor (TNF) and interleukins (ILs).[205] Under oxidative stress conditions, ROS leads to phosphorylation and degradation of IκBα, the negative regulator of NF-κB (P65). The released P65 can then translocate into the nucleus and regulate pro-inflammatory genes.[206] Interestingly, treatment with the antioxidant *N*-acetylcysteine (NAC) in rats attenuated endotoxin-induced NF-κB activation and decreased chemokine gene expression protecting the rats against neutrophilic lung inflammation.[188] Recently, NAC has shown promising results for treatment of many disorders including addiction, compulsive and grooming disorders, schizophrenia and bipolar disorder. The mechanism by which NAC modulates the symptoms of psychiatric disorders is potentially through the reductions in inflammatory cytokines in addition to its general activity as antioxidant by restoration of GSH level. Interestingly, NAC can further enhance expression of neurotrophic factors and modulate glutamate and dopamine levels by interacting with cysteine-glutamate antiporter and dopamine transporters, respectively.[207,208] Omega-3 fatty acids are effective antioxidants, it also has anti-inflammatory properties through the inhibition of NF-κB activation. In clinical trials, treatment with omega-3 fatty acid effectively reduced the positive symptoms in patients with schizophrenia when compared with placebo group.[209] Additionally, in bipolar disorder patients treated with omega-3 fatty acid showed significant delay in the period of remission as compared to the placebo group.[210] Interestingly, many other antioxidants such as, vitamin E, vitamin C, coenzyme Q10, GSH, curcumin, *ginkgo biloba*, and melatonin, among others, showed promising results in preclinical and clinical trials for treatment of psychiatric disorders.

9. Conclusion

Clearly, oxidative stress and inflammation have critical role in pathophysiology of stress-related disorders, hence, direct antioxidant intervention or manipulating cellular pathways that upregulate antioxidant gene expression (Nrf2, SOD, HO, etc.) and downregulate pro-oxidant genes (NF-κB, TNF-alpha, IL-6) are a promising therapeutic strategy for treatment/management/prevention of psychiatric symptoms, particularly, stress-related psychiatric disorders.

Disclosure of interest

The authors report no conflict of interest. Funding for this research was provided by an NIH grant awarded to Samina Salim (2R15MH093918-02).

References

1. Kiss A, Aguilera G. Regulation of the hypothalamic pituitary adrenal axis during chronic stress: responses to repeated intraperitoneal hypertonic saline injection. *Brain Res*. 1993;630(1–2):262–270.
2. LeDoux J. Fear and the brain: where have we been, and where are we going? *Biol Psychiatry*. 1998;44(12):1229–1238.

3. McEwen BS, Stellar E. Stress and the individual. Mechanisms leading to disease. *Arch Intern Med*. 1993;153(18):2093–2101.
4. Gunnar M, Quevedo K. The neurobiology of stress and development. *Annu Rev Psychol*. 2007;58:145–173.
5. McLaughlin KA, et al. Causal effects of the early caregiving environment on development of stress response systems in children. *Proc Natl Acad Sci USA*. 2015;112(18):5637–5642.
6. Chrousos GP, Gold PW. The concepts of stress and stress system disorders. Overview of physical and behavioral homeostasis. *JAMA*. 1992;267(9):1244–1252.
7. Semple BD, et al. Brain development in rodents and humans: identifying benchmarks of maturation and vulnerability to injury across species. *Prog Neurobiol*. 2013;106–107:1–16.
8. Pechtel P, Pizzagalli DA. Effects of early life stress on cognitive and affective function: an integrated review of human literature. *Psychopharmacology*. 2011;214(1):55–70.
9. Chen Y, Baram TZ. Toward understanding how early-life stress reprograms cognitive and emotional brain networks. *Neuropsychopharmacology*. 2016;41(1):197–206.
10. Brunson KL, et al. Mechanisms of late-onset cognitive decline after early-life stress. *J Neurosci*. 2005;25(41):9328–9338.
11. Graham YP, et al. The effects of neonatal stress on brain development: implications for psychopathology. *Dev Psychopathol*. 1999;11(3):545–565.
12. Oitzl MS, et al. Brain development under stress: hypotheses of glucocorticoid actions revisited. *Neurosci Biobehav Rev*. 2010;34(6):853–866.
13. Wohleb ES, et al. Monocyte trafficking to the brain with stress and inflammation: a novel axis of immune-to-brain communication that influences mood and behavior. *Front Neurosci*. 2015;8:447.
14. Sorrells SF, et al. The stressed CNS: when glucocorticoids aggravate inflammation. *Neuron*. 2009;64(1):33–39.
15. Salim S, et al. Moderate treadmill exercise prevents oxidative stress-induced anxiety-like behavior in rats. *Behav Brain Res*. 2010;208(2):545–552.
16. Salim S, Chugh G, Asghar M. Inflammation in anxiety. *Adv Protein Chem Struct Biol*. 2012;88:1–25.
17. Heim C, et al. The link between childhood trauma and depression: insights from HPA axis studies in humans. *Psychoneuroendocrinology*. 2008;33(6):693–710.
18. Heim C, Binder EB. Current research trends in early life stress and depression: review of human studies on sensitive periods, gene-environment interactions, and epigenetics. *Exp Neurol*. 2012;233(1):102–111.
19. Gee DG, Casey BJ. The impact of developmental timing for stress and recovery. *Neurobiol Stress*. 2015;1:184–194.
20. Entringer S, Buss C, Heim C. Early-life stress and vulnerability for disease in later life. *Bundesgesundheitsblatt Gesundheitsforschung Gesundheitsschutz*. 2016;59(10):1255–1261.
21. Wang Q, Shelton RC, Dwivedi Y. Interaction between early-life stress and FKBP5 gene variants in major depressive disorder and post-traumatic stress disorder: a systematic review and meta-analysis. *J Affect Disord*. 2018;225:422–428.
22. Heim C, Nemeroff CB. The role of childhood trauma in the neurobiology of mood and anxiety disorders: preclinical and clinical studies. *Biol Psychiatry*. 2001;49(12):1023–1039.
23. Teicher MH, et al. The neurobiological consequences of early stress and childhood maltreatment. *Neurosci Biobehav Rev*. 2003;27(1–2):33–44.
24. Koenen KC, Uddin M. FKBP5 polymorphisms modify the effects of childhood trauma. *Neuropsychopharmacology*. 2010;35(8):1623–1624.
25. Gilbert R, et al. Burden and consequences of child maltreatment in high-income countries. *Lancet*. 2009;373(9657):68–81.

26. Finkelhor D. The prevention of childhood sexual abuse. *Futur Child.* 2009;19(2):169–194.
27. Radford L, et al. The prevalence and impact of child maltreatment and other types of victimization in the UK: findings from a population survey of caregivers, children and young people and young adults. *Child Abuse Negl.* 2013;37(10):801–813.
28. Kessler RC, Davis CG, Kendler KS. Childhood adversity and adult psychiatric disorder in the US National Comorbidity Survey. *Psychol Med.* 1997;27(5):1101–1119.
29. Kendler KS, Gardner CO, Prescott CA. Toward a comprehensive developmental model for major depression in women. *Am J Psychiatry.* 2002;159(7):1133–1145.
30. Biederman J. Attention-deficit/hyperactivity disorder: a selective overview. *Biol Psychiatry.* 2005;57(11):1215–1220.
31. Johnson JG, et al. Childhood verbal abuse and risk for personality disorders during adolescence and early adulthood. *Compr Psychiatry.* 2001;42(1):16–23.
32. Bandelow B, et al. Early traumatic life events, parental attitudes, family history, and birth risk factors in patients with depressive disorder and healthy controls. *Int J Psychiatry Clin Pract.* 2013;17(1):56–63.
33. Carpenter LL, et al. Decreased adrenocorticotropic hormone and cortisol responses to stress in healthy adults reporting significant childhood maltreatment. *Biol Psychiatry.* 2007;62(10):1080–1087.
34. Elzinga BM, et al. Diminished cortisol responses to psychosocial stress associated with lifetime adverse events: a study among healthy young subjects. *Psychoneuroendocrinology.* 2008;33(2):227–237.
35. Lenroot RK, Giedd JN. Brain development in children and adolescents: insights from anatomical magnetic resonance imaging. *Neurosci Biobehav Rev.* 2006;30(6):718–729.
36. Rice D, Barone Jr S. Critical periods of vulnerability for the developing nervous system: evidence from humans and animal models. *Environ Health Perspect.* 2000; 108(Suppl. 3):511–533.
37. Andersen SL, Teicher MH. Stress, sensitive periods and maturational events in adolescent depression. *Trends Neurosci.* 2008;31(4):183–191.
38. Weiss MJ, Wagner SH. What explains the negative consequences of adverse childhood experiences on adult health? Insights from cognitive and neuroscience research. *Am J Prev Med.* 1998;14(4):356–360.
39. Fatemi SH, Folsom TD. The neurodevelopmental hypothesis of schizophrenia, revisited. *Schizophr Bull.* 2009;35(3):528–548.
40. Weaver IC, et al. Epigenetic programming by maternal behavior. *Nat Neurosci.* 2004;7(8):847–854.
41. Harlow HF, Suomi SJ. Social recovery by isolation-reared monkeys. *Proc Natl Acad Sci USA.* 1971;68(7):1534–1538.
42. Danese A, et al. Elevated inflammation levels in depressed adults with a history of childhood maltreatment. *Arch Gen Psychiatry.* 2008;65(4):409–415.
43. Heim C, et al. Effect of childhood trauma on adult depression and neuroendocrine function: sex-specific moderation by CRH receptor 1 gene. *Front Behav Neurosci.* 2009;3:41.
44. Gee DG, et al. Early developmental emergence of human amygdala–prefrontal connectivity after maternal deprivation. *Proc Natl Acad Sci USA.* 2013;110(39):15638–15643.
45. Grant MM, et al. Childhood trauma history differentiates amygdala response to sad faces within MDD. *J Psychiatr Res.* 2011;45(7):886–895.
46. Pollak SD, et al. Neurodevelopmental effects of early deprivation in post-institutionalized children. *Child Dev.* 2010;81(1):224–236.
47. Stein MB, et al. Hippocampal volume in women victimized by childhood sexual abuse. *Psychol Med.* 1997;27(4):951–959.

48. Tottenham N, et al. Elevated amygdala response to faces following early deprivation. *Dev Sci.* 2011;14(2):190–204.
49. Tottenham N, et al. Prolonged institutional rearing is associated with atypically large amygdala volume and difficulties in emotion regulation. *Dev Sci.* 2010;13(1):46–61.
50. van Harmelen A-L, et al. Reduceed medial prefrontal cortex volume in adults reporting childhood emotional maltreatment. *Biol Psychiatry.* 2010;68(9):832–838.
51. de Kloet ER, Joels M, Holsboer F. Stress and the brain: from adaptation to disease. *Nat Rev Neurosci.* 2005;6(6):463–475.
52. Ulrich-Lai YM, Herman JP. Neural regulation of endocrine and autonomic stress responses. *Nat Rev Neurosci.* 2009;10(6):397–409.
53. McEwen BS, Gianaros PJ. Stress- and allostasis-induced brain plasticity. *Annu Rev Med.* 2011;62:431–445.
54. Tsigos C, et al. Stress, endocrine physiology and pathophysiology. In: De Groot LJ, et al. eds. *Endotext.* South Dartmouth, MA: MDtextcom; 2000:102–221.
55. Smith SM, Vale WW. The role of the hypothalamic-pituitary-adrenal axis in neuroendocrine responses to stress. *Dialogues Clin Neurosci.* 2006;8(4):383–395.
56. Jacobson L, Sapolsky R. The role of the hippocampus in feedback regulation of the hypothalamic-pituitary-adrenocortical axis. *Endocr Rev.* 1991;12(2):118–134.
57. Watts AG. Glucocorticoid regulation of peptide genes in neuroendocrine CRH neurons: a complexity beyond negative feedback. *Front Neuroendocrinol.* 2005;26(3–4):109–130.
58. Feldman S, Conforti N, Weidenfeld J. Limbic pathways and hypothalamic neurotransmitters mediating adrenocortical responses to neural stimuli. *Neurosci Biobehav Rev.* 1995;19(2):235–240.
59. Herman JP, et al. Central mechanisms of stress integration: hierarchical circuitry controlling hypothalamo-pituitary-adrenocortical responsiveness. *Front Neuroendocrinol.* 2003;24(3):151–180.
60. Morrow BA, Roth RH. Serotonergic lesions alter cocaine-induced locomotor behavior and stress-activation of the mesocorticolimbic dopamine system. *Synapse.* 1996;23(3):174–181.
61. Chrousos GP. Stressors, stress, and neuroendocrine integration of the adaptive response. The 1997 Hans Selye memorial lecture. *Ann N Y Acad Sci.* 1998;851:311–335.
62. Gray TS, Carney ME, Magnuson DJ. Direct projections from the central amygdaloid nucleus to the hypothalamic paraventricular nucleus: possible role in stress-induced adrenocorticotropin release. *Neuroendocrinology.* 1989;50(4):433–446.
63. Nikolarakis KE, Almeida OF, Herz A. Stimulation of hypothalamic beta-endorphin and dynorphin release by corticotropin-releasing factor (in vitro). *Brain Res.* 1986;399(1):152–155.
64. Ben-Eliyahu S, et al. Stress increases metastatic spread of a mammary tumor in rats: evidence for mediation by the immune system. *Brain Behav Immun.* 1991;5(2):193–205.
65. Cohen S, Tyrrell DA, Smith AP. Psychological stress and susceptibility to the common cold. *N Engl J Med.* 1991;325(9):606–612.
66. Glaser R, Kiecolt-Glaser JK. Stress and immune function. *Clin Neuropharmacol.* 1986; 9(Suppl. 4):485–487.
67. Brindley DN, Rolland Y. Possible connections between stress, diabetes, obesity, hypertension and altered lipoprotein metabolism that may result in atherosclerosis. *Clin Sci (Lond).* 1989;77(5):453–461.
68. Kendler KS, Karkowski LM, Prescott CA. Causal relationship between stressful life events and the onset of major depression. *Am J Psychiatry.* 1999;156(6):837–841.
69. Finlay-Jones R, Brown GW. Types of stressful life event and the onset of anxiety and depressive disorders. *Psychol Med.* 1981;11(4):803–815.

70. Seifert AE, Polusny MA, Murdoch M. The association between childhood physical and sexual abuse and functioning and psychiatric symptoms in a sample of U.S. Army soldiers. *Mil Med.* 2011;176(2):176–181.
71. Shaw JA. Children exposed to war/terrorism. *Clin Child Fam Psychol Rev.* 2003;6(4):237–246.
72. Nader KO, et al. A preliminary study of PTSD and grief among the children of Kuwait following the Gulf crisis. *Br J Clin Psychol.* 1993;32(Pt. 4):407–416.
73. Macksoud MS, Aber JL. The war experiences and psychosocial development of children in Lebanon. *Child Dev.* 1996;67(1):70–88.
74. Tarullo AR, Gunnar MR. Child maltreatment and the developing HPA axis. *Horm Behav.* 2006;50(4):632–639.
75. Chrousos GP. Stress and disorders of the stress system. *Nat Rev Endocrinol.* 2009;5(7):374–381.
76. Heim C, et al. The dexamethasone/corticotropin-releasing factor test in men with major depression: role of childhood trauma. *Biol Psychiatry.* 2008;63(4):398–405.
77. Heim C, et al. Pituitary-adrenal and autonomic responses to stress in women after sexual and physical abuse in childhood. *JAMA.* 2000;284(5):592–597.
78. Kumari P, et al. An essential role for maternal control of Nodal signaling. *Elife.* 2013;2 e00683.
79. Branchi I, et al. Early social enrichment shapes social behavior and nerve growth factor and brain-derived neurotrophic factor levels in the adult mouse brain. *Biol Psychiatry.* 2006;60(7):690–696.
80. Lee J, Duan W, Mattson MP. Evidence that brain-derived neurotrophic factor is required for basal neurogenesis and mediates, in part, the enhancement of neurogenesis by dietary restriction in the hippocampus of adult mice. *J Neurochem.* 2002;82(6):1367–1375.
81. Bremner JD, et al. MRI and PET study of deficits in hippocampal structure and function in women with childhood sexual abuse and posttraumatic stress disorder. *Am J Psychiatry.* 2003;160(5):924–932.
82. Grassi-Oliveira R, et al. Low plasma brain-derived neurotrophic factor and childhood physical neglect are associated with verbal memory impairment in major depression—a preliminary report. *Biol Psychiatry.* 2008;64(4):281–285.
83. Daskalakis NP, et al. Early life stress effects on glucocorticoid—BDNF interplay in the hippocampus. *Front Mol Neurosci.* 2015;8:68.
84. Meaney MJ, et al. Molecular basis for the development of individual differences in the hypothalamic-pituitary-adrenal stress response. *Cell Mol Neurobiol.* 1993;13(4):321–347.
85. Watson JB, et al. Prenatal teratogens and the development of adult mental illness. *Dev Psychopathol.* 1999;11(3):457–466.
86. O'Connor TG, et al. Maternal antenatal anxiety and children's behavioural/emotional problems at 4 years. Report from the avon longitudinal study of parents and children. *Br J Psychiatry.* 2002;180:502–508.
87. O'Connor TG, et al. Maternal antenatal anxiety and behavioural/emotional problems in children: a test of a programming hypothesis. *J Child Psychol Psychiatry.* 2003;44(7):1025–1036.
88. Van den Bergh BR, Marcoen A. High antenatal maternal anxiety is related to ADHD symptoms, externalizing problems, and anxiety in 8- and 9-year-olds. *Child Dev.* 2004;75(4):1085–1097.
89. Van den Bergh BR, et al. Antenatal maternal anxiety is related to HPA-axis dysregulation and self-reported depressive symptoms in adolescence: a prospective study on the fetal origins of depressed mood. *Neuropsychopharmacology.* 2008;33(3):536–545.

90. Van den Bergh BR, et al. Antenatal maternal anxiety and stress and the neurobehavioural development of the fetus and child: links and possible mechanisms. A review. *Neurosci Biobehav Rev*. 2005;29(2):237–258.
91. Stein MB, et al. Genetic and environmental influences on trauma exposure and posttraumatic stress disorder symptoms: a twin study. *Am J Psychiatry*. 2002;159(10):1675–1681.
92. Rijsdijk FV, et al. Genetic and environmental influences on psychological distress in the population: general health questionnaire analyses in UK twins. *Psychol Med*. 2003;33(5):793–801.
93. Sullivan PF, Neale MC, Kendler KS. Genetic epidemiology of major depression: review and meta-analysis. *Am J Psychiatry*. 2000;157(10):1552–1562.
94. Gillespie CF, et al. Risk and resilience: genetic and environmental influences on development of the stress response. *Depress Anxiety*. 2009;26(11):984–992.
95. Egan MF, et al. The BDNF val66met polymorphism affects activity-dependent secretion of BDNF and human memory and hippocampal function. *Cell*. 2003;112(2):257–269.
96. Kemp AH, et al. Improving the prediction of treatment response in depression: integration of clinical, cognitive, psychophysiological, neuroimaging, and genetic measures. *CNS Spectr*. 2008;13(12):1066–1086.
97. Pezawas L, et al. The brain-derived neurotrophic factor val66met polymorphism and variation in human cortical morphology. *J Neurosci*. 2004;24(45):10099–10102.
98. Schofield P, et al. Disturbances in task-relevant information processing associated with the brain-derived neurotrophic factor Val66Met polymorphism: evidence from cognition, the P300 endophenotype and neuroimaging. *Biol Psychol*. 2009;80:176–188.
99. Gatt J, et al. Interactions between BDNF Val66Met polymorphism and early life stress predict brain and arousal pathways to syndromal depression and anxiety. *Mol Psychiatry*. 2009;14(7):681.
100. Edwards VJ, et al. Relationship between multiple forms of childhood maltreatment and adult mental health in community respondents: results from the adverse childhood experiences study. *Am J Psychiatry*. 2003;160(8):1453–1460.
101. Nemeroff CB. Neurobiological consequences of childhood trauma. *J Clin Psychiatry*. 2004;65(Suppl. 1):18–28.
102. Bradley RG, et al. Influence of child abuse on adult depression: moderation by the corticotropin-releasing hormone receptor gene. *Arch Gen Psychiatry*. 2008;65(2):190–200.
103. Licinio J, et al. Association of a corticotropin-releasing hormone receptor 1 haplotype and antidepressant treatment response in Mexican-Americans. *Mol Psychiatry*. 2004;9(12):1075–1082.
104. Polanczyk G, et al. Protective effect of CRHR1 gene variants on the development of adult depression following childhood maltreatment: replication and extension. *Arch Gen Psychiatry*. 2009;66(9):978–985.
105. Liu Z, et al. Association study of corticotropin-releasing hormone receptor1 gene polymorphisms and antidepressant response in major depressive disorders. *Neurosci Lett*. 2007;414(2):155–158.
106. Binder EB. The role of FKBP5, a co-chaperone of the glucocorticoid receptor in the pathogenesis and therapy of affective and anxiety disorders. *Psychoneuroendocrinology*. 2009;34(Suppl. 1):S186–S195.
107. Yehuda R, et al. Enhanced sensitivity to glucocorticoids in peripheral mononuclear leukocytes in posttraumatic stress disorder. *Biol Psychiatry*. 2004;55(11):1110–1116.
108. Kato T. Molecular genetics of bipolar disorder and depression. *Psychiatry Clin Neurosci*. 2007;61(1):3–19.

109. Binder EB, et al. Association of FKBP5 polymorphisms and childhood abuse with risk of posttraumatic stress disorder symptoms in adults. *JAMA*. 2008;299(11):1291–1305.
110. Ising M, et al. Polymorphisms in the FKBP5 gene region modulate recovery from psychosocial stress in healthy controls. *Eur J Neurosci*. 2008;28(2):389–398.
111. Binder EB, et al. Polymorphisms in FKBP5 are associated with increased recurrence of depressive episodes and rapid response to antidepressant treatment. *Nat Genet*. 2004;36(12):1319–1325.
112. Caspi A, et al. Influence of life stress on depression: moderation by a polymorphism in the 5-HTT gene. *Science*. 2003;301(5631):386–389.
113. Babenko O, Kovalchuk I, Metz GA. Stress-induced perinatal and transgenerational epigenetic programming of brain development and mental health. *Neurosci Biobehav Rev*. 2015;48:70–91.
114. Unternaehrer E, et al. Dynamic changes in DNA methylation of stress-associated genes (OXTR, BDNF) after acute psychosocial stress. *Transl Psychiatry*. 2012;2:e150.
115. Fuchikami M, et al. DNA methylation profiles of the brain-derived neurotrophic factor (BDNF) gene as a potent diagnostic biomarker in major depression. *PLoS One*. 2011;6(8):e23881.
116. Numata S, et al. Blood diagnostic biomarkers for major depressive disorder using multiplex DNA methylation profiles: discovery and validation. *Epigenetics*. 2015;10(2):135–141.
117. Elliott E, et al. Resilience to social stress coincides with functional DNA methylation of the Crf gene in adult mice. *Nat Neurosci*. 2010;13(11):1351–1353.
118. Caldji C, Diorio J, Meaney MJ. Variations in maternal care in infancy regulate the development of stress reactivity. *Biol Psychiatry*. 2000;48(12):1164–1174.
119. Caldji C, et al. Maternal care during infancy regulates the development of neural systems mediating the expression of fearfulness in the rat. *Proc Natl Acad Sci USA*. 1998;95(9):5335–5340.
120. Liu D, et al. Maternal care, hippocampal glucocorticoid receptors, and hypothalamic-pituitary-adrenal responses to stress. *Science*. 1997;277(5332):1659–1662.
121. Roth TL, et al. Lasting epigenetic influence of early-life adversity on the BDNF gene. *Biol Psychiatry*. 2009;65(9):760–769.
122. Murgatroyd C, et al. Dynamic DNA methylation programs persistent adverse effects of early-life stress. *Nat Neurosci*. 2009;12(12):1559–1566.
123. Smalheiser NR, Lugli G. microRNA regulation of synaptic plasticity. *NeuroMolecular Med*. 2009;11(3):133–140.
124. Schratt G. microRNAs at the synapse. *Nat Rev Neurosci*. 2009;10(12):842–849.
125. Aksoy-Aksel A, Zampa F, Schratt G. MicroRNAs and synaptic plasticity—a mutual relationship. *Philos Trans R Soc Lond Ser B Biol Sci*. 2014;369(1652):20130515.
126. He Y, et al. Genetic variations in microRNA processing genes are associated with susceptibility in depression. *DNA Cell Biol*. 2012;31(9):1499–1506.
127. Smalheiser NR, et al. MicroRNA expression is down-regulated and reorganized in prefrontal cortex of depressed suicide subjects. *PLoS One*. 2012;7(3):e33201.
128. Forstner AJ, et al. Genome-wide analysis implicates microRNAs and their target genes in the development of bipolar disorder. *Transl Psychiatry*. 2015;5:e678.
129. Haramati S, et al. MicroRNA as repressors of stress-induced anxiety: the case of amygdalar miR-34. *J Neurosci*. 2011;31(40):14191–14203.
130. Meerson A, et al. Changes in brain MicroRNAs contribute to cholinergic stress reactions. *J Mol Neurosci*. 2010;40(1–2):47–55.
131. Uchida S, et al. Characterization of the vulnerability to repeated stress in Fischer 344 rats: possible involvement of microRNA-mediated down-regulation of the glucocorticoid receptor. *Eur J Neurosci*. 2008;27(9):2250–2261.

132. Ledderose C, et al. Corticosteroid resistance in sepsis is influenced by microRNA-124-induced downregulation of glucocorticoid receptor-alpha. *Crit Care Med.* 2012;40(10):2745–2753.
133. Eendebak RJ, Lucassen PJ, Fitzsimons CP. Nuclear receptors and microRNAs: who regulates the regulators in neural stem cells? *FEBS Lett.* 2011;585(5):717–722.
134. Sober S, Laan M, Annilo T. MicroRNAs miR-124 and miR-135a are potential regulators of the mineralocorticoid receptor gene (NR3C2) expression. *Biochem Biophys Res Commun.* 2010;391(1):727–732.
135. Uchida S, et al. Early life stress enhances behavioral vulnerability to stress through the activation of REST4-mediated gene transcription in the medial prefrontal cortex of rodents. *J Neurosci.* 2010;30(45):15007–15018.
136. Schouten M, et al. microRNAs and the regulation of neuronal plasticity under stress conditions. *Neuroscience.* 2013;241:188–205.
137. Ader R. Effects of early experience and differential housing on behavior and susceptibility to gastric erosions in the rat. *J Comp Physiol Psychol.* 1965;60(2):233.
138. Coe CL, Rosenberg LT, Levine S. Effect of maternal separation on the complement system and antibody responses in infant primates. *Int J Neurosci.* 1988;40(3–4):289–302.
139. Hennessy MB, et al. Maternal separation produces, and a second separation enhances, core temperature and passive behavioral responses in guinea pig pups. *Physiol Behav.* 2010;100(4):305–310.
140. Wieck A, et al. Evidence for a neuroinflammatory mechanism in delayed effects of early life adversity in rats: relationship to cortical NMDA receptor expression. *Brain Behav Immun.* 2013;28:218–226.
141. Danese A, Lewis SJ. Psychoneuroimmunology of early-life stress: the hidden wounds of childhood trauma? *Neuropsychopharmacology.* 2017;42(1):99.
142. Pace TW, et al. Increased stress-induced inflammatory responses in male patients with major depression and increased early life stress. *Am J Psychiatry.* 2006;163(9):1630–1633.
143. Carpenter LL, et al. Association between plasma IL-6 response to acute stress and early-life adversity in healthy adults. *Neuropsychopharmacology.* 2010;35(13):2617.
144. Lam LL, et al. Factors underlying variable DNA methylation in a human community cohort. *Proc Natl Acad Sci USA.* 2012;109(Suppl. 2):17253–17260.
145. Guan Z, Fang J. Peripheral immune activation by lipopolysaccharide decreases neurotrophins in the cortex and hippocampus in rats. *Brain Behav Immun.* 2006;20(1):64–71.
146. Lapchak P, Araujo D, Hefti F. Systemic interleukin-1 beta decreases brain-derived neurotrophic factor messenger RNA expression in the rat hippocampal formation. *Neuroscience.* 1993;53(2):297–301.
147. Schnydrig S, et al. Peripheral lipopolysaccharide administration transiently affects expression of brain-derived neurotrophic factor, corticotropin and proopiomelanocortin in mouse brain. *Neurosci Lett.* 2007;429(1):69–73.
148. Cadenas E, Davies KJ. Mitochondrial free radical generation, oxidative stress, and aging. *Free Redic Biol Med.* 2000;29(3–4):222–230.
149. Serrano F, Klann E. Reactive oxygen species and synaptic plasticity in the aging hippocampus. *Ageing Res Rev.* 2004;3(4):431–443.
150. Son Y, et al. Mitogen-activated protein kinases and reactive oxygen species: how can ROS activate MAPK pathways? *J Signal Transduction.* 2011;2011 792639.
151. Arsenijevic D, et al. Differential roles of tumor necrosis factor-alpha and interferon-gamma in mouse hypermetabolic and anorectic responses induced by LPS. *Eur Cytokine Netw.* 2000;11(4):662–668.
152. Wang D, Malo D, Hekimi S. Elevated mitochondrial reactive oxygen species generation affects the immune response via hypoxia-inducible factor-1alpha in long-lived Mclk1 +/- mouse mutants. *J Immunol.* 2010;184(2):582–590.

153. Reczek CR, Chandel NS. ROS-dependent signal transduction. *Curr Opin Cell Biol.* 2015;33:8–13.
154. Schieber M, Chandel NS. ROS function in redox signaling and oxidative stress. *Curr Biol.* 2014;24(10):R453–R462.
155. Halliwell B. *Biochemistry of Oxidative Stress*. Portland Press Limited; 2007.
156. Pham-Huy LA, He H, Pham-Huy C. Free radicals, antioxidants in disease and health. *Int J Biomed Sci.* 2008;4(2):89.
157. Manni ML, Oury TD. Oxidative stress and pulmonary fibrosis. In: *Systems Biology of Free Radicals and Antioxidants*. 2014:1611–1631.
158. Matough FA, et al. The role of oxidative stress and antioxidants in diabetic complications. *Sultan Qaboos Univ Med J.* 2012;12(1):5–18.
159. Cichoż-Lach H, Michalak A. Oxidative stress as a crucial factor in liver diseases. *World J Gastroenterol.* 2014;20(25):8082.
160. Ng F, et al. Oxidative stress in psychiatric disorders: evidence base and therapeutic implications. *Int J Neuropsychopharmacol.* 2008;11(6):851–876.
161. Sarandol A, et al. Major depressive disorder is accompanied with oxidative stress: short-term antidepressant treatment does not alter oxidative—antioxidative systems. *Hum Psychopharmacol.* 2007;22(2):67–73.
162. Sorce S, et al. The NADPH oxidase NOX2 controls glutamate release: a novel mechanism involved in psychosis-like ketamine responses. *J Neurosci.* 2010;30(34):11317–11325.
163. Boskovic M, et al. Oxidative stress in schizophrenia. *Curr Neuropharmacol.* 2011;9(2):301–312.
164. Akyol M, et al. The effects of vitamin E on the skin lipid peroxidation and the clinical improvement in vitiligo patients treated with PUVA. *Eur J Dermatol.* 2002;12(1):24–26.
165. Dietrich-Muszalska A, et al. The oxidative stress may be induced by the elevated homocysteine in schizophrenic patients. *Neurochem Res.* 2012;37(5):1057–1062.
166. Nishioka N, Arnold SE. Evidence for oxidative DNA damage in the hippocampus of elderly patients with chronic schizophrenia. *Am J Geriatr Psychiatry.* 2004;12(2):167–175.
167. Stefanescu C, Ciobica A. The relevance of oxidative stress status in first episode and recurrent depression. *J Affect Disord.* 2012;143(1–3):34–38.
168. Guney E, et al. Oxidative stress in children and adolescents with anxiety disorders. *J Affect Disord.* 2014;156:62–66.
169. Altuntas I, et al. Erythrocyte superoxide dismutase and glutathione peroxidase activities, and malondialdehyde and reduced glutathione levels in schizophrenic patients. *Clin Chem Lab Med.* 2000;38(12):1277–1281.
170. Othmen LB, et al. Altered antioxidant defense system in clinically stable patients with schizophrenia and their unaffected siblings. *Prog Neuropsychopharmacol Biol Psychiatry.* 2008;32(1):155–159.
171. Ranjekar PK, et al. Decreased antioxidant enzymes and membrane essential polyunsaturated fatty acids in schizophrenic and bipolar mood disorder patients. *Psychiatry Res.* 2003;121(2):109–122.
172. Sivrioglu E, et al. The impact of ω-3 fatty acids, vitamins E and C supplementation on treatment outcome and side effects in schizophrenia patients treated with haloperidol: an open-label pilot study. *Prog Neuropsychopharmacol Biol Psychiatry.* 2007;31(7):1493–1499.
173. Mazza M, et al. Omega-3 fatty acids and antioxidants in neurological and psychiatric diseases: an overview. *Prog Neuropsychopharmacol Biol Psychiatry.* 2007;31(1):12–26.
174. Pandya CD, et al. Antioxidants as potential therapeutics for neuropsychiatric disorders. *Prog Neuropsychopharmacol Biol Psychiatry.* 2013;46:214–223.

175. Salim S. Oxidative stress and psychological disorders. *Curr Neuropharmacol.* 2014;12(2):140–147.
176. Patki G, et al. Depression, anxiety-like behavior and memory impairment are associated with increased oxidative stress and inflammation in a rat model of social stress. *Brain Res.* 2013;1539:73–86.
177. Masood A, et al. Reversal of oxidative stress-induced anxiety by inhibition of phosphodiesterase-2 in mice. *J Pharmacol Exp Ther.* 2008;326(2):369–379.
178. Patki G, et al. Tempol treatment reduces anxiety-like behaviors induced by multiple anxiogenic drugs in rats. *PLoS One.* 2015;10(3) e0117498.
179. Solanki N, et al. Modulating oxidative stress relieves stress-induced behavioral and cognitive impairments in rats. *Int J Neuropsychopharmacol.* 2017;20(7):550–561.
180. Patki G, Solanki N, Salim S. Witnessing traumatic events causes severe behavioral impairments in rats. *Int J Neuropsychopharmacol.* 2014;17(12):2017–2029.
181. Aruoma OI, Grootveld M, Bahorun T. Free radicals in biology and medicine: from inflammation to biotechnology. *Biofactors.* 2006;27(1–4):1–3.
182. Butterfield DA, et al. Vitamin E and neurodegenerative disorders associated with oxidative stress. *Nutr Neurosci.* 2002;5(4):229–239.
183. Hoshino Y, Mishima M. Redox-based therapeutics for lung diseases. *Antioxid Redox Signal.* 2008;10(4):701–704.
184. MacNee W. Oxidative stress and lung inflammation in airways disease. *Eur J Pharmacol.* 2001;429(1–3):195–207.
185. Prabhakar O. Cerebroprotective effect of resveratrol through antioxidant and anti-inflammatory effects in diabetic rats. *Naunyn Schmiedeberg's Arch Pharmacol.* 2013;386(8):705–710.
186. Haddad JJ, Olver RE, Land SC. Antioxidant/pro-oxidant equilibrium regulates HIF-1α and NF-κB redox sensitivity evidence for inhibition by glutathione oxidation in alveolar epithelial cells. *J Biol Chem.* 2000;275(28):21130–21139.
187. Schwartz G, et al. Early suppressive effects of chemotherapy and cytokine treatment on committed versus primitive haemopoietic progenitors in patient bone marrow. *Br J Haematol.* 1996;92(3):537–547.
188. Blackwell TS, et al. In vivo antioxidant treatment suppresses nuclear factor-kappa B activation and neutrophilic lung inflammation. *J Immunol.* 1996;157(4):1630–1637.
189. Dinkova-Kostova AT, et al. Extremely potent triterpenoid inducers of the phase 2 response: correlations of protection against oxidant and inflammatory stress. *Proc Natl Acad Sci USA.* 2005;102(12):4584–4589.
190. Ma Q, Battelli L, Hubbs AF. Multiorgan autoimmune inflammation, enhanced lymphoproliferation, and impaired homeostasis of reactive oxygen species in mice lacking the antioxidant-activated transcription factor Nrf2. *Am J Pathol.* 2006;168(6):1960–1974.
191. Aschbacher K, et al. Good stress, bad stress and oxidative stress: insights from anticipatory cortisol reactivity. *Psychoneuroendocrinology.* 2013;38(9):1698–1708.
192. Cabungcal J-H, et al. Early-life insults impair parvalbumin interneurons via oxidative stress: reversal by N-acetylcysteine. *Biol Psychiatry.* 2013;73(6):574–582.
193. Cohen S, et al. Chronic stress, glucocorticoid receptor resistance, inflammation, and disease risk. *Proc Natl Acad Sci USA.* 2012;109(16):5995–5999.
194. Greenberg MS, et al. Stress, PTSD, and dementia. *Alzheimers Dement.* 2014;10(3):S155–S165.
195. Irie M, et al. Psychosocial factors as a potential trigger of oxidative DNA damage in human leukocytes. *Jpn J Cancer Res.* 2001;92(3):367–376.
196. Metcalfe NB, Alonso-Alvarez C. Oxidative stress as a life-history constraint: the role of reactive oxygen species in shaping phenotypes from conception to death. *Funct Ecol.* 2010;24(5):984–996.

197. Powell N, et al. Psychosocial stress and inflammation in cancer. *Brain Behav Immun.* 2013;30:S41–S47.
198. Schiavone S, Colaianna M, Curtis L. Impact of early life stress on the pathogenesis of mental disorders: relation to brain oxidative stress. *Curr Pharm Des.* 2015;21(11):1404–1412.
199. Slavich GM, Irwin MR. From stress to inflammation and major depressive disorder: a social signal transduction theory of depression. *Psychol Bull.* 2014;140(3):774.
200. Adler A, et al. Gonadectomy in adult life increases tyrosine hydroxylase immuno-reactivity in the prefrontal cortex and decreases open field activity in male rats. *Neuroscience.* 1999;89(3):939–954.
201. Yamazaki H, et al. Role of the Keap1/Nrf2 pathway in neurodegenerative diseases. *Pathol Int.* 2015;65(5):210–219.
202. McGovern TK, et al. Neutrophilic oxidative stress mediates organic dust-induced pulmonary inflammation and airway hyperresponsiveness. *Am J Physiol Lung Cell Mol Physiol.* 2015;310:L155–L165.
203. Xiao L, et al. The mitochondria-targeted antioxidant MitoQ ameliorated tubular injury mediated by mitophagy in diabetic kidney disease via Nrf2/PINK1. *Redox Biol.* 2017;11:297–311.
204. Kang K, et al. Epigenetic modification of Nrf2 in 5-fluorouracil-resistant colon cancer cells: involvement of TET-dependent DNA demethylation. *Cell Death Dis.* 2014;5(4): e1183.
205. Baldwin AS. Series introduction: the transcription factor NF-κB and human disease. *J Clin Invest.* 2001;107(1):3–6.
206. Haddad JJ. Redox regulation of pro-inflammatory cytokines and IkappaB-alpha/NF-kappaB nuclear translocation and activation. *Biochem Biophys Res Commun.* 2003;296:625.
207. Bai O, et al. Expression of brain-derived neurotrophic factor mRNA in rat hippocampus after treatment with antipsychotic drugs. *J Neurosci Res.* 2003;71(1):127–131.
208. Dean O, et al. N-acetylcysteine in psychiatry: current therapeutic evidence and potential mechanisms of action. *J Psychiatry Neurosci.* 2011;36(2):78.
209. Peet M, Horrobin DF, E-E Multicenter Study Group. A dose-ranging exploratory study of the effects of ethyl-eicosapentaenoate in patients with persistent schizophrenic symptoms. *J Psychiatr Res.* 2002;36(1):7–18.
210. Stoll AL, et al. Omega 3 fatty acids in bipolar disorder: a preliminary double-blind, placebo-controlled trial. *Arch Gen Psychiatry.* 1999;56(5):407–412.

CHAPTER FOUR

Calpain in the cleavage of alpha-synuclein and the pathogenesis of Parkinson's disease

Ramsha Shams[a], Naren L. Banik[a,b,c], Azizul Haque[a,*]
[a]Department of Microbiology and Immunology, Hollings Cancer Center, Medical University of South Carolina, Charleston, SC, United States
[b]Department of Neurosurgery, Medical University of South Carolina, Charleston, SC, United States
[c]Ralph H. Johnson Veterans Administration Medical Center, Charleston, SC, United States
*Corresponding author: e-mail address: haque@musc.edu

Contents

Abstract

Parkinson's disease (PD) devastates 6.3 million people, ranking it as one of the most prevalent neurodegenerative motor disorders worldwide. PD patients may manifest symptoms of postural instability, bradykinesia, and resting tremors as a result of increasing α-synuclein aggregation and neuron death with disease progression. Therapy options are limited, and those available to patients may worsen their condition. Thus, investigations to understand disease progression may help develop therapeutic strategies for improvement of quality of life for patients suffering from PD. This review provides an overview of α-synuclein, a presynaptic neuronal protein whose function in the healthy brain and PD pathology remains a mystery. This review also focuses

Progress in Molecular Biology and Translational Science, Volume 167
ISSN 1877-1173
https://doi.org/10.1016/bs.pmbts.2019.06.007

on calcium-induced activation of calpain, a neutral protease, and the subsequent cascade of cellular processing of α-synuclein and emerging defense responses observed in experimental models of PD: microglial activation, dysregulation of T cells, and inflammatory responses in the brain. In addition, this review discusses the events of cross presentation of synuclein peptides by professional antigen presenting cells and microglia, induction of inflammatory responses in the periphery and brain, and emerging calpain-targeted therapeutic strategies to attenuate neuronal death in PD.

1. Introduction

Parkinson's disease (PD) is a complex, chronic progressive neurodegenerative disorder for which there is no cure.[1] Impacting approximately 6.3 million people in the world, this devastating disease is considered one of the most prevalent neurodegenerative motor disorders. During disease progression, neurons in the spinal cord (SC) and substantia nigra (SN) are damaged, leading to impaired motor function in the patient.[2,3] As a result, patients may manifest devastating symptoms including postural instability, bradykinesia, and resting tremors, which drastically lower the quality of life for the patient.[1,4] L-DOPA therapy is widely used but it shows temporary relief and can worsen the patients' symptoms if it is used for an extended period of time. Thus investigating the mechanisms involved in disease progression may help develop therapeutic strategies for the improvement of quality of life in PD.

Identified during disease prognosis are pathological features of PD, many of which are shared similarities among other neurodegenerative diseases, including rapid loss of dopaminergic neurons in the SN and, interestingly, accumulations of Lewy-bodies in the brain.[5] Lewy-bodies are toxic filaments and are composed of a small, lipid-binding protein called synuclein.[4] Alpha-synuclein (α-synuclein), a 140 amino acid-long protein, is abundant in the nervous system and is predominately expressed in neurons, even in healthy individuals, comprising 10% of cystolic protein.[6] When mutated, the α-synuclein protein may become truncated. Previous studies have correlated this truncated, mutated form of α-synuclein with the harmful Lewy-body aggregation and consequential cellular toxicity. A study using transgenic mice implicates α-synuclein C-terminus mutations with reduced dopamine in the striatum. These data, in turn, suggest toxicity of the Lewy-bodies in disease pathology.[7] It is not well known how the α-synuclein accumulations occur,[8,9] and the mechanisms promoting disease progression are also not well characterized.

α-Synuclein may promote a disruption of calcium homeostasis. Calcium is a key and versatile agent in many vital biological systems, and this powerful ion acts as a valuable physiological messenger for transport across the plasma membrane and activation of essential enzymes, among other functions. Ionic transport using Na^+ and K^+ utilizes an approximate 10- to 30-fold gradient across the cellular membrane. However, Ca^{2+}-powered transport is driven by a steep 20,000-fold concentration difference between the extracellular and intracellular space, affirming calcium's role as a valuable signal for responding to rapid-changing extracellular and intracellular conditions. By mediating calcium transport, cells can activate or inhibit Ca^{2+}-dependent signaling pathways, and neurons are especially sensitive to the effects provided by calcium for initiating response. Therefore, normal calcium homeostasis is necessary for improved biological function.[4]

Perturbation of Ca^{2+} homeostasis may activate calpain, a neutral protease, setting off a cascade of defense responses, as suggested by previous research.[4] Due to this dysregulation of Ca^{2+} homeostasis and calpain activation, the immune system, in turn, responds with the activation and migration of inflammatory T cells (CD4+ and CD8+), microglial activation, and astrocyte immobilization, features particularly studied in PD and other neurodegenerative diseases.[4,10,11]

Microglia activation is particularly interesting. Microglia are endogenous neural cells known as brain macrophages and are responsible for innate immunity in the brain. Classified under a family of immune cells, called the Major Histocompatibility Complex Class 2 (MHC-II), they can initiate responses leading to inflammation and neuron death.[12] Cells also relevant to inducing inflammation are chemokines and cytokines. Their primary function involves recruiting leukocytes to sites of inflammation. Their respective receptors are upregulated by calpain activation, therefore upregulating chemokine and cytokine molecules. The dysregulation of these molecules is often associated with PD.[12]

Recently, in addition to dopaminergic degeneration in the striatum as observed in PD,[1,5] damage to non-dopaminergic sites, namely, the spinal cord, has been demonstrated. The mechanisms describing how degeneration occurs both in the brain and in the spinal cord remain elusive. It has been suggested, however, that inflammation may be a key factor to this nervous system degeneration, and interestingly, neuro-inflammation has been implicated to promote the brain's susceptibility to neurodegeneration.[2,13]

The induced cell death following calpain activation is of particular interest to our laboratory. We suggest several hypotheses for the mechanisms following calpain activation and subsequent initiation of neuron death. The first hypothesis is that calpain will directly activate microglia. Microglial will secrete factors that will activate T cells. We also propose that calpain-activated T cells will produce neuroinflammatory factors, which may further activate microglia to produce factors killing neurons. Activated microglia may process α-synuclein generating peptides for T cell activation. Processing of α-synuclein by activated microglia may also produce toxic synuclein peptides causing neuronal cell death. Further, the damaged neurons by above processes may release detrimental factors that will activate and/or kill naïve immune cells and neurons, and thus perpetuating this vicious cycle for continued progression of the disease (Fig. 1).

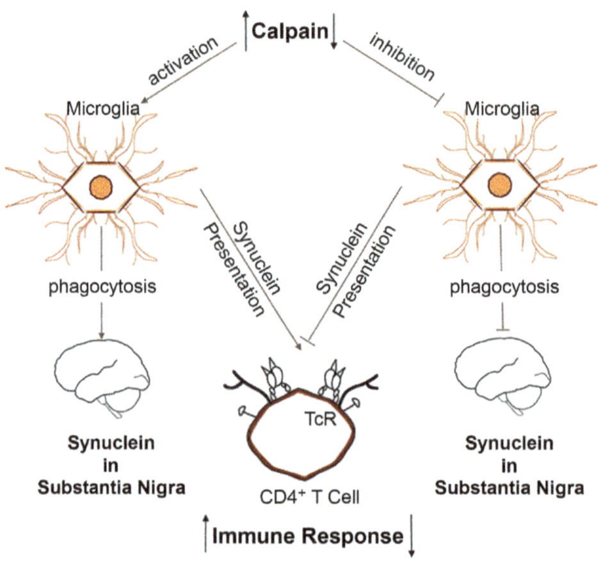

Fig. 1 Calpain activation, α-synuclein degradation, and microglial response in the brain. When there is increased calcium signaling in the brain, calpain is upregulated. Upregulation of calpain induces activation of microglia, which may promote multiple pathways of immune activation. Upon calpain activation, activated microglia can degrade α-synuclein and present immunogenic synuclein peptides to T cells, activating immune response, and inducing inflammation. Inhibition of calpain activation may downregulate microglia-mediated inflammatory responses in the brain, thereby attenuating inflammation. On the other hand, calpain activation can cause generation of toxic synuclein peptides and inflammatory T cells, which may induce neuronal cell death.

2. Calpain

Calpain is an intracellular Ca^{2+}-dependent cysteine protease that is widely distributed and show regulated protease activity at neutral pH.[14–16] Calpain is comprised of two subunits: an 80 kDa subunit called the "μ-calpain large" (μCl) subunit and a regulatory 30 kDa subunit.[16,17] The large subunit contains four domains. The functions of the first and third domains are not yet well characterized. It is known, however, that the fourth domain is a Ca^{2+}-binding domain. Moreover, as the main attributor to calpain's function as a protease, the second domain functions as cysteine protease. The small subunit is comprised of two domains. Domain 5 is an N-terminal glycine-clustering nonpolar region. Domain 6 (also called domain IV′) is a C-terminal Ca^{2+}-binding domain. In a cellular environment lacking calcium, the two subunits associate through these Ca^{2+}-binding domains.[18]

Calpain has a direct role in inducing inflammation since it cleaves the IκB-NFκB complex, and calpain promotes translocation of the transcription factor (NFκB) to the nucleus to enhance inflammation.[17,19] Previous studies have implicated calpain in the mechanisms of axon-myelin degeneration in Multiple Sclerosis (MS). The events leading to pathophysiology in MS are not clearly understood. Calpain activity is regulated by Ca^{2+}, calpastatin (an endogenous calpain inhibitor), growth factor-induced phosphorylation, and possibly by an activator(s) protein.[20–22] Besides, calpain activation appears to positively control further calpain translation since calpain protein expression was increased in EAE spinal cord but not calpain mRNA levels.[23] The pathophysiological role of calpain is also supported by mounting evidence of calpain's involvement in tissue degeneration in central nervous system (CNS) trauma,[23,24] cerebral ischemia,[25] muscular dystrophy,[26] Alzheimer's disease,[26] EAE,[27,28] and MS.[29–32] Greatly elevated activity of both neutral and acid proteases, including calpain, was found in the cerebrospinal fluid (CSF) of MS patients. The activity was significantly greater in acute cases than in the chronic stages or in remission. From these studies, there seems ample reason to stress the correlation between calpain activity and expression in relapse and remission stages in MS patients, as well as to examine further its value in this clinical context.

2.1 Calpain activation of T cells

One of the important transcription factors in T cell activation and cytokine production is NFAT. NFAT is a Ca^{2+}-dependent transcription factor that is involved in IL-2 gene expression.[33] Stimulation of CD4+ helper T cells requires two separate signals, which lead to production of Th1 or Th2 cytokines, depending on the local environment of the cell. The primary signal is given by the interaction of the T cell receptor (TCR) on the surface of the T cell with the major histocompatibility class II protein (MHCII) upon the surface of the APC.[34] Calpain expression may upregulate co-stimulatory molecules (CD80, CD86). CD40 expression in glial or bystander cells may modulate immune response shifts toward Th2-type response. One major secondary or co-stimulatory signal is given by the interaction of CD28 on the T cell surface with B7-1 (CD80) or B7-2 (CD86) upon the surface of the APC.[35] The result of TCR/MHCII ligation and CD28 co-stimulation, which can be mimicked by activating T cells with antibodies against CD3 and CD28, is an increase in the activity of several tyrosine kinases. Two of the more prominent tyrosine kinases that become activated are $p56^{lck}$ and ZAP70.[35] This cascade eventually leads to the tyrosine phosphorylation of an important linker protein, linker of activation of T cells (LAT).[36] LAT recruits phospholipase C (PLC), which then becomes activated by tyrosine phosphorylation,[37] and is dependent upon the presence of a non-membrane bound linker protein called SLP-76.[38] The result of activation of PLC is the hydrolysis of phosphatidylinositol-4,5-bisphosphate (PIP_2) into inositol-1,4,5-triphosphate (IP_3) and diacylglycerol (DAG).[39] IP_3 then binds to IP_3 receptors on the surface of the endoplasmic reticulum, causing an initial release of Ca^{2+} into the intracellular environment. This initial release of intracellular Ca^{2+} stores is responsible for a later sustained rise in Ca^{2+} levels via an influx of extracellular Ca^{2+} through the plasma membrane in an IP_3-independent manner. Increased Ca^{2+} levels due to release of intracellular stores and influx are at levels high enough to activate calpain. The rise in Ca^{2+}, liberation of DAG, and co-stimulation through CD28 are responsible for the activation of a number of signal transduction pathways that eventually lead to the synthesis of cytokines through activation of transcription factors including NF-kB[19,40] and NFAT.[41,42] While influx of Ca^{2+} in excitable cells is known to be mediated by voltage gated channels (Ca_v), current studies indicated that Ca_v are also present in non-excitable T cells.[37] This is one of the mechanisms responsible for Ca^{2+} influx in T lymphocytes.

Although Ca^{2+} plays many important roles in cell function, including activation of proteases and lipases, it also plays specific roles in T cell functions such as activation, proliferation, and cytokine production.[38] One of the many mechanisms involved in these processes is activation of calpain mediating activation and translocation of nuclear transcription factor NF-kB leading to IL-2 synthesis. We have shown that calpain inhibition can block IL-2 synthesis in PBMCs following stimulation with CD3/CD28 antibodies.[29] Calpain activation may occur due to elevated Ca^{2+} influx in T cells by increases in channel opening mediated by T cell receptor (TCR) stimulation.

3. α-Synuclein

α-Synuclein, a 140 amino acid-long protein, is translated from the SNCA gene.[43] The α-synuclein protein belongs to a family of proteins found in vertebrates called the synucleins, including also β-synuclein and γ-synuclein, classified for their small size, solubility, and having a highly-conserved α-helix lipid-binding motif.[44]

α-Synuclein protein contains three distinct regions. At residues 1–60, there is a six-repeat polyprotein lipid-binding motif at the amino terminus. At residues 61–95, there is a hydrophobic region called NAC, which promotes α-synuclein's β-sheet conformation and thereby promoting protein aggregation. At residues 61–95, there is a highly conserved acidic carboxyl terminus. This acidic tail is unstructured and shows a susceptibility to reacting.[43]

Even in healthy individuals, α-synuclein is expressed abundantly throughout the nervous system and body, and it has non-averse functions. Its expression is induced in neurogenesis after there is neuronal determination of pluripotent cells and after synaptic connections have been formed. In a previous study, α-synuclein was observed moving away from vesicles involved in synaptic transmission and gradually returning.[9,45] During this process, α-synuclein may be forming transient bonds with the synaptic vesicles and may be aiding in vesical synthesis. Interestingly, α-synuclein expression changes in conditions altering neural plasticity or injury, and therefore, the modulation of vesicle synthesis suggests α-synuclein may also have a function in modulating neurotransmitter production and release in response to stress. Similarly, α-synuclein may modulate cysteine string protein-α (CSP-α), a presynaptic protein whose deficiency is implicated with synaptic degeneration. In the mechanism for α-synuclein modulation of CSP-α, there is assembly of the SNARE complex when α-synuclein associates with CSP-α.[43]

Previous studies have implicated the polyprotein motif with promoting α-synuclein's harmful aggregation potential.[8] There are six repeats, containing point mutations implicated in the autosomal-dominant Parkinson's.[46–48] A point mutation may occur in the A30P gene in the third repeat. This mutation is also implicated in promoting protofibril formation.[49,50] In the A30P mutant, the rate of actin polymerization is increased and causes a disruption in the cytoskeleton. Cytoskeleton-associated functions, such as cell migration, may be inhibited, and endocytic/exocytic trafficking is perturbed.[51]

In repeat 5 of the SNCA gene, additional PD-associated point mutations are located at the E46K and A53T sites.[43] A53T mutation is especially interesting. Previous studies have found that the A53T mutation lacks a site essential for calpain 1 recognition. A proteolytic enzyme, calpain 1, uses repeat sequences to target and breakdown harmful α-synuclein. The absence of these recognition sequences means calpain 1 is unable to inhibit the formation of PD-associated Lewy-body aggregates.[52]

The SNCA gene transcript has two shorter variants, a product of alternative splicing, but their pathological and physiological impact has not been well studied.[46–48] Although the etiology of PD is unknown, susceptibility to synuclein may be associated with the genes encoding the human HLA class II molecules including MHC class II gene alleles DRB5*01 and DRB1*15:01.[53] Synuclein-derived peptides can elicit T cell responses in rats and mice, and it has been shown that neuronal death in the SN in α-synuclein overexpression model is absent in MHC II null mice. Synuclein proteins could be post-translationally modified in local microenvironment in which protein processing could be altered,[53] generating neoepitopes that may activate CD4+ T cells and aggravate the disease in PD. Thus, defining the role of microglia in presenting synuclein peptides to T cells under differential conditions of calpain activation will contribute to our knowledge of how synuclein presentation influences disease severity in PD. Synuclein is also expressed in cerebral cortex, bone marrow, colon, kidney, testis, soft tissue, and skin. Thus, peripheral T cells can also be activated by synuclein peptides and the activated T cells may migrate to the brain periphery secreting inflammatory cytokines and chemokines and contribute to the pathogenesis of PD.

4. Microglial activation, neuroinflammation, and neurodegeneration in PD

Microglia are neural immune cells, and they account for approximately 12% of all the cells in the brain.[54–56] They are structurally interesting. In healthy neural tissue, they have a small, circular cell body with branching projections, and this is considered to be its ramified state. Microglia in ramified morphology plays several crucial roles in the nervous system, including synaptic plasticity, axon growth, determination for neuron fate, and migration.[56–58] They are additionally essential for the maturation of the central nervous system for clearing cellular debris via phagocytosis, releasing neurotrophic cell signaling factors, and communicating with neurons.[12,59,60]

Additionally, microglia have the ability to sense homeostatic perturbations and respond accordingly.[58,61] Intense pathological traumas, such as inhibited neuronal function, injury, inflammation, infection, and ischemia, rapidly activate microglia, catalyzing a change in microglia morphology. No longer in the ramified state, the microglia respond to these stressors by manifesting an ameboid morphology to allow for rapid mobilization to the site of trauma. Subsequently, they can secrete cytotoxic factors including nitric oxide and reactive oxygen. They also secrete neurotrophins: nerve growth factors (NGF) and brain derived neurotrophic factors (BDNF), as well as anti-inflammatory/protective cytokines (e.g., IL-4, IL-5, IL-10, etc.), and chemokines (e.g., CXCL9, CXCL10, CXCL-12, etc.). Chemokine receptors are also expressed in the CNS by microglia, astrocytes, neurons, and endothelial cells in response to inflammatory T cells and immune activation. The systemic inflammatory reactions and responses can influence brain function.[59] Similarly, CNS reactions may affect distant lymphoid organs and immune functions by releasing neurohormones and neurotransmitters. Dopamine receptors are known to directly regulate neurotransmission of other neurotransmitters, release of cyclic adenosine monophosphate, cell proliferation, and differentiation.[62] Both human and mouse CD4+ and CD8+ T cells express dopamine receptor D3, and that dopamine plays a significant role in migration and homing of T cells via D3R, influencing the physiopathology of PD-like disease in MPTP mice.[63,64]

Activation of microglia can have dual effects. Short term activation plays a protective role in the CNS through the secretion of pro-inflammatory factors, leading to microglial phagocytosis of invading bacteria, for example. Inversely, long term activation of microglia plays a destructive role in the CNS through the release of pro-inflammatory cytokines: IL-1, IL-6, TNF-α, and IFN-γ. TNF-α signaling is mediated through TNFR1 and TNFR2 receptors, which promote apoptosis and consequential neuron death. Interestingly, factors released by damaged neurons perpetuate continued microglial activation. This will maintain neuroinflammation, thereby promoting death of dopamingeric neurons in the SN as well as neurons in the spinal cord. Neuroinflammation and the degeneration may then align with the pathology of progressive PD.[13,59,65]

5. T cell activation, neuroinflammation, and neurodegeneration in PD

A vital agent of the acquired immune system is the T cell. The CD4+ T cell is a variety of T cells heavily implicated with PD and the associated neurodegeneration. Once activated by antigen-presenting cells (APCs), the T cells will undergo phenotypic changes dependent on nearby mediating factors. For example, T cells may differentiate into a T-helper 1 phenotype when in the presence of IL-12. The diverse phenotypes that the T cells can possess suggest a specialization for target pathogens. Of interest, CD4+ T cells and Th17 T cells can manifest pro-inflammatory functions in a number of neurological disorders including MS,[30,66] optic neuritis,[67,68] and PD.[69,70]

Recent evidence suggests that cytokines activate dopaminergic neurons of the SN, causing them to express MHC class I cells. CD8+ T cells kill neurons that present the appropriate combination of peptide and MHC class I and peptide. Native peptides and nitrated synuclein-derived peptides elicit T cell responses in rats and mice.[11] Inflammatory CD4+ T cells in the CNS can induce neuroinflammation, and depending on the microenvironment and antigenic peptides, these activated T cells may contribute to neuronal death. Microglial presentation of pathogenic synuclein peptides can activate CD4+ T cells in the presence of diverse mediators, producing various cytokines and chemokines which may further activate microglia and maintain a pathogenic environment in PD. Calpain processing of degraded synuclein peptides into fine antigenic epitopes can also occur in the brain although the precise nature of these peptides and their neuroinflammatory and neurodegenerative functions in PD and other neurodegenerative diseases remain unclear.

In the presence of cytokines such as IL-12, the differentiation of CD4+ T cells toward the Th1 is favored,[71,72] and this inflammatory phenotype could be associated with neuroinflammation and neuronal damage. While the role of Th1 CD4+ T cells in PD is well documented, Th17 CD4+ cells have been associated with neuroinflammation and neurodegeneration in the presence of IL-23 in the microenvironment.[72,73] α-Synuclein processing by APCs and microglia, and dysregulation of immune responses in the presence various growth factors can modulate PD-associated neurodegeneration.[74] We have previously shown that calpain activation promotes Th1/Th17 cells in mouse models of human MS.[30] Calpain inhibition can downregulate these inflammatory Th1/Th17 responses and favors Th2 differentiation and production of IL-4, which may contribute to attenuation of multiple CNS diseases.

6. Calpain cleavage of α-synuclein and presentation by microglia and professional APCs

Both calpain and endogenous calpain inhibitor calpastatin are localized in the cytosol, and take part in selective protein cleavage in response to calcium signaling.[75] Alterations in calcium homeostasis may lead to calpain activation and contribute to CNS disorders. Under abnormal calcium signaling, calpastatin can be degraded by calpain, allowing calpain to escape from inhibition by calpastatin. This could be problematic because of insufficient regulation of calpain-mediated proteolysis in various disease situations, which needs to be addressed.

T Cells often require stimulation by professional APCs, like macrophages, B-Cells, and dendritic cells. APCs are a class of cells having antigens (Ag) on its surface side with which MHC complexes can bind. APCs have proteolytic functions, degrading target antigenic proteins into shorter peptides, in an internal, endosomal compartment with proteolytic enzymes. These fragments are subsequently loaded on MHC molecules.[76] Microglial cells have similar processing machinery like professional APCs, and should be able to process synuclein to yield antigenic peptides to stimulate CD4+ T cells in the brain, and present immunogenic peptides to T cells. APCs can also target α-synuclein outside of the brain. The molecular mechanisms between MHC-loading of toxic synuclein peptides and presentation to inflammatory CD4+ T cells have not yet been characterized.

There are substantial gaps in our knowledge of antigen processing and presentation, and the role of APCs in regulating immunodominance in the CNS. Processing reactions in microglia versus professional APCs may modulate the number of peptide: MHC class II complexes by enhancing epitope formation and display, or alternatively catalyzing peptide destruction or editing. Our previous studies have observed a hierarchy of naturally derived peptide epitopes from an autoantigen human IgG, bound to HLA class II DR4 molecules.[77,78] Following immunization of HLA-DR4 transgenic mice, T cells responsive to an immunodominant IgG-derived epitope were almost 20-fold more abundant than cells reactive with a minor or subdominant epitope from this same Ag. Although the binding affinity for HLA-DR4 of both dominant and subdominant IgG epitopes was equal, T cell responses to those epitopes were different. In functional antigen assays utilizing T cell hybridomas, the hierarchy of dominant epitope presentation over subdominant epitope was confirmed using human or murine B cells with either exogenous or endogenous IgG as a source of antigen.

A different hierarchy of epitopes was bound to HLA-DR4 in human a macrophage cell line, THP-1.DR4.[77] In inflammatory autoimmune diseases, it has been hypothesized that initially immune responses are directed against immunodominant epitopes of select Ags with later activation of inflammatory CD4+ T cell responses to subdominant or cryptic epitopes within the same or other Ags. The molecular events which lead to this change in CD4+ T cell responses in the host remain unclear. Thus, Ag processing, synuclein processing in particular, and presentation may be key steps in the development and progression of inflammatory diseases like synucleinopathy and PD in humans. Whether similar steps are involved in synuclein processing and presentation by microglia and astrocytes in the brain remains unclear.

7. Cross presentation of synuclein peptides to T cells

The brain's blood brain barrier is a border protecting the brain from invasion of foreign proteins. For this reason, it was previously understood that the blood brain barrier plays as a separation between the brain and the rest of the body. However, it has been suggested that the brain directly communicates with the immune and endocrine systems.[79] Therefore, when these systems elicit systemic inflammatory reactions and responses, the brain is influenced and can change brain function. Similarly, the brain can communicate and influence processes occurring in the rest of the body via neurohormones and neurotransmitters release, thereby affecting distant endocrine and lymphoid organs and immune functions.[80]

Both human and mouse CD4+ and CD8+ T cells express dopamine receptor D3, and that dopamine plays a significant role in migration and homing of T cells via D3R, influencing the physiopathology of PD-like disease in MPTP mice. Our recent study suggests that both CD4+ and CD8+ T cell populations are activated in MPTP mice during acute and chronic phases of the disease, and that the increased number of inflammatory T cells as well as the severity of the disease, are significantly reduced by calpain inhibition, indicating the role of calpain in the pathology of PD or PD-like disease. Regulatory T cells (CD4+ CD25+, Tregs) are known to suppress immune activation and maintain immune homeostasis and tolerance.[81] While dysfunctions of Tregs are observed in MPTP mice, administration of Tregs into MPTP mice has been shown to suppress microglial inflammatory responses as well as pathogenic functions of Th1/Th17 cells, leading to robust nigrostriatal protection.

Bone marrow, kidney, colon, testis, soft tissue, and skin tissue demonstrate abundant expression of α-synuclein. Because B-cells, macrophages, dendritic cells can efficiently present Ags to T cells eliciting an immune response, it is proposed that macrophages may also process α-synuclein in the body outside of the brain. These synuclein peptide fragments may be captured and loaded onto both macrophages and microglia to activate T cells, and influence immunity in the periphery and the brain (Fig. 2).

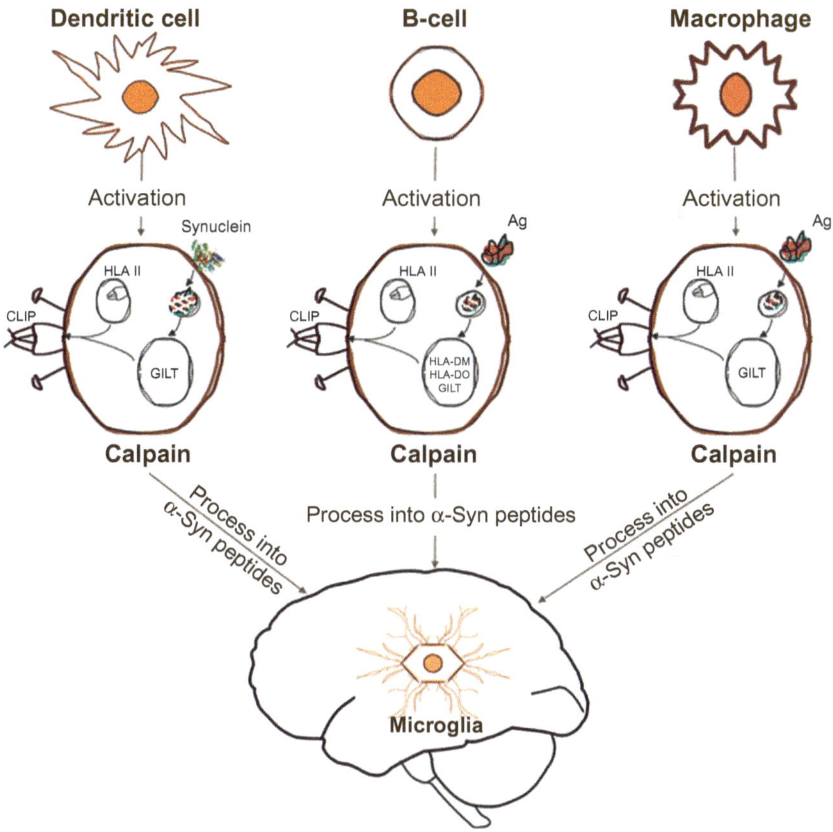

Fig. 2 Cross presentation of α-synuclein peptides to microglia in the brain. When the body elicits an immune response, dendritic cells, B-cells, and macrophages can be activated as a part of this immune response. Upon activation, these APCs can process and present synuclein to activate T cells. Calpain processing of synuclein may occur in the endolysosomal compartments producing MHC-peptide complexes to be captured by other non-professional APCs for immune activation. Gamma-interferon-inducible lysosomal thiol reductase (GILT), and non-classical class II molecules (HLM-DO and HLA-DM) in APCs can modulate immune responses in the host. The processed synuclein peptides can also be transported across the blood brain barrier and presented to microglia for T cell activation and induction of neuroinflammation.

Microglia can also pick up processed synuclein peptides in the brain and can cross present them to T cells in the brain. After activation of T cells, inflammatory cytokines and chemokines may travel to the brain, crossing the blood brain barrier, and thus contributing to the pathogenesis of PD as well as other degenerative diseases.

8. Conclusions

Alterations in calcium homeostasis may lead to calpain activation. We believe differential conditions of calpain activation can contribute to our knowledge of PD disease progression. Particularly interesting is calpin's role in microglial activation. The duality of microglia activation, promoting protection with short term activation and degradation with long term activation, gives hints for understanding the disease's complexity. As one of its damaging roles, microglia cells maintain neuroinflammation, which promote dopamingeric neuron death in the SN and in the spinal cord. In further, microglia can process secreted synuclein and present peptides to T cells, promoting neuroinflammation. Interestingly, bone marrow, kidney, colon, testis, soft tissue, and skin tissue abundantly express α-synuclein. Because antigen presenting cells, such as B-cells, macrophages, dendritic cells, can efficiently present Ags to T cells and therefore elicit an immune response, it is proposed that macrophages may also process α-synuclein in the body outside of the brain. These synuclein peptide fragments may be loaded onto macrophages and microglia for T cell activation in both the periphery and, after crossing the blood brain barrier, in the CNS. Thus, T cell activation may occur in the brain, releasing neuroinflammatory factors, promoting neuronal cell death, and thus aggravating the pathogenesis of PD.

Acknowledgments

This study was made possible by grant from the South Carolina Spinal Cord Injury Research Funds (SCIRF#2016 I-03 and SCIRF #2018 I-01) to A.H. Contents do not necessarily represent the policy of the SCIRF and do not imply endorsement by the funding agency. This work was also supported by grants from the Ralph H. Johnson Veterans Administration Medical Center, Charleston (BX004269, 1I01BX002349-01) to N.L.B.

Conflicts of interest

The authors have no financial conflicts of interest.

References

1. Ball N, et al. Parkinson's disease and the environment. *Front Neurol.* 2019;10:218.
2. Hayes MT. Parkinson's disease and parkinsonism. *Am J Med.* 2019;.
3. Zhou ZD, et al. Molecular targets for modulating the protein translation vital to proteostasis and neuron degeneration in Parkinson's disease. *Transl Neurodegener.* 2019;8:6.
4. Zaichick SV, McGrath KM, Caraveo G. The role of Ca(2+) signaling in Parkinson's disease. *Dis Model Mech.* 2017;10(5):519–535.
5. Bove C, Travagli RA. Neurophysiology of the brainstem in Parkinson's disease. *J Neurophysiol.* 2019;121:1856–1864.
6. Kahle PJ. alpha-Synucleinopathy models and human neuropathology: similarities and differences. *Acta Neuropathol.* 2008;115(1):87–95.
7. Mishizen-Eberz AJ, et al. Cleavage of alpha-synuclein by calpain: potential role in degradation of fibrillized and nitrated species of alpha-synuclein. *Biochemistry.* 2005;44(21):7818–7829.
8. Mehra S, Sahay S, Maji SK. alpha-Synuclein misfolding and aggregation: implications in Parkinson's disease pathogenesis. *Biochim Biophys Acta Proteins Proteomics.* 2019;.
9. Huang M, et al. alpha-Synuclein: a multifunctional player in exocytosis, endocytosis, and vesicle recycling. *Front Neurosci.* 2019;13:28.
10. Dufty BM, et al. Calpain-cleavage of alpha-synuclein: connecting proteolytic processing to disease-linked aggregation. *Am J Pathol.* 2007;170(5):1725–1738.
11. Sulzer D, et al. T cells from patients with Parkinson's disease recognize alpha-synuclein peptides. *Nature.* 2017;546(7660):656–661.
12. Lull ME, Block ML. Microglial activation and chronic neurodegeneration. *Neurotherapeutics.* 2010;7(4):354–365.
13. Caggiu E, et al. Inflammation, infectious triggers, and Parkinson's disease. *Front Neurol.* 2019;10:122.
14. Pontremoli S, Melloni E. Extralysosomal protein degradation. *Annu Rev Biochem.* 1986;55:455–481.
15. Sorimachi H, Saido TC, Suzuki K. New era of calpain research. Discovery of tissue-specific calpains. *FEBS Lett.* 1994;343(1):1–5.
16. Banik NL, et al. Purification of a calcium-activated neutral proteinase from bovine brain. *Neurochem Res.* 1983;8(11):1389–1405.
17. Ono Y, Saido TC, Sorimachi H. Calpain research for drug discovery: challenges and potential. *Nat Rev Drug Discov.* 2016;15(12):854–876.
18. Sorimachi H, Ishiura S, Suzuki K. Structure and physiological function of calpains. *Biochem J.* 1997;328(Pt. 3):721–732.
19. Schaecher K, Goust JM, Banik NL. The effects of calpain inhibition on IkB alpha degradation after activation of PBMCs: identification of the calpain cleavage sites. *Neurochem Res.* 2004;29(7):1443–1451.
20. Mikosik A, et al. Roles of calpain-calpastatin system (CCS) in human T cell activation. *Oncotarget.* 2016;7(47):76479–76495.
21. Ray SK, Hogan EL, Banik NL. Calpain in the pathophysiology of spinal cord injury: neuroprotection with calpain inhibitors. *Brain Res Brain Res Rev.* 2003;42(2):169–185.
22. Wingrave JM, et al. Higher calpastatin levels correlate with resistance to calpain-mediated proteolysis and neuronal apoptosis in juvenile rats after spinal cord injury. *J Neurotrauma.* 2004;21(9):1240–1254.
23. Guyton MK, et al. Calpeptin attenuated inflammation, cell death, and axonal damage in animal model of multiple sclerosis. *J Neurosci Res.* 2010;88(11):2398–2408.
24. Shields DC, et al. A putative mechanism of demyelination in multiple sclerosis by a proteolytic enzyme, calpain. *Proc Natl Acad Sci USA.* 1999;96(20):11486–11491.

25. Ahmad S, et al. Bromine inhalation mimics ischemia-reperfusion cardiomyocyte injury and calpain activation in rats. *Am J Physiol Heart Circ Physiol.* 2019;316(1):H212–H223.
26. Martinez-Thompson JM, Moore SA, Liewluck T. A novel CAPN3 mutation in late-onset limb-girdle muscular dystrophy with early respiratory insufficiency. *J Clin Neurosci* 2018;53:229–231.
27. Schaecher K, et al. Calpain expression and infiltration of activated T cells in experimental allergic encephalomyelitis over time: increased calpain activity begins with onset of disease. *J Neuroimmunol.* 2002;129(1–2):1–9.
28. Shields DC, Banik NL. Pathophysiological role of calpain in experimental demyelination. *J Neurosci Res.* 1999;55(5):533–541.
29. Imam SA, et al. Increased calpain correlates with Th1 cytokine profile in PBMCs from MS patients. *J Neuroimmunol.* 2007;190(1–2):139–145.
30. Smith AW, et al. Regulation of Th1/Th17 cytokines and IDO gene expression by inhibition of calpain in PBMCs from MS patients. *J Neuroimmunol.* 2011;232(1–2):179–185.
31. Trager N, et al. Effects of a novel orally administered calpain inhibitor SNJ-1945 on immunomodulation and neurodegeneration in a murine model of multiple sclerosis. *J Neurochem.* 2014;130(2):268–279.
32. Trager N, et al. The involvement of calpain in CD4(+) T helper cell bias in multiple sclerosis. *J Clin Cell Immunol.* 2013;4(4):1000153.
33. Letavernier E, et al. Critical role of the calpain/calpastatin balance in acute allograft rejection. *Eur J Immunol.* 2011;41(2):473–484.
34. t Hart BA, et al. The major histocompatibility complex influences the ethiopathogenesis of MS-like disease in primates at multiple levels. *Hum Immunol.* 2001;62(12):1371–1381.
35. Laurent L, et al. Prevention of lupus nephritis development in NZB/NZW mice by selective blockade of CD28. *Eur J Immunol.* 2017;47(8):1368–1376.
36. Raab M, et al. LFA-1 activates focal adhesion kinases FAK1/PYK2 to generate LAT-GRB2-SKAP1 complexes that terminate T-cell conjugate formation. *Nat Commun.* 2017;8:16001.
37. Saul S, et al. How ORAI and TRP channels interfere with each other: interaction models and examples from the immune system and the skin. *Eur J Pharmacol.* 2014;739:49–59.
38. Valimaki E, et al. Calpain activity is essential for ATP-driven unconventional vesicle-mediated protein secretion and inflammasome activation in human macrophages. *J Immunol.* 2016;197(8):3315–3325.
39. Carrasco S, Merida I. Diacylglycerol-dependent binding recruits PKCtheta and RasGRP1 C1 domains to specific subcellular localizations in living T lymphocytes. *Mol Biol Cell.* 2004;15(6):2932–2942.
40. Rios R, et al. Solanum paniculatum L. decreases levels of inflammatory cytokines by reducing NFKB, TBET and GATA3 gene expression in vitro. *J Ethnopharmacol.* 2017;209:32–40.
41. Park H, et al. Activation of channel catfish (Ictalurus punctatus) T cells involves NFAT-like transcription factors. *Dev Comp Immunol.* 2002;26(9):775–784.
42. Lee JU, Kim LK, Choi JM. Revisiting the concept of targeting NFAT to control T cell immunity and autoimmune diseases. *Front Immunol.* 2018;9:2747.
43. Stefanis L. alpha-Synuclein in Parkinson's disease. *Cold Spring Harb Perspect Med.* 2012;2(2):a009399.
44. George JM. The synucleins. *Genome Biol.* 2002;3(1):1, REVIEWS3002.
45. Sarafian TA, et al. Stimulation of synaptoneurosome glutamate release by monomeric and fibrillated alpha-synuclein. *J Neurosci Res.* 2017;95(9):1871–1887.
46. Maroteaux L, Scheller RH. The rat brain synucleins; family of proteins transiently associated with neuronal membrane. *Brain Res Mol Brain Res.* 1991;11(3–4):335–343.

47. Ueda K, et al. Molecular cloning of cDNA encoding an unrecognized component of amyloid in Alzheimer disease. *Proc Natl Acad Sci USA*. 1993;90(23):11282–11286.
48. Maroteaux L, Campanelli JT, Scheller RH. Synuclein: a neuron-specific protein localized to the nucleus and presynaptic nerve terminal. *J Neurosci*. 1988;8(8):2804–2815.
49. Conway KA, et al. Acceleration of oligomerization, not fibrillization, is a shared property of both alpha-synuclein mutations linked to early-onset Parkinson's disease: implications for pathogenesis and therapy. *Proc Natl Acad Sci USA*. 2000;97(2):571–576.
50. Fredenburg RA, et al. The impact of the E46K mutation on the properties of alpha-synuclein in its monomeric and oligomeric states. *Biochemistry*. 2007;46(24):7107–7118.
51. Sousa VL, et al. {alpha}-synuclein and its A30P mutant affect actin cytoskeletal structure and dynamics. *Mol Biol Cell*. 2009;20(16):3725–3739.
52. Mishizen-Eberz AJ, et al. Distinct cleavage patterns of normal and pathologic forms of alpha-synuclein by calpain I in vitro. *J Neurochem*. 2003;86(4):836–847.
53. Garretti F, et al. Autoimmunity in Parkinson's disease: the role of alpha-Synuclein-specific T cells. *Front Immunol*. 2019;10:303.
54. Jarry U, et al. Efficiently stimulated adult microglia cross-prime naive CD8+ T cells injected in the brain. *Eur J Immunol*. 2013;43(5):1173–1184.
55. Thion MS, Ginhoux F, Garel S. Microglia and early brain development: an intimate journey. *Science*. 2018;362(6411):185–189.
56. Filipello F, et al. The microglial innate immune receptor TREM2 is required for synapse elimination and normal brain connectivity. *Immunity*. 2018;48(5):979–991 e8.
57. Cunningham C. Microglia and neurodegeneration: the role of systemic inflammation. *Glia*. 2013;61(1):71–90.
58. Lee WH, et al. Reverse signaling of tumor necrosis factor superfamily proteins in macrophages and microglia: superfamily portrait in the neuroimmune interface. *Front Immunol*. 2019;10:262.
59. Nasrolahi A, et al. Immune system and new avenues in Parkinson's disease research and treatment. *Rev Neurosci*. 2019;.
60. Aldana BI. Microglia-specific metabolic changes in neurodegeneration. *J Mol Biol*. 2019;431:1830–1842.
61. Paolicelli RC, Ferretti MT. Function and dysfunction of microglia during brain development: consequences for synapses and neural circuits. *Front Synaptic Neurosci*. 2017;9:9.
62. Mishra A, Singh S, Shukla S. Physiological and functional basis of dopamine receptors and their role in neurogenesis: possible implication for Parkinson's disease. *J Exp Neurosci*. 2018;12. 1179069518779829.
63. Lao CL, et al. Intranasal and subcutaneous administration of dopamine D3 receptor agonists functionally restores nigrostriatal dopamine in MPTP-treated mice. *Neurotox Res*. 2013;24(4):523–531.
64. Gonzalez H, et al. Dopamine receptor D3 expressed on CD4+ T cells favors neurodegeneration of dopaminergic neurons during Parkinson's disease. *J Immunol*. 2013;190(10):5048–5056.
65. Smith JA, et al. Role of pro-inflammatory cytokines released from microglia in neurodegenerative diseases. *Brain Res Bull*. 2012;87(1):10–20.
66. Acharya S, et al. Amelioration of experimental autoimmune encephalomyelitis and DSS induced colitis by NTG-A-009 through the inhibition of Th1 and Th17 cells differentiation. *Sci Rep*. 2018;8(1):7799.
67. Wu GF, Parker Harp CR, Shindler KS. Optic neuritis: a model for the immunopathogenesis of central nervous system inflammatory demyelinating diseases. *Curr Immunol Rev*. 2015;11(2):85–92.
68. Das A, et al. Calpain inhibitor attenuated optic nerve damage in acute optic neuritis in rats. *J Neurochem*. 2013;124(1):133–146.

69. Samantaray S, et al. Inhibition of calpain activation protects MPTP-induced nigral and spinal cord neurodegeneration, reduces inflammation, and improves gait dynamics in mice. *Mol Neurobiol.* 2015;52(2):1054–1066.
70. Kustrimovic N, et al. Parkinson's disease patients have a complex phenotypic and functional Th1 bias: cross-sectional studies of CD4+ Th1/Th2/T17 and Treg in drug-naive and drug-treated patients. *J Neuroinflammation.* 2018;15(1):205.
71. Vacaflores A, et al. Exposure of human CD4 T cells to IL-12 results in enhanced TCR-induced cytokine production, altered TCR signaling, and increased oxidative metabolism. *PLoS One.* 2016;11(6):e0157175.
72. Zhou J, et al. BLT1 in dendritic cells promotes Th1/Th17 differentiation and its deficiency ameliorates TNBS-induced colitis. *Cell Mol Immunol.* 2018;15(12):1047–1056.
73. Prots I, Winner B. Th17 cells: a promising therapeutic target for Parkinson's disease? *Expert Opin Ther Targets.* 2019;23(4):309–314.
74. Solleiro-Villavicencio H, Rivas-Arancibia S. Effect of chronic oxidative stress on neuroinflammatory response mediated by CD4(+)T cells in neurodegenerative diseases. *Front Cell Neurosci.* 2018;12:114.
75. De Tullio R, et al. Differential regulation of the calpain-calpastatin complex by the L-domain of calpastatin. *Biochim Biophys Acta.* 2014;1843(11):2583–2591.
76. Haverson K, et al. Professional and non-professional antigen-presenting cells in the porcine small intestine. *Immunology.* 2000;101(4):492–500.
77. Haque MA, Hawes JW, Blum JS. Cysteinylation of MHC class II ligands: peptide endocytosis and reduction within APC influences T cell recognition. *J Immunol.* 2001;166(7):4543–4551.
78. Doonan BP, Haque A. Prostate cancer immunotherapy: exploiting the HLA class II pathway in vaccine design. *J Clin Cell Immunol.* 2015;6(4):351.
79. Tsygan NV, et al. Autoimmunity in acute ischemic stroke and the role of blood-brain barrier: the dark side or the light one? *Front Med.* 2019;1–7.
80. Fuzzati-Armentero MT, Cerri S, Blandini F. Peripheral-central neuroimmune crosstalk in Parkinson's disease: what do patients and animal models tell us? *Front Neurol.* 2019;10:232.
81. Lane PJ, et al. Lymphoid tissue inducer cells: pivotal cells in the evolution of CD4 immunity and tolerance? *Front Immunol.* 2012;3:24.

CHAPTER FIVE

Calcineurin signaling as a target for the treatment of alcohol abuse and neuroinflammatory disorders

Patrick J. Ronan[a,b,c,*], **Sarah A. Flynn**[b], **Thomas P. Beresford**[c,d]
[a]Sioux Falls VA Research Service, Sioux Falls, SD, United States
[b]Department of Psychiatry and Division of Basic Biomedical Sciences, Sanford USD School of Medicine, Sioux Falls, SD, United States
[c]Laboratory for Clinical and Translational Research in Psychiatry, Research Service and Psychiatry, Rocky Mountain Regional VA Medical Center (116), Aurora, CO, United States
[d]Department of Psychiatry, University of Colorado Denver School of Medicine, Anschutz Medical Campus, Aurora, CO, United States
*Corresponding author: e-mail address: patrick.ronan@usd.edu

Contents

Abstract

Converging lines of evidence point to a significant role of neuroinflammation in a host of psychiatric conditions, including alcohol use disorder, TBI, and PTSD. A complex interaction of both peripheral and central signaling underlies processes involved in neuroinflammation. Calcineurin is a molecule that sits at the nexus of these processes and has been clearly linked to a number of psychiatric disorders including alcohol use disorder (AUD). Like its role in regulating peripheral immune cells, calcineurin (CN) plays an integral role in processes regulating neuroimmune function and neuroinflammatory processes. Targeting CN or elements of its signaling pathways at critical

Progress in Molecular Biology and Translational Science, Volume 167
ISSN 1877-1173
https://doi.org/10.1016/bs.pmbts.2019.06.008

points may aid in the functional recovery from neuroinflammatory related disorders. In this review we will highlight the role of neuroinflammation and calcineurin signaling in AUD, TBI and stress-induced disorders and discuss recent findings demonstrating a therapeutic effect of immunosuppressant-induced calcineurin inhibition in a pre-clinical model of binge alcohol drinking.

1. Introduction

Converging lines of evidence point to a role of neuroinflammation in a host of psychiatric conditions, including alcohol use disorder, TBI, and PTSD. A complex interaction of both peripheral and central signaling underlies processes involved in neuroinflammation. Calcineurin is a molecule that sits at the nexus of these processes and has been clearly linked to a number of psychiatric disorders including alcohol use disorder (AUD). Like its role in regulating peripheral immune cells, calcineurin (CN) plays an integral role in processes regulating neuroimmune function and neuroinflammatory processes. Calcineurin is a "fundamental link" between morphological changes in astrocytes and microglia and immune/inflammatory signaling in both acute and chronic neuroinflammation (for review see Ref.[1]). Calcineurin regulates glial cell activation and cytokine expression and is induced during disease, injury, and aging.[1,2] As such, targeting CN or elements of its signaling pathways at critical points may aid in the functional recovery from neuroinflammatory related disorders.[2,3]

2. Calcineurin

Calcineurin regulates immune/inflammatory processes in both peripheral T cells and brain glial cells through activation of the transcription factors nuclear factor of activated T-cells (NFAT) and nuclear factor kappa-light-chain-enhancer of activated B cells (NFκB). Calcineurin is a Ca^{2+} and calmodulin-dependent serine-threonine phosphatase also known as protein phosphatase 3 (PPP3).[4] Dephosphorylation of NFAT by CN causes activation and nuclear translocation where it stimulates transcription of several proinflammatory genes including interleukin-2 (IL-2).[5–7] Calcineurin mediates activation and translocation of NFκB not through direct dephosphorylation of the transcription factor but through activation of a kinase that causes the removal of inhibitory subunits.[8] A host of pro-inflammatory cytokines strongly activate CN in primary astrocyte culture which leads to the expression of other immune/inflammatory molecules. Through its actions on NFAT and NFκB-dependent transcription, CN seems to be perfectly suited to drive the self-perpetuating "cytokine cycles" that have been implicated in chronic neuroinflammation (for review see Ref.[1]).

3. Alcohol and neuroinflammation

It is now widely accepted that immune mechanisms, and specifically neuroimmune mechanisms, play important, if poorly understood, roles in the etiology and progression of alcohol abuse.[3,9] The hypothesis that an overactive neuroimmune system promotes alcohol consumption is gaining widespread support among scientists.[10–12] The first mention of a possible connection between immunosuppressants and alcohol consumption was published as a brief clinical report in 1990.[13] It was noted that AUD liver graft patients were more abstinent while on immunosuppressants.

Conventional treatment for alcoholism results in a 35–45% abstinence rate after 1 year.[14] Liver transplant for AD patients is controversial because of the concern that the patients may relapse and not only reject the organ, but also exhaust medical and financial resources.[15] However, over the past 15 years, AD liver graft recipients have been noted as a clinically distinct group with much higher abstinence rates than AUD individuals taking FDA approved treatment.[16–22] Most transplant patients 1 year following transplant showed about 90% abstinence rate and about 75–80% after 3 years.[23] These abstinence rates are significantly higher than the abstinent rates seen in non-liver graft patients using conventional treatment. At first, it was assumed that selection of patients, environmental and psychosocial factors contributed to such high abstinence rates. However, detailed studies showed that none of these factors corresponded with the long-term abstinence rates shown in AD liver graft recipients.[24] However, these studies did not rule out the effect of medications frequently used after a transplant. High abstinence rates in AD liver graft recipients following treatment with the immunosuppressant cyclosporine-A (CsA) were noted but did not get much attention.[13]

Our group was the first to test a preclinical strategy to investigate the use of immunosuppressants to treat excessive drinking.[25] It has taken many years for the research community to embrace and investigate this possible connection. But now multiple lines of converging evidence point to neuroinflammation as playing a fundamental role in the development and maintenance of alcohol abuse. In our initial preclinical investigation, mice treated (IP) with the immunosuppressant cyclosporine A (CsA) drank significantly less ethanol in a controlled, continuous-access, two-bottle choice model while there was no effect on total liquid intake.[25]

We then sought to determine the mechanism of this effect by comparing three different immunosuppressants acting though different modes of action in a modified "drinking in the dark" (DID) paradigm of excessive drinking.[26] Again, immunosuppressants were highly effective in decreasing alcohol consumption, *but only those acting through inhibition of the phosphatase calcineurin (CN) and those that crossed the blood-brain barrier.* CsA and tacrolimus (TRL or FK-506) inhibit CN activity through different mechanisms and both decreased ethanol consumption. A third immunosuppressant, sirolimus (SRL), that acts through a CN-independent pathway had no effect on ethanol consumption. Though inhibition of CN was clearly implicated, it was still unclear whether the anti-drinking effects were due to CN inhibition-mediated suppression of peripheral immune responses or direct inhibition of CN in brain or via another mechanism in brain. Both CsA and TRL crossed the blood brain barrier in this study. They also have known inhibitory effects on CN in brain where immunophilins and calcineurin are abundant and widespread.[27] Sirolimus did not effectively penetrate brain[26,28,29] suggesting that a general suppression of peripheral immune function was not enough to decrease excessive alcohol consumption.

We next sought to clarify if the CN-mediated inhibitory effect on drinking is acting via suppression of peripheral immune response or acting directly in the central nervous system (CNS). We hypothesized that central administration of CsA via intracerebroventricular (ICV) injection would be sufficient to inhibit excessive drinking behavior similar to what was observed after peripheral administration (IP). This was the case, implicating neuroregulatory and neuroinflammatory processes in binge drinking behaviors.[3] These findings add to a growing body of work implicating neuroinflammation in alcoholism.

Alcohol use leads to a range of neuroinflammatory responses and multiple compounds or genetic approaches that inhibit neuroimmune function have been shown to limit ethanol intake to varying degrees in animal models.[10–12] These studies target different specific mechanisms, but all converge on the common outcome of reducing neuroimmune activation. Of these compounds, CsA has one of the strongest effects on attenuation of binge drinking; perhaps because along with reducing neuroinflammation, CsA also plays a key role in complementary neuronal signaling processes related to brain reward and addiction.

4. Calcineurin and extended brain reward pathways

The effectiveness of calcineurin inhibitors in reducing ethanol consumption[3,25,26] may be due to their complementary roles in both neuroinflammatory pathways and directly in neuronal signaling pathways related to brain reward and addiction. During the past few decades, the diverse and important roles CN plays in neuronal signaling processes have been more and more appreciated. Calcineurin is one of the most abundant phosphatases in brain and is known to play important roles in a wide variety of signal transduction pathways. Not only involved in neuroimmune processes, CN is also expressed at especially high levels in brain neurons.[1] There are particularly high concentrations of CN in the hippocampus, neocortex, striatum, and amygdala.[1] where CN modulates diverse cellular functions including receptor and ion channel trafficking, ion channel function, apoptosis, and gene regulation.[30] Calcineurin expression is very similar to that of calmodulin kinase II (CamKII) and it is abundantly expressed and in all of the regions of the brain where CaMKII is found including the dopaminergic reward pathway of the ventral tegmental area (VTA) and the nucleus accumbens (NAc).[31]

5. Alcohol, calcineurin, and corticotropin releasing factor

Calcineurin signaling regulates a host of pathways known to be involved in addictive behaviors. One key substrate that has been implicated in ethanol consumption and the development of alcohol abuse is corticotropin releasing factor.[32–34] A primary mediator of central and peripheral stress responses, CRF signaling in brain through CRF1 receptors has been consistently demonstrated to play a key role in many aspects alcohol abuse including binge-like ethanol drinking in the DID model in mice.[34–36] Particularly, the CRF system has been shown to be central to the processes driving higher levels of intake, withdrawal-induced anxiety, relapse after withdrawal, and sensitivity to stress. We have also recently demonstrated that increased conditioned fear cue-induced ethanol consumption is blocked by a CRF1 antagonist (unpublished results).

Calcineurin regulates CRF gene expression and CRF-mediated signaling. Transcription of CRF is mediated via cAMP response element binding protein (CREB) phosphorylation. Phosphorylation of CREB is necessary but not sufficient for CRF transcription.[37] A CN-regulated coactivator called CRTC (cAMP-regulated transcriptional co-activators, formerly known as "TORC," transducer of regulated CREB activity, not to be confused with "TORC"—mTOR signaling complex) is also required for CREB-induced CRF transcription.[38,39] Dephosphorylation of CRTC by CN facilitates translocation to the nucleus where it interacts with the dimerization and DNA binding domain of CREB and enables the recruitment of the pre-initiation complex. This facilitates transcriptional activation of CREB-dependent genes including CRF. Stress alone induces nuclear translocation of CRTC and transcription of CRF in rodents and knockdown of CRTC isoforms completely blocks cyclic AMP-stimulated CRH promoter activity without affecting CREB phosphorylation.[38,40] This CRTC mediated mechanism of stress induced CRF expression may underlie the enhanced sensitivity to stressors seen in models of ethanol abuse. Calcineurin is also involved in CRF receptor mediated signaling through a CN-NFAT pathway. Cyclosporine blocks CRF-receptor mediated catecholamine synthesis and release through inhibition of NFAT activation in cell culture.[41]

6. Calcineurin and reward signaling

Monoamine activity is also integral to processes related to reward, stress and addiction and CN has been shown to be involved in monoamine signaling. Along with direct effects on monoamine synthesis and release it has been shown that CN activity regulates an important dopamine and addiction related signal transduction protein—dopamine and cAMP-regulated neuronal phosphoprotein (DARPP-32). This protein regulates both dopaminergic and glutamatergic (NMDA) receptor responses in extended brain reward pathways[42–44] and is critically involved in regulating responses to ethanol and other drugs of abuse.[45,46] DARPP-32 is a direct CN substrate and is a critical factor for dopamine dependent striatal synaptic plasticity.[47] Dopamine causes increased phosphorylation and CN acts to maintain low levels of phosphorylated DARPP in neostriatum.[48] It has been hypothesized that CN inhibition (i.e., CsA) acts as a dopamine mimetic via dopamine-like increases in phosphorylated DARPP-32.[49] In this way, CN inhibition could potentially dampen temporally linked dopamine reward cues elicited by alcohol and lessen reinforcement.

Calcineurin also plays a regulatory role in other neurotransmitter systems related to alcohol reward and addiction, including NMDA and GABAergic neurons. We have demonstrated that quantitative NMR metabolomics can reliably determine brain levels of major neurotransmitters (such as glutamate, GABA, aspartate) and their changes upon CN inhibition.[50] We demonstrated that CsA, but not SRL, reduces brain levels of glutamate and GABA.[28,29,51] We have also shown that CsA and TRL treatment in the brain led to a specific decrease in mitochondrial metabolism of neurons, while their requirements for carbohydrate consumption (glucose) were elevated.[28,52] Interestingly, we and other have shown that only CN-inhibitors (but not SRL) are known to produce a unique protective effects against ischemia, so-called ischemic preconditioning, by preserving/slowing down mitochondrial metabolism and glutamate release in the brain, heart and kidney.[53,54] The proposed protective mechanisms of CsA are directly linked to its CN-inhibition including increased phosphorylation of ERK and activation of AMP-activated protein kinase AMPK, specifically in rat hippocampus.[55–57] AMPK signaling has long been recognized as an important regulator of brain, heart and body metabolism and energy homeostasis. There is a need to further establish the precise metabolic crosstalk between alcohol-induced changes in the brain and CN-inhibition using a metabolomic approach.

7. Corticotropin releasing factor, ethanol and neuroimmune processes

There is a reciprocal functional interaction between neuroimmune and CRF systems in brain. CRF is a critical mediator of neuroimmune responses including those seen after chronic ethanol administration. Enhanced cytokine expression (CCL2, IL1β, and TNFα mRNA) following acute withdrawal from chronic ethanol treatment is induced by CRF through CRF1 activation[58] and cytokines drive anxiety during ethanol withdrawal via CRF signaling in the CeA.[58,59] There is also a significant complex interaction between proinflammatory cytokines and the HPA axis. Cytokines such as IL-1β and IL-6 stimulate hypothalamic CRF transcription, CRF-induced ACTH release, and subsequent glucocorticoid release that functions to dampen further inflammatory responses.[60,61]

8. Calcineurin signaling in other neuroinflammatory disorders

Along with neurocognitive impairments, other persistent debilitating effects of TBI are common. These include chronic relapsing conditions such as PTSD and substance abuse, especially alcohol.[62] These problems have been perhaps best identified in military populations. Alcohol has been recognized in a recent Institute of Medicine report as the key substance use problem in need of intervention and/or treatment among military personnel.[63] The rate of alcohol use disorder (AUD) among veterans is 32%, significantly higher than the general population.[64] and there is evidence that Veterans with TBI have almost twice that as Veterans without TBI.[65] Comorbid mental health conditions occur in 37% of people with a drinking problem.[66] One of the most debilitating comorbid conditions is PTSD. Estimates of PTSD rates in Veterans who served in OEF/OIF range from double to almost 10 times that of community rates.[67,68] Veterans from other conflicts also have a higher incidence. Vietnam era Veterans have lifetime rates estimated as high as 30%.[69] Complicating the matter, individuals with PTSD, in the absence of TBI, have very high rates of comorbid AUD and substance abuse ranging from 28% to 75%.[69,70] These conditions are inextricably intertwined and the causal mechanisms underlying them are likely shared, at least in part.

One key mechanism that connects these conditions is dysregulated neuroinflammatory signaling (Fig. 1). Neuroinflammation is a key feature associated with TBI, stress, and alcohol abuse. As such this is an area ripe for the investigation and development of therapies. As we have recently shown, an anti-inflammatory immunosuppressant acting in brain is sufficient to decrease binge alcohol drinking.[3] Evidence supports the notion that this general strategy targeting various aspects of neuroimmune signaling will be effective in treating some TBI comorbid conditions. Presently, there are no FDA approved medications for the treatment of comorbid TBI/AUD/PTSD.

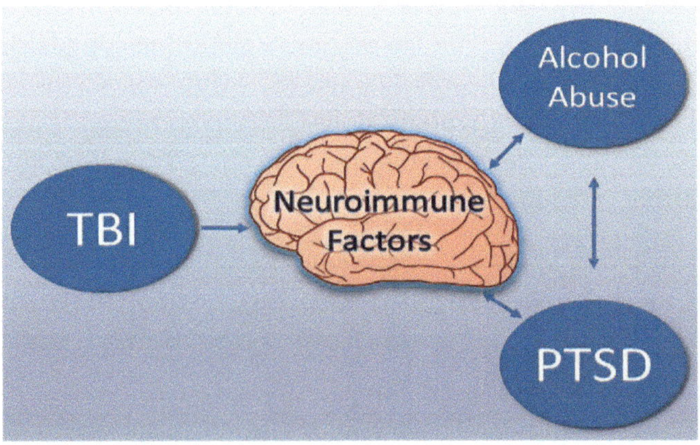

Fig. 1 Dysregulated neuroinflammatory signaling is a common factor linking multiple comorbid conditions.

9. Traumatic brain injury and alcohol abuse

By far, most studies investigating a link between alcohol abuse and TBI center on alcohol as a risk factor not the opposite. It is established that alcohol abuse is a risk factor for TBI but, importantly, alcohol abuse after TBI has been linked to increased seizures, psychiatric disorders including mood and anxiety.[71] There is not a clear answer to the question of whether TBI increases alcohol use due to many factors. First of all, many TBI patients abused alcohol before their injury which is highly predictive of abuse after injury. Second, after an acute decline in drinking found in many studies many return to drinking, or start drinking, so long-term follow-up studies are needed, and are lacking. Finally, other comorbid conditions such as PTSD and not TBI may account for increased use. There is an enhanced risk of increased alcohol abuse after TBI in military populations (for review see Ref.[71]). Given the high incidence of TBI among Veterans in Iraq and Afghanistan, post-TBI drinking has been the subject of many recent studies. A link between combat brain injury and increased alcohol abuse has been reported by numerous studies.[72–76] However, in some other studies, comorbid PTSD was suggested as the cause.[77,78] Regardless, TBI increases the risk for PTSD which, in turn, increases the risk for alcohol abuse. Even when PTSD is accounted for, TBI has been found to be a significant risk factor for increased drinking.[72,73] In one large study of over 13,000 Veterans who served in Afghanistan and/or Iraq, confirmed active duty TBI was associated with a threefold increase in substance abuse.[79] Other studies of soldiers have also found increased association between TBI and alcohol abuse. A study of over 4000 British soldiers who served in the same theaters found that TBI more than doubled the risk of later alcohol abuse.[76] A third study of active duty soldiers also found an association between TBI and increased binge-drinking (odds ratio = 1.48).[72,73]

There is not a clear understanding why there is a greater link between TBI and alcohol abuse in military personnel but certainly, there is an interaction with stressors and a link with enhanced PTSD. It is likely that increased neuroinflammatory processes caused by chronic stressors may exacerbate the effects of TBI and that stressors after TBI may not only prevent healing but may act on a sensitized neuroinflammatory system yielding greater negative outcomes. There is evidence of this from a recent study. A single exposure to a predator odor delivered 8 months after blast-induced TBI produces elevated, persistent anxiety-like behavior compared to controls.[80]

10. Traumatic brain injury and neuroinflammation

There is evidence that TBI injuries cause chronic dysregulated neuroinflammatory processes (for review see Refs.[81,82]). Much evidence suggests that after well-documented acute increases in neuroinflammatory processes after TBI insult, there is a more protracted increase as well. Inflammation following TBI involves a complex interaction between and within peripheral and central systems. There is early activation of microglia and peripheral neutrophil recruitment followed by infiltration of lymphocytes and monocyte-derived macrophages along with a complex array of both pro- and anti-inflammatory cytokines. Post-TBI inflammation can be beneficial as it promotes both clearance of debris and regeneration. Acute inflammatory activity likely is also involved in secondary injury but later, more delayed sub-acute inflammatory responses may be beneficial for healing. Broad spectrum immunosuppressants are effective in preventing injury in models of ischemia but may not be as effective in the stages after insult. In fact, neuroinflammation may subside in the later stages of TBI. In rats 40 weeks after exposure to a model of low-energy blast-induced TBI chronic neuroinflammation (40 weeks) does not occur in the brain regions examined though they did find that focal hemorrhage may trigger chronic focal neuroinflammation.[83] It is likely that while TBI does not necessarily lead to long-term overt measures of neuroinflammation, it sensitizes neuroinflammatory systems such that responses to future insults such as those caused by stress and alcohol are more robust.

11. Stress and neuroinflammation

Stress also has a major effect on neuroinflammatory processes. Both acute and chronic stressors stimulate neuroinflammatory systems. Pro-inflammatory activation of microglia and increased infiltrated monocyte-derived brain macrophages result from psychosocial stressors.[84–86] There is also an interaction between anxiety and neuroinflammation. High anxiety is associated with higher pro-inflammatory activation of microglia, especially after a lipopolysaccharide (LPS) challenge that drives a pro-inflammatory polarization of macrophages and microglia.[87–89] Microglial changes in morphology, indicative of functional responses, are induced by both classic inflammatory signals such as LPS, but many of these same morphological changes are induced by psychosocial stress (for review see Ref.[90]).

Overall, a variety of rodent models of stress have demonstrated increased microglial activity as determined by increased Iba-1 expression in key limbic and brain reward regions that regulate neural responses to both stress and substance abuse such as amygdala, nucleus accumbens, paraventricular nucleus, hippocampus, and prefrontal cortex.[90]

Evidence to date suggests an important role of neuroinflammatory processes in the development and expression of depression and PTSD. Different animal models of mild TBI have shown have been shown to increase anxiety and PTSD-like phenotypes.[91,92] Recently, we have shown that inducing neuroinflammation with LPS (ICV) in lieu of a traumatic stressor dose-dependently causes a depression/PTSD-like phenotype (unpublished data). Together these studies suggest that traumatic stressors are acting, at least in part, through neuroinflammatory mechanisms to cause PTSD. Stressors also cause increased cytokine expression such as IL-1β in brain which is thought to, at least in part, mediate the neurochemical and behavioral consequences of stress such as PTSD.[93]

Given the complexity of variables surrounding the connection between TBI and alcohol abuse there is a clear need for good animal models to gain better insight into the complex connections between these disorders. Most rodent studies have focused solely on relatively acute alcohol use after TBI, none have investigated the interaction of TBI and stressors on drinking behaviors. We need to understand the persistent, more protracted effects of TBI most commonly seen in our VA clinics to better develop strategies for functional recovery. To accomplish this, researchers will have to utilize models of TBI, combined with stress and PTSD models and alcohol paradigms to better understand their relationship and to decipher interactive mechanisms as well as test putative pre-clinical treatment strategies.

12. Genetic approach to understanding calcineurin function

Identification of the specific pathways and substrates involved immunosuppressant effects on drinking and other neuroinflammatory disorders needs to be studied taking a variety of approaches. One line of inquiry among many that we are pursuing is to identify specific CLN mediated effects using a genetic approach. We are using Cre-LoxP conditional knockouts to determine the role of specific calcineurin-mediated pathways. We have mouse line with a floxed regulatory subunit of calcineurin (a generous gift of S. Tonegawa, MIT) that we are using to determine cell types and regions in brain contributing to this response.[94] For example, to test the hypothesis that calcineurin pathways in the dopamine reward system are regulating rewarding properties of ethanol, we are knocking out CLN specifically in dopaminergic neurons and testing mice in drinking models. To accomplish this, floxed CLN mice are crossed with another line expressing cre recombinase in tyrosine hydroxylase expressing neurons. Likewise, to determine in general if CLN in glia or neurons mediate the anti-drinking effect of CsA we have crossed our floxed mice with mice expressing cre recombinase only in glia or neurons. Another strategy we are employing involves the use of viral vectors expressing cre or vectors to induce RNA interference mediated silencing of CLN.[95,96] This allows for the focal knockout/silencing of CLN in specific brain regions. These studies and others will yield information that will allow us to target specific aspects of CLN signaling to better enable us to develop effective treatments for AUD and other neuroinflammatory disorders.

13. Conclusions

Calcineurin signaling provides a rich avenue for the pursuit of therapeutic targets for treatment of AUD, TBI, and PTSD in humans. Calcineurin is key mediator of neuroinflammation and is directly involved in neuronal signaling pathways known to be associated with a host of psychiatric disorders. A growing body of evidence suggests that various calcineurin mediated neuro-immuno-modulatory pathways could provide novel targets for the development of pharmacotherapies.[3,12,97] Calcineurin acts in both glia and neurons to control signaling related to neuroinflammation, reward, and stress. These pathways include NFAT, DARPP-32, CREB via its co-activator CRTC (TORC), monoamine synthesis and release, CRF, GABA, and NMDA signaling, among others. Given the relative ubiquity of calcineurin regulated signaling in brain it is imperative to understand mechanisms of its signaling in order to determine effective targets and strategies for precision therapies. Because calcineurin signaling is so diffuse in brain, global inhibition with antagonists, though effective as an anti-drinking agent, is associated significant side effects, both neurological and peripheral. The ultimate goal is to understand the specific calcineurin-mediated pathways involved in psychiatric disorders and provide novel treatments.

Acknowledgments

This work was supported by a Merit Review Award from the Department of Veterans Affairs (1 I01 BX004712-01, T.B. and P.R.), an Advanced VA Research Fellowship (PR), Great Plains Veterans Research Foundation (PR). It also received support from the USD Center for Brain and Behavior Research and the U. Discover Summer Research Scholars Program (P.R.). The views expressed in this article are those of the authors and do not necessarily reflect the position or policy of the Department of Veterans Affairs or the United States government.

References

1. Furman JL, Norris CM. Calcineurin and glial signaling: neuroinflammation and beyond. *J Neuroinflammation*. 2014;11:158.
2. Kipanyula MJ, Kimaro WH, Seke Etet PF. The emerging roles of the calcineurin-nuclear factor of activated T-lymphocytes pathway in nervous system functions and diseases. *J Aging Res*. 2016;2016:5081021.
3. Ronan PJ, Strait SA, Palmer GM, Beresford TP. Central administration of cyclosporine a decreases ethanol drinking. *Alcohol Alcohol*. 2018;53(2):193–199.
4. Liu J, Farmer Jr JD, Lane WS, Friedman J, Weissman I, Schreiber SL. Calcineurin is a common target of cyclophilin-cyclosporin A and FKBP-FK506 complexes. *Cell*. 1991;66(4):807–815.
5. Musson RE, Smit NP. Regulatory mechanisms of calcineurin phosphatase activity. *Curr Med Chem*. 2011;18(2):301–315.
6. Clipstone NA, Crabtree GR. Calcineurin is a key signaling enzyme in T lymphocyte activation and the target of the immunosuppressive drugs cyclosporin A and FK506. *Ann N Y Acad Sci*. 1993;696:20–30.
7. Ho S, Clipstone N, Timmermann L, et al. The mechanism of action of cyclosporin A and FK506. *Clin Immunol Immunopathol*. 1996;80(3 Pt. 2):S40–S45.
8. Palkowitsch L, Marienfeld U, Brunner C, Eitelhuber A, Krappmann D, Marienfeld RB. The Ca2+−dependent phosphatase calcineurin controls the formation of the Carma1-Bcl10-Malt1 complex during T cell receptor-induced NF-kappaB activation. *J Biol Chem*. 2011;286(9):7522–7534.
9. Mayfield J, Harris RA. The neuroimmune basis of excessive alcohol consumption. *Neuropsychopharmacology*. 2017;42(1):376.
10. Mayfield J, Ferguson L, Harris RA. Neuroimmune signaling: a key component of alcohol abuse. *Curr Opin Neurobiol*. 2013;23(4):513–520.
11. Crews FT, Lawrimore CJ, Walter TJ, Coleman Jr LG. The role of neuroimmune signaling in alcoholism. *Neuropharmacology*. 2017;122:56–73.
12. Cui C, Shurtleff D, Harris RA. Neuroimmune mechanisms of alcohol and drug addiction. *Int Rev Neurobiol*. 2014;118:1–12.
13. Giles HG, Orrego H, Sandrin S, Saldivia V. The influence of cyclosporine on abstinence from alcohol in transplant patients. *Transplantation*. 1990;49(6):1201–1202.
14. Yates WR, Booth BM, Reed DA, Brown K, Masterson BJ. Descriptive and predictive validity of a high-risk alcoholism relapse model. *J Stud Alcohol*. 1993;54(6):645–651.
15. Zhou M, Wagner LM, Diflo T, Naegle M. Implementation of the high-risk alcoholism relapse scale in a liver transplant clinic. *Gastroenterol Nurs*. 2015;38(6):447–454.
16. Beresford TP, Lucey MR. Alcoholics and liver transplantation: facts, biases, and the future. *Addiction*. 1994;89(9):1043–1048.

17. Campbell Jr DA, Magee JC, Punch JD, Merion RM, Turcotte JG, Bromberg JS. One center's experience with liver transplantation: alcohol use relapse over the long-term. *Liver Transpl Surg*. 1998;4(5 Suppl 1):S58–S64.
18. Pageaux GP, Michel J, Coste V, et al. Alcoholic cirrhosis is a good indication for liver transplantation, even for cases of recidivism. *Gut*. 1999;45(3):421–426.
19. Cowling T, Jennings LW, Jung GS, et al. Comparing quality of life following liver transplantation for Laennec's versus non-Laennec's patients. *Clin Transpl*. 2000;14(2): 115–120.
20. Mackie J, Groves K, Hoyle A, et al. Orthotopic liver transplantation for alcoholic liver disease: a retrospective analysis of survival, recidivism, and risk factors predisposing to recidivism. *Liver Transpl*. 2001;7(5):418–427.
21. Pereira SP, Williams R. Alcohol relapse and functional outcome after liver transplantation for alcoholic liver disease. *Liver Transpl*. 2001;7(3):204–205.
22. DiMartini A, Weinrieb R, Fireman M. Liver transplantation in patients with alcohol and other substance use disorders. *Psychiatr Clin North Am*. 2002;25(1):195–209.
23. Berlakovich GA, Steininger R, Herbst F, Barlan M, Mittlbock M, Muhlbacher F. Efficacy of liver transplantation for alcoholic cirrhosis with respect to recidivism and compliance. *Transplantation*. 1994;58(5):560–565.
24. Lucey MR, Carr K, Beresford TP, et al. Alcohol use after liver transplantation in alcoholics: a clinical cohort follow-up study. *Hepatology*. 1997;25(5):1223–1227.
25. Beresford HF, Deitrich R, Beresford TP. Cyclosporine-A discourages ethanol intake in C57bl/6j mice: a preliminary study. *J Stud Alcohol*. 2005;66(5):658–662.
26. Beresford T, Fay T, Serkova NJ, Wu PH. Immunophyllin ligands show differential effects on alcohol self-administration in C57BL mice. *J Pharmacol Exp Ther*. 2012;341(3): 611–616.
27. Dawson TM, Steiner JP, Lyons WE, Fotuhi M, Blue M, Snyder SH. The immunophilins, FK506 binding protein and cyclophilin, are discretely localized in the brain: relationship to calcineurin. *Neuroscience*. 1994;62(2):569–580.
28. Serkova N, Jacobsen W, Niemann CU, et al. Sirolimus, but not the structurally related RAD (everolimus), enhances the negative effects of cyclosporine on mitochondrial metabolism in the rat brain. *Br J Pharmacol*. 2001;133(6):875–885.
29. Gottschalk S, Cummins CL, Leibfritz D, Christians U, Benet LZ, Serkova NJ. Age and sex differences in the effects of the immunosuppressants cyclosporine, sirolimus and everolimus on rat brain metabolism. *Neurotoxicology*. 2011;32(1):50–57.
30. Baumgartel K, Mansuy IM. Neural functions of calcineurin in synaptic plasticity and memory. *Learn Mem*. 2012;19(9):375–384.
31. Sola C, Tusell JM, Serratosa J. Comparative study of the distribution of calmodulin kinase II and calcineurin in the mouse brain. *J Neurosci Res*. 1999;57(5):651–662.
32. Heilig M, Koob GF. A key role for corticotropin-releasing factor in alcohol dependence. *Trends Neurosci*. 2007;30(8):399–406.
33. Ronan PJ, Summers CH. Molecular signaling and translational significance of the corticotropin releasing factor system. *Prog Mol Biol Transl Sci*. 2011;98:235–292.
34. Phillips TJ, Reed C, Pastor R. Preclinical evidence implicating corticotropin-releasing factor signaling in ethanol consumption and neuroadaptation. *Genes Brain Behav*. 2015;14(1):98–135.
35. Sparta DR, Hopf FW, Gibb SL, et al. Binge ethanol-drinking potentiates corticotropin releasing factor R1 receptor activity in the ventral tegmental area. *Alcohol Clin Exp Res*. 2013;37(10):1680–1687.
36. Giardino WJ, Ryabinin AE. CRF1 receptor signaling regulates food and fluid intake in the drinking-in-the-dark model of binge alcohol consumption. *Alcohol Clin Exp Res*. 2013;37(7):1161–1170.

37. Liu Y, Kamitakahara A, Kim AJ, Aguilera G. Cyclic adenosine 3′,5′-monophosphate responsive element binding protein phosphorylation is required but not sufficient for activation of corticotropin-releasing hormone transcription. *Endocrinology*. 2008;149(7): 3512–3520.
38. Liu Y, Coello AG, Grinevich V, Aguilera G. Involvement of transducer of regulated cAMP response element-binding protein activity on corticotropin releasing hormone transcription. *Endocrinology*. 2010;151(3):1109–1118.
39. Spencer RL, Weiser MJ. TORC: a new twist on corticotropin-releasing hormone gene expression. *Endocrinology*. 2010;151(3):855–858.
40. Liu Y, Knobloch HS, Grinevich V, Aguilera G. Stress induces parallel changes in corticotrophin-releasing hormone (CRH) transcription and nuclear translocation of transducer of regulated cAMP response element-binding activity 2 in hypothalamic CRH neurones. *J Neuroendocrinol*. 2011;23(3):216–223.
41. Dermitzaki E, Tsatsanis C, Gravanis A, Margioris AN. The calcineurin-nuclear factor of activated T cells signaling pathway mediates the effect of corticotropin releasing factor and urocortins on catecholamine synthesis. *J Cell Physiol*. 2012; 227(5):1861–1872.
42. Greengard P, Allen PB, Nairn AC. Beyond the dopamine receptor: the DARPP-32/ protein phosphatase-1 cascade. *Neuron*. 1999;23(3):435–447.
43. Svenningsson P, Nishi A, Fisone G, Girault JA, Nairn AC, Greengard P. DARPP-32: an integrator of neurotransmission. *Annu Rev Pharmacol Toxicol*. 2004; 44:269–296.
44. Bibb JA, Snyder GL, Nishi A, et al. Phosphorylation of DARPP-32 by Cdk5 modulates dopamine signalling in neurons. *Nature*. 1999;402(6762):669–671.
45. Svenningsson P, Nairn AC, Greengard P. DARPP-32 mediates the actions of multiple drugs of abuse. *AAPS J*. 2005;7(2):E353–E360.
46. Donohue T, Hoffman PL, Tabakoff B. Effect of ethanol on DARPP-32 phosphorylation in transgenic mice that express human type VII adenylyl cyclase in brain. *Alcohol Clin Exp Res*. 2005;29(3):310–316.
47. Yagishita S, Hayashi-Takagi A, Ellis-Davies GC, Urakubo H, Ishii S, Kasai H. A critical time window for dopamine actions on the structural plasticity of dendritic spines. *Science*. 2014;345(6204):1616–1620.
48. Nishi A, Snyder GL, Nairn AC, Greengard P. Role of calcineurin and protein phosphatase-2A in the regulation of DARPP-32 dephosphorylation in neostriatal neurons. *J Neurochem*. 1999;72(5):2015–2021.
49. Wera S, Neyts J. Calcineurin as a possible new target for treatment of Parkinson's disease. *Med Hypotheses*. 1994;43(3):132–134.
50. Serkova N, Brand A, Christians U, Leibfritz D. Evaluation of the effects of immunosuppressants on neuronal and glial cells in vitro by multinuclear magnetic resonance spectroscopy. *Biochim Biophys Acta*. 1996;1314(1–2):93–104.
51. Klawitter J, Gottschalk S, Hainz C, Leibfritz D, Christians U, Serkova NJ. Immunosuppressant neurotoxicity in rat brain models: oxidative stress and cellular metabolism. *Chem Res Toxicol*. 2010;23(3):608–619.
52. Serkova NJ, Christians U, Benet LZ. Biochemical mechanisms of cyclosporine neurotoxicity. *Mol Interv*. 2004;4(2):97–107.
53. Serkova N, Donohoe P, Gottschalk S, et al. Comparison of the effects of cyclosporin a on the metabolism of perfused rat brain slices during normoxia and hypoxia. *J Cereb Blood Flow Metab*. 2002;22(3):342–352.
54. Christians U, Gottschalk S, Miljus J, et al. Alterations in glucose metabolism by cyclosporine in rat brain slices link to oxidative stress: interactions with mTOR inhibitors. *Br J Pharmacol*. 2004;143(3):388–396.

55. Ikeda Y, Miura T, Sakamoto J, et al. Activation of ERK and suppression of calcineurin are interacting mechanisms of cardioprotection afforded by delta-opioid receptor activation. *Basic Res Cardiol.* 2006;101(5):418–426.
56. Park HG, Yi H, Kim SH, et al. The effect of cyclosporine A on the phosphorylation of the AMPK pathway in the rat hippocampus. *Prog Neuro-Psychopharmacol Biol Psychiatry.* 2011;35(8):1933–1937.
57. Jeanneteau FD, Lambert WM, Ismaili N, et al. BDNF and glucocorticoids regulate corticotrophin-releasing hormone (CRH) homeostasis in the hypothalamus. *Proc Natl Acad Sci U S A.* 2012;109(4):1305–1310.
58. Whitman BA, Knapp DJ, Werner DF, Crews FT, Breese GR. The cytokine mRNA increase induced by withdrawal from chronic ethanol in the sterile environment of brain is mediated by CRF and HMGB1 release. *Alcohol Clin Exp Res.* 2013;37(12):2086–2097.
59. Knapp DJ, Whitman BA, Wills TA, et al. Cytokine involvement in stress may depend on corticotrophin releasing factor to sensitize ethanol withdrawal anxiety. *Brain Behav Immun.* 2011;25(Suppl 1):S146–S154.
60. Venihaki M, Dikkes P, Carrigan A, Karalis KP. Corticotropin-releasing hormone regulates IL-6 expression during inflammation. *J Clin Invest.* 2001;108(8):1159–1166.
61. Kageyama K, Tamasawa N, Suda T. Signal transduction in the hypothalamic corticotropin-releasing factor system and its clinical implications. *Stress.* 2011;14(4): 357–367.
62. Department of Defense Alcohol and Substance Abuse Disorder Research Program. *The Alcohol Use Disorder and Comorbid PTSD/TBI Landscape.* Department of Defense; 2017, ed2017.
63. Army Center for Substance Abuse Program. *Report.* Department of Defense; 2011.
64. Lan CW, Fiellin DA, Barry DT, et al. The epidemiology of substance use disorders in US veterans: a systematic review and analysis of assessment methods. *Am J Addict.* 2016;25(1):7–24.
65. Taylor BC, Hagel EM, Carlson KF, et al. Prevalence and costs of co-occurring traumatic brain injury with and without psychiatric disturbance and pain among Afghanistan and Iraq war veteran V.A. users. *Med Care.* 2012;50(4):342–346.
66. Kessler RC, Crum RM, Warner LA, Nelson CB, Schulenberg J, Anthony JC. Lifetime co-occurrence of DSM-III-R alcohol abuse and dependence with other psychiatric disorders in the National Comorbidity Survey. *Arch Gen Psychiatry.* 1997;54(4):313–321.
67. Milliken CS, Auchterlonie JL, Hoge CW. Longitudinal assessment of mental health problems among active and reserve component soldiers returning from the Iraq war. *JAMA.* 2007;298(18):2141–2148.
68. Hoge CW, McGurk D, Thomas JL, Cox AL, Engel CC, Castro CA. Mild traumatic brain injury in U.S. soldiers returning from Iraq. *N Engl J Med.* 2008;358(5): 453–463.
69. Kulka RA. *Trauma and the Vietnam War Generation: Report of Findings From the National Vietnam Veterans Readjustment Study.* New York: Brunner/Mazel; 1990.
70. Baker DG, Heppner P, Afari N, et al. Trauma exposure, branch of service, and physical injury in relation to mental health among U.S. veterans returning from Iraq and Afghanistan. *Mil Med.* 2009;174(8):773–778.
71. Weil ZM, Corrigan JD, Karelina K. Alcohol abuse after traumatic brain injury: experimental and clinical evidence. *Neurosci Biobehav Rev.* 2016;62:89–99.
72. Adams RS, Corrigan JD, Larson MJ. Alcohol use after combat-acquired traumatic brain injury: what we know and don't know. *J Soc Work Pract Addict.* 2012;12(1):28–51.
73. Adams RS, Larson MJ, Corrigan JD, Horgan CM, Williams TV. Frequent binge drinking after combat-acquired traumatic brain injury among active duty military personnel with a past year combat deployment. *J Head Trauma Rehabil.* 2012;27(5): 349–360.

74. Adams RS, Larson MJ, Corrigan JD, et al. Combat-acquired traumatic brain injury, posttraumatic stress disorder, and their relative associations with postdeployment binge drinking. *J Head Trauma Rehabil.* 2016;31(1):13–22.
75. Johnson LA, Eick-Cost A, Jeffries V, Russell K, Otto JL. Risk of alcohol use disorder or other drug use disorder among U.S. service members following traumatic brain injury, 2008–2011. *Mil Med.* 2015;180(2):208–215.
76. Rona RJ, Jones M, Fear NT, et al. Mild traumatic brain injury in UK military personnel returning from Afghanistan and Iraq: cohort and cross-sectional analyses. *J Head Trauma Rehabil.* 2012;27(1):33–44.
77. Miles SR, Graham DP, Teng EJ. Examining the influence of mild traumatic brain injury and posttraumatic stress disorder on alcohol use disorder in OEF/OIF veterans. *Mil Med.* 2015;180(1):45–52.
78. Polusny MA, Kehle SM, Nelson NW, Erbes CR, Arbisi PA, Thuras P. Longitudinal effects of mild traumatic brain injury and posttraumatic stress disorder comorbidity on postdeployment outcomes in national guard soldiers deployed to Iraq. *Arch Gen Psychiatry.* 2011;68(1):79–89.
79. Carlson KF, Nelson D, Orazem RJ, Nugent S, Cifu DX, Sayer NA. Psychiatric diagnoses among Iraq and Afghanistan war veterans screened for deployment-related traumatic brain injury. *J Trauma Stress.* 2010;23(1):17–24.
80. Perez-Garcia G, Gama Sosa MA, De Gasperi R, et al. Exposure to a predator scent induces chronic behavioral changes in rats previously exposed to low-level blast: implications for the relationship of blast-related TBI to PTSD. *Front Neurol.* 2016;7:176.
81. Simon DW, McGeachy MJ, Bayir H, Clark RS, Loane DJ, Kochanek PM. The far-reaching scope of neuroinflammation after traumatic brain injury. *Nat Rev Neurol.* 2017;13(3):171–191.
82. Quillinan N, Herson PS, Traystman RJ. Neuropathophysiology of brain injury. *Anesthesiol Clin.* 2016;34(3):453–464.
83. Gama Sosa MA, De Gasperi R, Perez Garcia GS, et al. Lack of chronic neuroinflammation in the absence of focal hemorrhage in a rat model of low-energy blast-induced TBI. *Acta Neuropathol Commun.* 2017;5(1):80.
84. McKim DB, Niraula A, Tarr AJ, Wohleb ES, Sheridan JF, Godbout JP. Neuroinflammatory dynamics underlie memory impairments after repeated social defeat. *J Neurosci.* 2016;36(9):2590–2604.
85. Wohleb ES, Fenn AM, Pacenta AM, Powell ND, Sheridan JF, Godbout JP. Peripheral innate immune challenge exaggerated microglia activation, increased the number of inflammatory CNS macrophages, and prolonged social withdrawal in socially defeated mice. *Psychoneuroendocrinology.* 2012;37(9):1491–1505.
86. Wohleb ES, Powell ND, Godbout JP, Sheridan JF. Stress-induced recruitment of bone marrow-derived monocytes to the brain promotes anxiety-like behavior. *J Neurosci.* 2013;33(34):13820–13833.
87. Piirainen S, Youssef A, Song C, et al. Psychosocial stress on neuroinflammation and cognitive dysfunctions in Alzheimer's disease: the emerging role for microglia? *Neurosci Biobehav Rev.* 2017;77:148–164.
88. Li Z, Ma L, Kulesskaya N, Voikar V, Tian L. Microglia are polarized to M1 type in high-anxiety inbred mice in response to lipopolysaccharide challenge. *Brain Behav Immun.* 2014;38:237–248.
89. Orihuela R, McPherson CA, Harry GJ. Microglial M1/M2 polarization and metabolic states. *Br J Pharmacol.* 2016;173(4):649–665.
90. Calcia MA, Bonsall DR, Bloomfield PS, Selvaraj S, Barichello T, Howes OD. Stress and neuroinflammation: a systematic review of the effects of stress on microglia and the implications for mental illness. *Psychopharmacology.* 2016;233(9):1637–1650.

91. Elder GA, Dorr NP, De Gasperi R, et al. Blast exposure induces post-traumatic stress disorder-related traits in a rat model of mild traumatic brain injury. *J Neurotrauma.* 2012;29(16):2564–2575.
92. Davies DR, Olson D, Meyer DL, et al. Mild traumatic brain injury with social defeat stress alters anxiety, contextual fear extinction, and limbic monoamines in adult rats. *Front Behav Neurosci.* 2016;10:71.
93. Mendoza C, Barreto GE, Avila-Rodriguez M, Echeverria V. Role of neuroinflammation and sex hormones in war-related PTSD. *Mol Cell Endocrinol.* 2016;434: 266–277.
94. Zeng H, Chattarji S, Barbarosie M, et al. Forebrain-specific calcineurin knockout selectively impairs bidirectional synaptic plasticity and working/episodic-like memory. *Cell.* 2001;107(5):617–629.
95. Scammell TE, Arrigoni E, Thompson MA, Ronan PJ, Saper CB, Greene RW. Focal deletion of the adenosine A1 receptor in adult mice using an adeno-associated viral vector. *J Neurosci.* 2003;23(13):5762–5770.
96. Callahan LB, Tschetter KE, Ronan PJ. Inhibition of corticotropin releasing factor expression in the central nucleus of the amygdala attenuates stress-induced behavioral and endocrine responses. *Front Neurosci.* 2013;7:195.
97. Coller JK, Hutchinson MR. Implications of central immune signaling caused by drugs of abuse: mechanisms, mediators and new therapeutic approaches for prediction and treatment of drug dependence. *Pharmacol Ther.* 2012;134(2):219–245.

CHAPTER SIX

Emerging evidence for the role of pituitary adenylate cyclase-activating peptide in neuropsychiatric disorders

Kabirullah Lutfy[a,*], Gollapudi Shankar[b]
[a]College of Pharmacy, Western University of Health Sciences, Pomona, CA, United States
[b]Outpatient Clinic, VA Affairs-VCB, Laredo, TX, United States
*Corresponding author: e-mail address: klutfy@westernu.edu

Contents

Abstract

Stress activates many brain nuclei and causes acute changes in several physiological and behavioral responses to restore homeostasis in affected organisms. While this response is protective, chronic stress exposure causes the sustained activation of these nuclei, leading to maladaptive physiological changes that underlie pathological mood and affective states. Hence, chronic stress may produce anxiety and mood disorders by promoting neuronal plasticity within these stress-responsive nuclei. A growing body of evidence attributes neuropeptide systems in mediating not only the physiological stress response but also pathological states that develop following chronic stress exposure. Recent preclinical data suggest that pituitary adenylyl cyclase-activating polypeptide (PACAP) and its receptors (PAC_1, $VPAC_1$, and $VPAC_2$) play an important role in the behavioral and endocrine responses to stress, as well as in mood and affective disorders. Human studies also point out the significance of the PACAP/PAC_1 receptor system in these disorders. For instance, PACAP through PAC_1 receptor up-regulates the expression of DISC1 (disrupted in schizophrenia 1) and impedes its association with its interacting protein. Interestingly, the DISC1 gene mutation is linked to schizophrenia and depression. Moreover, a link between PACAP blood titer and fear physiology,

Progress in Molecular Biology and Translational Science, Volume 167
ISSN 1877-1173
https://doi.org/10.1016/bs.pmbts.2019.06.009

post-traumatic stress disorder (PTSD) diagnosis and symptoms has been reported in heavily traumatized female patients. Additionally, in the peripheral blood, methylation of the gene encoding the PAC_1 receptor is also associated with PTSD. This book chapter describes the emerging evidence that entails PACAP in the stress response and stress-mediated neuropsychiatric disorders.

1. Introduction

Pituitary adenylyl cyclase activating polypeptide (PACAP), a pleiotropic polypeptide, is related to the secretin/glucagon/vasoactive intestinal polypeptide (VIP) family and is present in two biologically active forms, PACAP-38 and PACAP-27.[1] PACAP and VIP have a high resemblance in their primary structure, and both peptides exert a variety of functions through activating G-protein coupled receptors. VIP and PACAP share two common receptors ($VPAC_1$ and $VPAC_2$ receptors) while only PACAP binds with high affinity to PAC_1 receptors.[2]

High to moderate expression of PACAP and its selective (PAC_1) receptor have been reported in the hypothalamic and limbic structures and in other brain regions as well as in peripheral tissues.[3] The broader distribution of PACAP in and outside the central nervous system suggests its involvement in several biological processes, such as regulation of food intake, energy metabolism, body temperature, neuronal survival, and the neuroendocrine and behavioral response to stress.[3–7]

Earlier studies showed expression of PACAP and its receptors in the hypothalamus and adrenal glands. Predominantly abundant innervations of hypophysiotropic neurons by nerve fibers comprising of the neuropeptide PACAP suggests a prominent role of this peptide/receptor system in the control of the hypothalamic-pituitary-adrenal (HPA) axis and stress response (reviewed in Ref. 8). Along this line, PACAP increases the expression of CRH in the paraventricular nucleus (PVN) of the hypothalamus[9–11] and intra-PVN administration of PACAP mimics behavioral changes result from the activation of PVN by CRH or a stressor.[12] Likewise, restraint stress-induced secretion of corticosterone is attenuated in mice lacking PACAP,[13] suggesting that PACAP plays a functional role in the stress response.

While activation of the stress circuit acutely promotes a “fight or flight” response, the prolonged activation of the HPA axis can lead to maladaptive neuroendocrine and behavioral changes. The negative behavioral consequences of chronic stress result from continuous or inappropriate activation of the stress circuit and continued secretion of stress mediators, leading to improper activation of the fear circuit and the development of anxiety. These mediators also exert anti-rewarding actions, leading to diminished mood and anhedonia as observed in depressed patients. These mediators even overwhelm the sleep-wake cycle that causes insomnia as well as daytime somnolence which are frequently reported in patients with post-traumatic stress disorder (PTSD). Interruption of executive and cognitive functions also could occur as an outcome of continuous activation of the stress circuit.[14] Considering that stress is one of the leading causes of neuropsychiatric disorders, we first discuss the role of PACAP in the stress response followed by its involvement in the pathophysiology of stress-induced mental illnesses.

2. PACAP and stress response

Stress can be defined as a condition when homeostasis is threatened, or the organism observes a situation to be threatening. The HPA axis mediates the stress response. Corticotropin releasing factor/hormone (CRF/CRH) has long been known as the key regulator of the HPA axis and plays a crucial role in stress responses.[15] Briefly, activation of the HPA axis by a stressor involves the release of CRH from neurosecretory cells in the PVN of the hypothalamus. CRH travels through the portal circulation to the pituitary and stimulates the release of adrenocorticotropin hormone (ACTH) from the anterior pituitary. ACTH, in turn, reaches its site of action and acts on the adrenal glands to cause the release of glucocorticoids, which not only mediate the stress response but also act as negative feedback to reduce further activation of the HPA axis.[16] The latter is a biologically protective mechanism that promotes a "fight or flight" response to a stressor. However, the prolonged activation of the HPA axis can lead to maladaptive behavioral changes, such as anxiety and/or depression, in which the negative feedback mechanism may be impaired. Thus, the HPA axis can be the target not only for the stressors but also that of endogenous chemicals that serve as regulators of the stress response and could contribute to the pathophysiology of stress-related neuropsychiatric disorders.

PACAP and its receptors are found in the hypothalamus and adrenal glands. Predominantly abundant innervation of hypophysiotropic neurons by nerve fibers comprising of the neuropeptide PACAP suggests a prominent role of this peptide/receptor system in the control of the HPA axis and stress response (reviewed in Ref. 8). Indeed, PACAP increases the expression of CRH in the PVN[9–11] and local administration of PACAP into the PVN mimics behavioral changes result from the activation of PVN by CRH or a stressor,[12] showing that PACAP is involved in the stress response. Along this line, restraint stress causes secretion of corticosterone in wild-type mice, a response that is attenuated in mice lacking PACAP.[13] Moreover, repeated stress (2–3 h per day for 7 days) leads to unhabituated corticosterone secretion and this response is reduced in mice lacking PACAP.[17]

Chronic variable stress increase PACAP and PAC_1 receptor transcript in the bed nucleus of stria terminalis (BNST), raising the possibility that PACAP may regulate the stress response and could contribute to the pathophysiology of mood and affective disorders via acting not only on the HPA axis but also on extrahypothalamic brain regions. Indeed, local administration of PACAP into the BNST increases corticosterone secretion[18,19] and induces anxiety-like behaviors.[18] Furthermore, PACAP injection into the central nucleus of amygdala (CeA) mimics the behavioral and neuroendocrine manifestations of stress.[20]

Epinephrine and norepinephrine release from the adrenal medulla represent the cardinal feature of acute stress responses. Neuropeptides are active as transmitters in this circuitry. Remarkably, recent findings in the preceding decade have enhanced our understanding of the perception of the process of neuropeptide–catecholamine interactions in the brain and periphery. A new picture of stress circuitry has emerged, in which catecholamine and neuropeptide systems are intimately interconnected, both centrally and peripherally, during responses to both systemic and psychogenic stress. PACAP modulates the HPA axis in response to acute psychogenic but not systemic stressors, through activation of CRH (reviewed in Ref. 8). PACAP facilitates the biosynthesis of catecholamines in the adrenal medulla.[21] Thus, one other target could be via a local action at the level of the adrenal gland which has not been fully explored and not covered in this chapter.

Recent studies have suggested that action of antidepressants on synaptic plasticity is mediated by their regulatory influence not only upon small-molecule neurotransmitters but also via neuropeptides which may act both as neurotransmitters and as neuromodulators. Neuropeptides and catecholamines act on several circuits in the central and peripheral nervous systems to facilitate systemic and psychological stress responses, as well as enduring adaptive and maladaptive changes to stress. These could lead to survival with resilience, or survival with vulnerability, as manifested in mood and affective disorders, such as anxiety, depression, PTSD, obesity and eating disorders (reviewed in Ref. 22). A growing body of literature suggests that the PACAP/PAC_1 receptor system may be a potential contributor to neuropsychiatric disorders, which is described below.

3. PACAP and anxiety disorders

Anxiety disorders are the most common psychiatric disorder in the USA. Anxiety disorders usually are long-lasting and debilitating and characterized by thoughts and feelings of uncontrollable fear and accompanied by somatic, affective, and behavioral symptoms. PACAP is present in most of the brainstem areas including autonomic sensory and motor system related areas, but also in the centers involved in stress adaptation, such as the ventral tegmental area and locus coeruleus.[7] As stated above, local administration of PACAP in the PVN influences the CRF expression[9] and mimics behavioral changes result from the activation of PVN by CRH or a stressor.[12] Furthermore, intracerebroventricular PACAP administration produces anxiety- and depressive-like behaviors in rodents, effects produced by CRF.[23] On the other hand, PACAP deficiency seems to be protective against the deleterious effects and preventing the chronic activation of the HPA axis.[24]

PACAP and PAC_1 receptors are widely expressed in the CeA and BNST, raising the possibility that this peptide/receptor system is not only involved in the stress response but also in anxiety disorders. Interestingly, fear conditioning and estrogen replacement therapy each is shown to increase PAC_1 receptor gene expression.[25] Furthermore, local administration of PACAP in the CeA is found to be anxiogenic in rats and that the melanocortin receptor 4 (MC4R) system mediates the anxiogenic effects of PACAP.[20] These authors suggested that dysregulation of this neuropeptide system may contribute to the pathophysiology of the anxiety group of disorders.[20] Interestingly, chronic variable stress, which leads to anxiety and depression in rodents, increases PACAP and PAC_1 receptor transcript expression in the BNST and local administration of PACAP in the BNST induces anxiety-like behaviors in rats.[18] On the other hand, continuous infusion of PACAP antagonist (PACAP6–38) in the BNST reduces anxiety-like behaviors induced by chronic variable stress,[18] suggesting that PACAP alterations in the BNST may mediate stress-induced BNST neuroplasticity and that this may be the underlying mechanism of some forms of human anxiety disorders. In line with these observations, mice lacking

PACAP spend more time on the open arms of the elevated plus maze compared to their wild-type controls,[26–28] suggesting that the lack of PACAP is protective against the development of anxiety disorders.

Human studies have also suggested the possible involvement of the PACAP/receptor system in the pathophysiology of anxiety and PTSD.[25] Interestingly, a single nucleotide polymorphism in the PAC_1 receptor gene leads to decreases in amygdala and hippocampus connectivity and may contribute to aberrant fear and anxiety disorders.[29] Together, these studies assign a significant role to neuropeptide systems in the mediation of the stress response, making them potential drug targets for the treatment of stress and anxiety disorders.

4. PACAP and major depression

Major depressive disorder (MDD) is an enduring and life-threatening psychiatric condition characterized by low mood, psychomotor changes, and marked decreased interest or pleasure in most activities, also known as anhedonia (American Psychiatric Association 2013). Major depression is among the leading causes of disability around the globe according to the World Health Organization (WHO, 2012). The lifetime prevalence of depression in the industrialized population ranges between 15% and 25%.[30] It is important to note that patients diagnosed with depression show somewhat different symptoms and the diagnostic criteria themselves are already somewhat contradictory as opposing phenomena might support the same "major depressive disorder" diagnosis (like hyperphagia vs. hypophagia, insomnia vs. hypersomnia or weight gain vs. weight loss). Considering this heterogeneous pathophysiological background, it is not surprising that 30% of patients do not respond to the currently available medications influencing monoaminergic neurotransmission in the brain. The need for new strategies to treat the disease is evident.

PACAP has been a critical neurotransmitter/neuropeptide mediating activation of the HPA and hormonal sympathetic adrenal axes. CRH not only initiates activation of the HPA axis via the release of ACTH from the pituitary but also is released from hypothalamic and extrahypothalamic neurons to feedback on noradrenergic systems driving central to peripheral stress "executive" programs (reviewed in Ref. 22). PACAP and its receptor PAC_1 are widely expressed in several nuclei of the hypothalamus as well as in various extra-hypothalamic regions including the amygdala, hippocampus, and nucleus accumbens.[7,31,32] Given the importance of these brain regions in motivated behaviors, reward, and the pathophysiology of mental illnesses, PACAP and its receptors may be involved in the pathogenesis of depressive disorders and other neuropsychiatric disorders. Indeed, there is accumulating evidence that the PACAP-ergic signaling plays a vital role in the behavioral and endocrine responses to stress, as well as in synaptic plasticity and neuroprotection.[24,28,33–35]

Intracranial self-stimulation (ICSS) is a useful approach in which electrical activation of specific brain regions is associated with reward. In rodents, it is achieved by the animal performing an operant response, such as lever pressing. An increase in the ICSS threshold is considered as a reflection of deficits in brain reward circuitries, used as a sign of anhedonia, that is a core symptom of major depression and other mood disorders (American Psychiatric Association 2013). Chronic treatment with antidepressant drugs attenuate increases in the ICSS threshold induced by cocaine withdrawal.[36] PACAP, on the other hand, increases ICSS threshold, suggesting that PACAP promotes depressive-like behaviors in rodents.[37] PACAP administration also reduces social interaction and preference for a saccharine solution, which is commonly used as a model of anhedonia. However, it fails to alter immobility time in the forced swim test, an animal model of despair.[37] The changes observed are blocked by the PACAP receptor antagonist, PACAP6-38, suggesting that the PACAP/PAC_1 receptor system may play a functional role in the pathogenesis of depression, particularly the anhedonia and social dysfunction associated with the disease.[37]

While exogenous PACAP appears to promote depression-like behaviors in rodents, conflicting results are reported regarding the role of endogenous PACAP in these behaviors in mice lacking PACAP. Hattori and colleagues[27] reported a slight reduction in immobility time in the forced swim test in mice lacking PACAP, indicating attenuation of depressive-like behaviors in the absence of endogenous PACAP. However, other groups show that mice lacking PACAP express increased immobility time in the forced swim test,[34,38–40] suggesting that endogenous PACAP may serve to protect against the development of depressive-like behaviors in rodents.

Endogenous PACAP, however, promotes depressive-like behaviors in subjects exposed to chronic stress. Thus, early life stress (180-min maternal deprivation) in combination with chronic variable stress induces depressive-like behaviors in wild-type mice, and this behavioral change is found to be reduced in mice lacking PACAP.[41] Additionally, chronic variable stress is also found to cause reduced depressive-like behaviors in adult mice lacking PACAP compared to their wild-type controls.[42] Moreover, Lehmann and colleagues[24] demonstrate that a 14-day social defeat stress exposure caused a significant increase in PVN *c*-fos expression as well as plasma corticosterone secretion and depression-like behaviors in wild-type mice. These stress-mediated changes are found to be reduced in mice lacking PACAP, suggesting that prolonged activation of the PACAP/PAC_1 receptor system in response to chronic stress may lead to the enduring changes observed in depression.

Human studies underpin the significance of the PACAP/PAC_1 receptor system in mood disorders. For instance, PACAP through PAC_1 receptor up-regulates the expression of the protein DISC1 (disrupted in schizophrenia 1) and radically but momentarily interferes with the interaction of DISC1 with its interacting protein.[43] Interestingly, the DISC1 gene mutation is linked not only to schizophrenia[44] but also to depression.[45,46] Furthermore, Hashimoto and colleagues[40] report on a possible linkage between the PACAP gene and MDD, in which they compare 637 MDD patients versus 967 healthy controls and find that a variant in the PACAP gene is associated with MDD. Additionally, a SNP in the PAC_1 receptor gene leads to a significant interaction between this variant and neighborhood crime for patients suffering from major depression.[47] Interestingly, the PACAP variant gene (rs2856966) is found to be associated with a better treatment outcome to venlafaxine.[48] Taken together, the above studies suggest that the PACAP system may represent a potential target for the development of novel pharmacotherapies to treat depression and possibly other stress-mediated neuropsychiatric disorders.

5. PACAP and post-traumatic stress disorder (PTSD)

Post-traumatic stress disorder (PTSD) is a pervasive and well-known of all the trauma- and stress-related disorders with a twofold greater prevalence in women than men. It affects about 5–10% of the US population and is sparked by exposure to an acute or recurring traumatic event(s). Patients with PTSD, in comparison to trauma-exposed controls, not only exhibit psychological symptoms but also abnormally high conditioned fear responses. This high level of fear may result from a lack of ability to habituate to aversive stimuli, to extinguish or inhibit fear memories, and/or possibly strengthening of the original fear memory.[49–51]

The cardinal feature of PTSD is the dysregulation of the HPA axis, and that the system is altered and hyper-responsive to cortisol feedback. Recent research in PTSD suggests that increased activity of CRH-containing circuits is involved in the pathophysiology of the disease.[52] The neuropeptide PACAP is involved in regulating CRH, a key mediator of the HPA axis,[9] raising the possibility that the PACAP system may be involved in the pathophysiology of PTSD and other stress-related neuropsychiatric disorders.

Available genetic data suggest that PACAP/PAC_1 receptor expression and signaling may be integrally involved in regulating the psychological and physiological responses to traumatic stress and variants of PACAP or PAC_1 receptor gene may be potential contributors of PTSD. For example, in highly traumatized female but not male subjects, there seems to be a strong sex-specific association of PACAP blood levels and PAC_1 receptor gene variants with fear discrimination, PTSD diagnosis, and symptoms.[25] The sex-specific association occurs within a single-nucleotide polymorphism (rs2267735) that is located on the so-called estrogen response element within the PAC_1 receptor gene that mediates the regulation of PAC_1 receptor gene by the sex hormones.[25] In line with these human data, Mercer and colleagues report that fear conditioning and estradiol additively increase the expression of the PAC_1 receptor gene in mice.[53] In line with this, in women with low levels of estradiol, the CC allele is associated with reduced PAC_1 receptor gene expression which is linked to higher signs and symptoms of PTSD.[53] These data suggest that perturbations in the PACAP/PAC_1 receptor system are regulated by estrogen and are involved in abnormal fear responses underlying PTSD. Additionally, these sex-related differences in PACAP signaling during emotional learning could provide novel targets for the treatment of PTSD.[54]

Functional magnetic resonance imaging of patients with moderate to high trauma revealed a higher reactivity of limbic structures, such as the amygdala and hippocampus, in patients with this genotype. Likewise, there was decreased connectivity of the amygdala and hippocampus in these individuals,[55] suggesting that this gene alteration may contribute to aberrant fear, which is a hallmark of PTSD and anxiety disorders.[29] Moreover, decreased hippocampal activity is observed in female subjects carrying the PAC_1 variant.[56] Considering that genetic risk factors may account for up to 30–40% of the heritability of PTSD, the above findings suggest that perturbations in the PACAP/PAC_1 receptor system are regulated by estrogen and are involved in abnormal fear responses underlying PTSD. Additionally, these sex-related differences in PACAP signaling during emotional learning could provide novel targets for the treatment of PTSD.[54]

6. PACAP and schizophrenia

Schizophrenia is a severe and debilitating neuropsychiatric brain disorder that affects 1% of the population. It is equally prevalent across people irrespective of ethnicity, race, and nationality across the globe and is equally prevalent in men and women. Its cause is due to the interaction of several abnormal genes with environmental factors. The course of schizophrenia is variable, with some individuals experiencing exacerbations and others remaining stable but chronically ill.[57] Schizophrenia is not only a neurodevelopmental disorder but also neurodegenerative disorder.[58] There have been several reviews recently of various aspects of PACAP as well as its role in neurodevelopmental disorders.[59] It has been proposed that PACAP may play a role in mental diseases and that regulation of the PACAPergic signals could be a potential treatment for schizophrenia considering schizophrenia is not only neurodegenerative but also neurodegenerative disorder.[60] PAC_1 receptor gene alternative splicing could contribute to the risk of developing schizophrenia, not only by affecting embryonic neurogenesis but also by impairing proliferation and differentiation of adult stem cells.[61] Accordingly, a reduction in the amount of hippocampal neural stem cells is reported in post-mortem schizophrenic brains,[62] raising the possibility that this could explain the decrease in hippocampal volume and the correlation with several cognitive deficits observed in schizophrenic patients.

PAC_1 is a G-protein-coupled receptor encoded by *Adcyap1*, which is a potential schizophrenia susceptibility gene in rodents and humans. Variations of the PACAP and/or PACAP receptor genes may be a genetic risk factor for psychiatric disorders including schizophrenia-like phenotype in mice.[63] Studies have shown that PACAP-deficient mice exhibit behavioral and neurophysiological abnormalities including emotional lability, depression-like behaviors, and memory impairments that are ameliorated with atypical antipsychotic drugs. For example, Ago and colleagues[38] show that mice lacking PACAP display reduced performance in the object recognition test. Likewise, PACAP heterozygous mice exhibit deficits in sensorimotor gating, as measured in the prepulse inhibition paradigm following the administration of an aversive stimulus, i.e., administration of a $5HT_2$ receptor agonist.[64] However, an earlier study reports no change in prepulse inhibition in mice lacking PACAP compared to their wild-type controls.[27] These authors also show more social interaction, unlike previous studies, with a mild performance deficit in working memory in mice lacking PACAP.[27]

Human studies, however, have implicated the PACAP/PAC_1 receptor system in schizophrenia. For instance, the PACAP gene allele that is over-expressed in patients with schizophrenia is found to be associated with reduced hippocampal volume and poorer memory performance.[63] Likewise, PACAP through PAC_1 receptor up-regulates the production of the protein DISC1 and drastically but briefly reduces interaction of DISC1 with its interacting protein[43] and may alter neurite outgrowth. Interestingly, DISC1 gene mutation was linked to schizophrenia.[44] Likewise, stathmin1, which induces abnormal axonal arborization, may be a linkage for the involvement of PACAP in schizophrenia, as the expression of stathamin 1 is upregulated in the brain of mice lacking PACAP as well as in patients with schizophrenia (reviewed in Ref. 60). These findings suggest that it is possible that neural development via these two proteins is disturbed in schizophrenic patients and PACAP may be involved in this process.

Not only the PACAP/PAC_1 receptor system but also the PACAP/$VPAC_2$ signaling is suggested to be associated with mental illnesses. Indeed, a duplication of the chromosome that encodes the $VPAC_2$ receptor is found to be linked to schizophrenia.[65] Thus, restoring the changes induced by the PACAP/PAC_1/$VPAC_2$ signaling may be a novel approach to treat schizophrenia.

7. Concluding remarks

The PACAP/PAC_1 receptor system has been implicated in several neuropsychiatric disorders. Previous studies have shown alterations in anxiety and depression in mice lacking PACAP or its receptor compared to their wild-type controls. However, considering that compensatory developmental changes may occur in knockout mice, further studies using pharmacological tools and other approaches are needed to fully characterize the role of the PACAP/PAC1 receptor system in these disorders. Currently, the lack of selective small-molecule PAC1 receptor antagonists represents a challenge. However, attempts have been made to develop such small molecules.[66]

PACAP has the potential to be neuroprotective and regulate mood and affective states via multiple mechanisms. Earlier studies have demonstrated that PACAP is associated with the expression of tyrosine hydroxylase, which is the rate-limiting enzyme in the catecholamine synthesis. This can be significant in the monoamine theory of depression. Likewise, studies have shown that PACAP can increase the expression of brain-derived neurotrophic factor (BDNF), which is involved in neurogenesis and neuronal survival.

PACAP is thus well positioned to alter mood and be involved in the pathophysiology of depression and related disorders. Furthermore, studies have shown that PACAP receptors are also found on macrophages and immune cells. Thus, PACAP has the potential to impact neurological and immunological pathways via altering the function of these cells. Future studies can be directed to address these unexplored areas of research.

Acknowledgment

This chapter was supported by the Tobacco Related Disease Research Program (TRDRP) Grant # 24RT-0023 (funded to K.L.).

References

1. Miyata A, Arimura A, Dahl RR, et al. Isolation of a novel 38 residue-hypothalamic polypeptide which stimulates adenylate cyclase in pituitary cells. *Biochem Biophys Res Commun*. 1989;164(1):567–574.
2. Cazillis M, Gonzalez BJ, Billardon C, et al. VIP and PACAP induce selective neuronal differentiation of mouse embryonic stem cells. *Eur J Neurosci*. 2004;19(4): 798–808.
3. Ghatei MA, Takahashi K, Suzuki Y, Gardiner J, Jones PM, Bloom SR. Distribution, molecular characterization of pituitary adenylate cyclase-activating polypeptide and its precursor encoding messenger RNA in human and rat tissues. *J Endocrinol*. 1993;136(1):159–166.
4. Reglodi D, Kiss P, Horvath G, et al. Effects of pituitary adenylate cyclase activating polypeptide in the urinary system, with special emphasis on its protective effects in the kidney. *Neuropeptides*. 2012;46(2):61–70.
5. Reglodi D, Kiss P, Szabadfi K, et al. PACAP is an endogenous protective factor-insights from PACAP-deficient mice. *J Mol Neurosci*. 2012;48(3):482–492.
6. Reglodi D, Tamas A, Koppan M, Szogyi D, Welke L. Role of PACAP in female fertility and reproduction at gonadal level—recent advances. *Front Endocrinol (Lausanne)*. 2012;3:155.
7. Vaudry D, Falluel-Morel A, Bourgault S, et al. Pituitary adenylate cyclase-activating polypeptide and its receptors: 20 years after the discovery. *Pharmacol Rev*. 2009;61(3): 283–357.
8. Mustafa T. Pituitary adenylate cyclase-activating polypeptide (PACAP): a master regulator in central and peripheral stress responses. *Adv Pharmacol*. 2013;68:445–457.
9. Agarwal A, Halvorson LM, Legradi G. Pituitary adenylate cyclase-activating polypeptide (PACAP) mimics neuroendocrine and behavioral manifestations of stress: evidence for PKA-mediated expression of the corticotropin-releasing hormone (CRH) gene. *Brain Res Mol Brain Res*. 2005;138(1):45–57.
10. Grinevich V, Fournier A, Pelletier G. Effects of pituitary adenylate cyclase-activating polypeptide (PACAP) on corticotropin-releasing hormone (CRH) gene expression in the rat hypothalamic paraventricular nucleus. *Brain Res*. 1997;773(1–2): 190–196.
11. Stroth N, Liu Y, Aguilera G, Eiden LE. Pituitary adenylate cyclase-activating polypeptide controls stimulus-transcription coupling in the hypothalamic-pituitary-adrenal axis to mediate sustained hormone secretion during stress. *J Neuroendocrinol*. 2011;23(10): 944–955.

12. Norrholm SD, Das M, Legradi G. Behavioral effects of local microinfusion of pituitary adenylate cyclase activating polypeptide (PACAP) into the paraventricular nucleus of the hypothalamus (PVN). *Regul Pept.* 2005;128(1):33–41.
13. Tsukiyama N, Saida Y, Kakuda M, et al. PACAP centrally mediates emotional stress-induced corticosterone responses in mice. *Stress.* 2011;14(4):368–375.
14. Charmandari E, Tsigos C, Chrousos G. Endocrinology of the stress response. *Annu Rev Physiol.* 2005;67:259–284.
15. Vale W, Spiess J, Rivier C, Rivier J. Characterization of a 41-residue ovine hypothalamic peptide that stimulates secretion of corticotropin and beta-endorphin. *Science.* 1981;213(4514):1394–1397.
16. Vyas S, Rodrigues AJ, Silva JM, et al. Chronic stress and glucocorticoids: from neuronal plasticity to neurodegeneration. *Neural Plast.* 2016;2016:6391686.
17. Jiang SZ, Eiden LE. Activation of the HPA axis and depression of feeding behavior induced by restraint stress are separately regulated by PACAPergic neurotransmission in the mouse. *Stress.* 2016;19(4):374–382.
18. Roman CW, Lezak KR, Hartsock MJ, et al. PAC1 receptor antagonism in the bed nucleus of the stria terminalis (BNST) attenuates the endocrine and behavioral consequences of chronic stress. *Psychoneuroendocrinology.* 2014;47:151–165.
19. Lezak KR, Roelke E, Harris OM, et al. Pituitary adenylate cyclase-activating polypeptide (PACAP) in the bed nucleus of the stria terminalis (BNST) increases corticosterone in male and female rats. *Psychoneuroendocrinology.* 2014;45:11–20.
20. Iemolo A, Seiglie M, Blasio A, Cottone P, Sabino V. Pituitary adenylate cyclase-activating polypeptide (PACAP) in the central nucleus of the amygdala induces anxiety via melanocortin receptors. *Psychopharmacology (Berl).* 2016;233(17):3269–3277.
21. Stroth N, Kuri BA, Mustafa T, Chan SA, Smith CB, Eiden LE. PACAP controls adrenomedullary catecholamine secretion and expression of catecholamine biosynthetic enzymes at high splanchnic nerve firing rates characteristic of stress transduction in male mice. *Endocrinology.* 2013;154(1):330–339.
22. Eiden LE. Neuropeptide-catecholamine interactions in stress. *Adv Pharmacol.* 2013;68: 399–404.
23. Dore R, Iemolo A, Smith KL, Wang X, Cottone P, Sabino V. CRF mediates the anxiogenic and anti-rewarding, but not the anorectic effects of PACAP. *Neuropsychopharmacology.* 2013;38(11):2160–2169.
24. Lehmann ML, Mustafa T, Eiden AM, Herkenham M, Eiden LE. PACAP-deficient mice show attenuated corticosterone secretion and fail to develop depressive behavior during chronic social defeat stress. *Psychoneuroendocrinology.* 2013;38(5):702–715.
25. Ressler KJ, Mercer KB, Bradley B, et al. Post-traumatic stress disorder is associated with PACAP and the PAC1 receptor. *Nature.* 2011;470(7335):492–497.
26. Hashimoto H, Shintani N, Tanaka K, et al. Altered psychomotor behaviors in mice lacking pituitary adenylate cyclase-activating polypeptide (PACAP). *Proc Natl Acad Sci USA.* 2001;98(23):13355–13360.
27. Hattori S, Takao K, Tanda K, et al. Comprehensive behavioral analysis of pituitary adenylate cyclase-activating polypeptide (PACAP) knockout mice. *Front Behav Neurosci.* 2012;6:58.
28. Otto C, Martin M, Wolfer DP, Lipp HP, Maldonado R, Schutz G. Altered emotional behavior in PACAP-type-I-receptor-deficient mice. *Brain Res Mol Brain Res.* 2001;92(1–2):78–84.
29. Stevens JS, Jovanovic T, Fani N, et al. Disrupted amygdala-prefrontal functional connectivity in civilian women with posttraumatic stress disorder. *J Psychiatr Res.* 2013;47(10):1469–1478.
30. Kessler RC, Berglund P, Demler O, et al. The epidemiology of major depressive disorder: Results from the National Comorbidity Survey Replication (NCS-R). *JAMA.* 2003;289(23):3095–3105.

31. Joo KM, Chung YH, Kim MK, et al. Distribution of vasoactive intestinal peptide and pituitary adenylate cyclase-activating polypeptide receptors (VPAC1, VPAC2, and PAC1 receptor) in the rat brain. *J Comp Neurol.* 2004;476(4):388–413.
32. Piggins HD, Stamp JA, Burns J, Rusak B, Semba K. Distribution of pituitary adenylate cyclase activating polypeptide (PACAP) immunoreactivity in the hypothalamus and extended amygdala of the rat. *J Comp Neurol.* 1996;376(2):278–294.
33. Hammack SE, May V. Pituitary adenylate cyclase activating polypeptide in stress-related disorders: data convergence from animal and human studies. *Biol Psychiatry.* 2015;78(3):167–177.
34. Hashimoto H, Hashimoto R, Shintani N, et al. Depression-like behavior in the forced swimming test in PACAP-deficient mice: amelioration by the atypical antipsychotic risperidone. *J Neurochem.* 2009;110(2):595–602.
35. Stroth N, Holighaus Y, Ait-Ali D, Eiden LE. PACAP: a master regulator of neuroendocrine stress circuits and the cellular stress response. *Ann N Y Acad Sci.* 2011;1220:49–59.
36. Markou A, Hauger RL, Koob GF. Desmethylimipramine attenuates cocaine withdrawal in rats. *Psychopharmacology (Berl).* 1992;109(3):305–314.
37. Seiglie MP, Smith KL, Blasio A, Cottone P, Sabino V. Pituitary adenylate cyclase-activating polypeptide induces a depressive-like phenotype in rats. *Psychopharmacology (Berl).* 2015;232(20):3821–3831.
38. Ago Y, Hiramatsu N, Ishihama T, et al. The selective metabotropic glutamate 2/3 receptor agonist MGS0028 reverses psychomotor abnormalities and recognition memory deficits in mice lacking the pituitary adenylate cyclase-activating polypeptide. *Behav Pharmacol.* 2013;24(1):74–77.
39. Gaszner B, Kormos V, Kozicz T, Hashimoto H, Reglodi D, Helyes Z. The behavioral phenotype of pituitary adenylate-cyclase activating polypeptide-deficient mice in anxiety and depression tests is accompanied by blunted c-Fos expression in the bed nucleus of the stria terminalis, central projecting Edinger-Westphal nucleus, ventral lateral septum, and dorsal raphe nucleus. *Neuroscience.* 2012;202:283–299.
40. Hashimoto R, Hashimoto H, Shintani N, et al. Possible association between the pituitary adenylate cyclase-activating polypeptide (PACAP) gene and major depressive disorder. *Neurosci Lett.* 2010;468(3):300–302.
41. Farkas J, Kovacs LA, Gaspar L, et al. Construct and face validity of a new model for the three-hit theory of depression using PACAP mutant mice on CD1 background. *Neuroscience.* 2017;354:11–29.
42. Kormos V, Gaspar L, Kovacs LA, et al. Reduced response to chronic mild stress in PACAP mutant mice is associated with blunted FosB expression in limbic forebrain and brainstem centers. *Neuroscience.* 2016;330:335–358.
43. Hattori T, Baba K, Matsuzaki S, et al. A novel DISC1-interacting partner DISC1-binding zinc-finger protein: implication in the modulation of DISC1-dependent neurite outgrowth. *Mol Psychiatry.* 2007;12(4):398–407.
44. Millar JK, Wilson-Annan JC, Anderson S, et al. Disruption of two novel genes by a translocation co-segregating with schizophrenia. *Hum Mol Genet.* 2000;9(9):1415–1423.
45. Blackwood DH, Muir WJ. Clinical phenotypes associated with DISC1, a candidate gene for schizophrenia. *Neurotox Res.* 2004;6(1):35–41.
46. Blackwood DH, Pickard BJ, Thomson PA, Evans KL, Porteous DJ, Muir WJ. Are some genetic risk factors common to schizophrenia, bipolar disorder and depression? Evidence from DISC1, GRIK4 and NRG1. *Neurotox Res.* 2007;11(1):73–83.
47. Lowe SR, Pothen J, Quinn JW, et al. Gene-by-social-environment interaction (GxSE) between ADCYAP1R1 genotype and neighborhood crime predicts major depression symptoms in trauma-exposed women. *J Affect Disord.* 2015;187:147–150.

48. Cooper AJ, Narasimhan S, Rickels K, Lohoff FW. Genetic polymorphisms in the PACAP and PAC1 receptor genes and treatment response to venlafaxine XR in generalized anxiety disorder. *Psychiatry Res.* 2013;210(3):1299–1300.
49. Jovanovic T, Norrholm SD, Blanding NQ, et al. Impaired fear inhibition is a biomarker of PTSD but not depression. *Depress Anxiety.* 2010;27(3):244–251.
50. Jovanovic T, Norrholm SD, Blanding NQ, et al. Fear potentiation is associated with hypothalamic-pituitary-adrenal axis function in PTSD. *Psychoneuroendocrinology.* 2010;35(6):846–857.
51. Jovanovic T, Ressler KJ. How the neurocircuitry and genetics of fear inhibition may inform our understanding of PTSD. *Am J Psychiatry.* 2010;167(6):648–662.
52. Dunlop BW, Rothbaum BO, Binder EB, et al. Evaluation of a corticotropin releasing hormone type 1 receptor antagonist in women with posttraumatic stress disorder: study protocol for a randomized controlled trial. *Trials.* 2014;15:240.
53. Mercer KB, Dias B, Shafer D, et al. Functional evaluation of a PTSD-associated genetic variant: estradiol regulation and ADCYAP1R1. *Transl Psychiatry.* 2016;6(12):e978.
54. Kirry AJ, Herbst MR, Poirier SE, et al. Pituitary adenylate cyclase-activating polypeptide (PACAP) signaling in the prefrontal cortex modulates cued fear learning, but not spatial working memory, in female rats. *Neuropharmacology.* 2018;133:145–154.
55. Grimm O, Pohlack S, Cacciaglia R, et al. Amygdalar and hippocampal volume: a comparison between manual segmentation, Freesurfer and VBM. *J Neurosci Methods.* 2015;253:254–261.
56. Pohlack ST, Nees F, Ruttorf M, et al. Neural mechanism of a sex-specific risk variant for posttraumatic stress disorder in the type I receptor of the pituitary adenylate cyclase activating polypeptide. *Biol Psychiatry.* 2015;78(12):840–847.
57. Yanos PT, Moos RH. Determinants of functioning and well-being among individuals with schizophrenia: an integrated model. *Clin Psychol Rev.* 2007;27(1):58–77.
58. Owen MJ, O'Donovan MC. Schizophrenia and the neurodevelopmental continuum: evidence from genomics. *World Psychiatry.* 2017;16(3):227–235.
59. Hill JM. Vasoactive intestinal peptide in neurodevelopmental disorders: therapeutic potential. *Curr Pharm Des.* 2007;13(11):1079–1089.
60. Katayama T, Hattori T, Yamada K, Matsuzaki S, Tohyama M. Role of the PACAP-PAC1-DISC1 and PACAP-PAC1-stathmin1 systems in schizophrenia and bipolar disorder: novel treatment mechanisms? *Pharmacogenomics.* 2009;10(12):1967–1978.
61. Yan Y, Zhou X, Pan Z, Ma J, Waschek JA, DiCicco-Bloom E. Pro- and anti-mitogenic actions of pituitary adenylate cyclase-activating polypeptide in developing cerebral cortex: potential mediation by developmental switch of PAC1 receptor mRNA isoforms. *J Neurosci.* 2013;33(9):3865–3878.
62. Reif A, Fritzen S, Finger M, et al. Neural stem cell proliferation is decreased in schizophrenia, but not in depression. *Mol Psychiatry.* 2006;11(5):514–522.
63. Hashimoto R, Hashimoto H, Shintani N, et al. Pituitary adenylate cyclase-activating polypeptide is associated with schizophrenia. *Mol Psychiatry.* 2007;12(11):1026–1032.
64. Hazama K, Hayata-Takano A, Uetsuki K, et al. Increased behavioral and neuronal responses to a hallucinogenic drug in PACAP heterozygous mutant mice. *PLoS One.* 2014;9(2):e89153.
65. Shen S, Gehlert DR, Collier DA. PACAP and PAC1 receptor in brain development and behavior. *Neuropeptides.* 2013;47(6):421–430.
66. Takasaki I, Watanabe A, Yokai M, et al. In silico screening identified novel small-molecule antagonists of PAC1 receptor. *J Pharmacol Exp Ther.* 2018;365(1):1–8.

CHAPTER SEVEN

Glial mechanisms underlying major depressive disorder: Potential therapeutic opportunities

Shafiqur Rahman*, Sami Alzarea
Department of Pharmaceutical Sciences, College of Pharmacy, South Dakota State University, Brookings, SD, United States
*Corresponding author: e-mail address: shafiqur.rahman@sdstate.edu

Contents

Abstract

Major depressive disorder is a severe and the most prevalent form of neuropsychiatric disorder that affects millions of individuals worldwide. The current treatment focuses on brain monoamine hypothesis and conventional antidepressants that inhibit the reuptake of monoamine neurotransmitters to elevate synaptic monoamine concentrations to interact with postsynaptic receptors. However, many people with major depressive disorder fail to respond to conventional therapies experiencing relapse and significant functional impairment. It is increasingly recognized that glia-associated inflammatory mechanisms contribute to the pathophysiology of major depressive disorder. Furthermore, microglial nicotinic cholinergic mechanisms were recently proposed to play a critical role in major depressive disorder-linked inflammatory processes. In this review, we will provide recent advances on glial dysfunctions, the role of inflammatory

Progress in Molecular Biology and Translational Science, Volume 167
ISSN 1877-1173
https://doi.org/10.1016/bs.pmbts.2019.06.010

cytokines, transcription factors, and brain derived neurotrophic factor underlying the pathophysiology of major depressive disorder. We will also evaluate and discuss the anti-inflammatory effects of nicotinic acetylcholine receptor positive allosteric modulators and glial nicotinic receptor-mediated mechanisms that might be responsive to potential pharmacologic interventions. Overall, these important, but underappreciated, roles of brain glial targets and mechanisms might offer new therapeutic opportunities for major depressive disorder.

1. Introduction

Major depressive disorder (MDD) is the most prevalent form of neuropsychiatric disorder affecting approximately 400 million people worldwide and is the leading cause of disability in many developed nations.[1–3] Epidemiological data indicate that MDD will rank in the top three disorders responsible for global disease burden causing lost disability-adjusted life and untimely death by the year 2030.[2,3] Emerging evidence indicates that approximately 30% of individuals with MDD are addicted to alcohol or nicotine, indicating a significant development of co-morbid substance use disorder.[4–6] MDD is also widely recognized as a risk factor for suicidal behavior.[7] The current treatment focuses on brain monoamine hypothesis of MDD and conventional antidepressants such as serotonin and norepinephrine reuptake inhibitors, selective serotonin reuptake inhibitors, tricyclic antidepressants, and monoamine oxidase inhibitors enhance the synaptic level of monoamine neurotransmitters in the brain.[8] However, these medications exhibit limited effects, and only a third of depressed patients benefit from this therapeutic strategy. The onset of action for these drugs takes 3–4 weeks to produce limited benefits for MDD symptoms.[9] Therefore, understanding the emerging cellular and molecular mechanisms underlying the pathophysiology of MDD is important for better treatment opportunities.

Previous studies have demonstrated that specific brain regions and circuitry are critical in regulating emotion, mood and reward.[10] Malfunction of these brain regions and circuits has been suggested to contribute to MDD.[11] Current understanding regarding specific neuronal circuitry or pathways in MDD is still limited. Individuals with MDD exhibited abnormal changes in blood flow or related measures in several brain regions such as the prefrontal cortex, cingulate cortex, hippocampus, striatum, amygdala and thalamus.[12] Previous studies have shown that post-mortem samples in depressed patients reveal serious cytoarchitectural changes in some of these brain regions.[13] For example, depressed patients showed a decrease in gray-matter volume in their prefrontal cortex and hippocampus. Preclinical data from animal studies also indicate that the hippocampus and prefrontal cortex are involved in MDD,[10] indicating the important role of these specific brain regions in behaviors associated with MDD.[13]

A growing body of evidence indicates that inflammation is critically involved in the pathophysiology of MDD.[14,15] For instance, treatments for enhancing immune system activity such as interferon alpha immunotherapy resulted in developing depression-related symptoms in patients with hepatitis C or malignant melanoma.[16] Previous studies have found that serum proinflammatory cytokines such as interleukin 1 beta (IL-1β) and tumor necrosis factor alpha (TNF-α) increased in patients with MDD.[17,18] Interestingly, healthy human volunteers have shown some depression-related symptoms such as anxiety, mood changes, and cognitive disruption associated with upregulation of proinflammatory cytokines after receiving endotoxin or vaccines.[19–21] In addition, animal models were found to develop MDD-like behaviors after receiving proinflammatory cytokines.[22–24] Furthermore, attenuation of proinflammatory cytokines prevents induction of depression-like behaviors and reduce other MDD-related behaviors such as anxiety and cognitive deficit in preclinical models.[25–27] Thus, there is a growing appreciation from clinical and preclinical data that inflammation plays an essential role in the pathophysiology of MDD.

2. Astroglial dysfunction and major depressive disorder

Astroglial cells or astrocytes are the most abundant glial cell type and are traditionally known to regulate the homeostatic environment involving glucose metabolism, glutamate reuptake, synaptic development and maturation, and blood-brain barrier maintenance in the central nervous system (CNS).[28–30] Astrocytes also play an essential role in immune cell trafficking and activation.[31] During inflammation, neuroinflammatory cytokines are upregulated in the CNS.[32] Emerging evidence has indicated that inflammatory signaling pathway, such as nuclear factor-κB (NF-κB) activation, exhibits higher activity in astrocytes upon immune response.[31] This downstream signaling has an important role in the transcription and production of proinflammatory cytokines and these cytokines were shown to have pathogenic influence in MDD.[32,33]

Previous studies suggest that astrocytes have neuroprotective function through synthesis and release of kynurenic acid, an NMDA receptor antagonist, at normal physiological condition.[34,35] However, astrocytes lose their neuroprotective functions during inflammation and neurotoxic transmission becomes more abundant.[34,36] Consequently, neurotoxic transmission and disabled neuroprotective function of astrocytes may harm the neurons and likely develop MDD-related symptoms.[15,37] Human postmortem studies indicate that astroglial changes occur in several brain regions in individuals with MDD.[38] For example, depressed patients exhibited larger size of nuclei of astrocytes in the prefrontal cortex.[38] In addition, suicide victims with major depression showed hypertrophic astrocytes in the anterior cingulate cortex.[39] Thus, hypertrophy of astrocytes in depressed patients indicates the presence of neuroinflammation in MDD.[40]

Activation of astrocytes can be identified using several common markers.[38] Glial fibrillary acidic protein (GFAP), 50 kilodaltons (kDa), is expressed by mature and reactive astrocytes to form the structure of intermediate cytoskeletal filaments.[41] GFAP was believed to help astrocytes in the maintenance of mechanical strength and shape.[42] Astroglial movement and structure can be regulated by GFAP that controls the communication between astrocytes and neurons.[43] Older individuals with MDD showed increased GFAP immunoreactivity in orbitofrontal cortex compared to younger patients.[44] Older patients with late-onset of depression exhibited an elevation in GFAP immunoreactivity and cell density in dorsolateral prefrontal cortex.[45] Preclinical studies have also shown that GFAP mRNA is upregulated in the hippocampus and prefrontal cortex in inflammatory animal models of MDD.[36,46] S100 calcium binding protein B (S100B), 10 kDa protein, is another astroglial marker is increased in MDD.[38] Reactive astrocytes were found to increase mRNA and protein expressions of S100B in intracellular and extracellular fluids.[47] Leaked S100B from cerebrospinal fluid to bloodstream of depressed patients increased as an indication of the astroglial response.[48] An increased S100B was found in cerebrospinal fluid of subjects with MDD.[49] Overall, astrocytes were shown to have important inflammatory roles in the pathophysiology of MDD.[38] Increased astroglial markers may indicate abnormal functions of astrocytes during neuroinflammation. Therefore, reduction of dysfunctional states of astrocytes might have potential therapeutic implications for MDD.

3. Microglial dysfunction and major depressive disorder

Microglia are part of the innate immune system and are known resident macrophages that maintain immunological functions in the CNS.[29,50,51]

Microglia are derived from hematopoietic progenitors to become macrophage-lineage cells and are activated during pathological conditions such as immune challenge by inflection, injury, and neurodegeneration.[50] Upon activation, chronic activations of microglia, including inflammatory, protective, recuperative, and toxic processes may result in serious effects in the neurons and other glial cells.[52] Furthermore, the primary source of inflammatory cytokines in the CNS is produced by microglia and therefore, microglia play a major role in regulating the inflammation in the brain and spinal cord.[51,53] In addition, recent findings have shown that microglia have a role in developing mature synapses upon embryogenesis.[50] Microglia contribute to the regulation of numerous synaptic functions, for example, microglia have a regulatory function in neurogenesis via apoptosis-coupled phagocytosis.[54,55] These functions are regulated at normal physiological conditions.[56] Microglial structure and function determine the important roles based on resting or dormant conditions of the cells.[57] Thus, there is a difference between microglial morphology and function under pathological challenges and normal physiological response. However, the normal physiological function may be disrupted by pathological changes. For instance, chronic infections, stroke, trauma, neurodegenerative diseases, or chronic psychological stress are the main pathological states that cause abnormal regulatory functions of microglia. Consequently, these complex effects of microglia lead to behavioral changes in cognition and mood. Therefore, microglia are considered microgliopathy in abnormal situations.[57,58]

Immune challenges by bacterial and viral infections were believed to correlate with MDD-related symptoms and these infections are susceptible to activate microglia in the brain.[59,60] Production of proinflammatory cytokines or upregulation in depressed patients is induced by the infections.[61] Immune challenge by administration endotoxin [lipopolysaccharide (LPS)], microglial activator, lead to development of depression symptoms were reported in human studies.[21,62] Moreover, suicide patients with depression showed morphological changes in microglia in several studies.[63,64]

Inflammatory animal model of MDD exhibits microglial activation in the rodent brain and peripheral inflammation produced by LPS resulted in depression-like behavior induction.[24,35] Traumatic brain injury in mice was associated with depression-like behavior and microglial activation.[65] Moreover, inhibition of microglial activation based on genetic and pharmacological studies leads to a reduction in depression-like behavior.

For example, minocycline, a microglial inhibitor, abrogated LPS-induced depression-like behavior.[24,66] Attenuation of microglia by blocking their ATP-gated P2X7 ion channel resulted in reduction of stress and LPS-induced depression-like behavior in animal models.[67]

Microglial activation can be identified using known common markers.[68] Previous studies have shown that several microglial markers are upregulated in patients with MDD or animal models of MDD.[69] For example, ionized calcium-binding adapter molecule 1 (Iba-1), 17 kDa protein, is one of the microglial markers, upregulated during microglial activation and this marker was suggested to participate in the cytoskeleton of microglia.[70] Immunoreactivity studies indicated that Iba-1 expression in dorsal anterior cingulate cortex is upregulated in suicide patients with MDD.[71] Increased expression of Iba-1 in the brain was reported in stress model of MDD.[72] Furthermore, recent findings have shown the upregulation of CD11b, 126 kDa protein, in the hippocampus, amygdule, and striatum of mice following LPS administration.[36] Therefore, elevations of these important markers indicate dysfunction of microglia associated with MDD pathophysiology in animal models which might have important therapeutic significance.

3.1 Role of inflammatory cytokines

Immune cells synthesize and produce important mediators during the immune response and the most common mediators released by immune cells are cytokines. These proteins have several functions that allow interacting or communicating between the cells. The cytokines induce immune cells to produce similar or different mediators that augment or antagonize their functions.[73,74] The primary function of these cytokines, also known as proinflammatory cytokines, is to induce inflammation in the organs or tissues. The primary cytokines in this category includes IL-1β, IL-6, IL-15, IL-17, IL-18, TNF-α, and interferon-γ (IFN-γ).[74] Additional cytokines are known as anti-inflammatory cytokines and the main anti-inflammatory cytokines include IL-4, IL10, and IL13.[15,74] Therefore, the imbalance between proinflammatory and anti-inflammatory cytokines may lead to a pathogenic condition. Previous studies have indicated that MDD is associated with upregulation of proinflammatory cytokines.[32] The widely known proinflammatory cytokines that could contribute to MDD-related symptoms include IL-1β and TNF-α, which are produced primarily by microglia in the CNS.[32] Serum samples from depressed patients show an elevated level of IL-1β and TNF-α.[17,18] LPS administration in healthy human volunteers' results in increased of these proinflammatory cytokines associated with mood disorders.[21,62] In addition, several studies reveal increased expression of IL-1β and TNF-α in inflammatory animal models of MDD.[36,46]

The IL-1β and TNF-α were investigated to determine their roles in MDD based on genetic and pharmacological manipulations in animal models. For example, IL-1β and TNF-α knockout mice exhibited less MDD-like behavior compared to wild type mice.[22] Intracerebroventricular administration of IL-1β and TNF-α results in an induction of MDD-like behavior in animal models.[23] Overall, the evidence indicates the involvement of IL-1β and TNF-α in MDD pathophysiology. Thus, decreased production of IL-1β and TNF-α might have therapeutic benefits in MDD-related symptoms.

3.2 Role of transcription factors

Evidence suggests that the IL-1β, TNF-α, and other proinflammatory cytokines are synthesized by activation of inflammatory downstream signaling such as NF-κB.[14,33] The NF-κB was reported to have a role in mediating blood-brain interface allowing for transferring peripheral inflammatory signals to the CNS. Central inhibition of NF-κB results in attenuation of peripheral administration of IL-1β-induced behavioral alterations.[75] Healthy human volunteers exposed to stressor show elevation in NF-κB DNA interaction in their blood mononuclear cells, suggesting the role of NF-κB in MDD.[76] One of the mediators reported for NF-κB activation is IκB mRNA and without immune challenge, IκB protein prevents NF-κB to be active by interacting with NF-κB in the cytoplasm.[77] Following induction of the inflammatory process, kinase enzyme phosphorylates IκB to release NF-κB.[78] The NF-κB release allows transferring to the nucleus for transcription of several genes including IκB mRNA.[79]

Nuclear peroxisome proliferator-activated receptor type-α (PPAR-α) regulates several important cellular functions, including anti-inflammatory effects.[80] The PPAR-α was found to play an important role in inhibiting NF-κB activation.[81] Moreover, pharmacological activation of PPAR-α caused reduction of LPS-induced depression-like behavior.[82] PPAR-α knockout mice show increased depression-like behavior compared to wild type mice.[83] Therefore, the regulation of anti-inflammatory effects by targeting PPAR-α and NF-κB could represent a therapeutic strategy in modulating MDD-related symptoms.

3.3 Role of brain derived neurotrophic factor

Brain-derived neurotrophic factor (BDNF) was proposed to play a critical role in MDD.[84] The BDNF was shown to be dysregulated during inflammation associated with MDD.[85] Activation of NF-κB was found to be stimulated by interacting tyrosine receptor kinase B (TrkB) receptor via BDNF.[86] The expression of BDNF is upregulated in the hippocampus associated with LPS-induced depression-like behavior as shown in animal studies.[87] As an inflammatory mediator, BDNF also can be released by microglia to regulate inflammatory processes.[85] Blockade of TrkB receptor by ANA12 was found to attenuate LPS induced-depression-like behavior in preclinical models.[88] These findings suggest that increased BDNF expression could be an important factor in MDD and therefore, modulation of BDNF expression might represent a therapeutic approach in MDD.

As described previously, neuronal excitability can be induced by microglial BDNF or this occurs due to dysregulation of intracellular chloride concentration.[89] There are two neuronal co-transporters that regulate chloride concentration: sodium-potassium-chloride co-transporter 1 (NKCC1) and potassium-chloride co-transporter 2 (KCC2).[89–91] The NKCC1 allows efflux of chloride, whereas KCC2 allows influx of chloride. Interruption of these co-transporter leads to a reduction of intracellular chloride concentration and consequently to neuronal excitability.[89] Moreover, γ-aminobutyric acid (GABA) was found to regulate MDD by enhancing intracellular chloride via stimulating GABAA receptors to reduce neuronal excitability.[90] There is a crosstalk between NKCC1 and GABAA receptor in the mediation of intracellular chloride.[91] A recent study indicated that bumetanide, an NKCC1 antagonist, reduced brain trauma-induced depression-like behavior associated with inflammation.[92] To this end, decreasing NKCC1 expression and increasing KCC2 expression associated with MDD might have a unique therapeutic significance.

While microglial dysfunction discussed above represents important glial targets and mechanisms to modulate neuron-glia interactions which have potential implications in the pathophysiology of MDD, the following sections will review and discuss the essential role of microglial nicotinic cholinergic receptor mediated mechanisms associated with MDD and potential therapeutic opportunities.

4. Microglial nicotinic cholinergic mechanisms and major depressive disorder

There is growing evidence that brain nicotinic acetylcholine receptors (nAChRs) are important therapeutic target for several CNS disorders, including drug addiction and MDD.[5,93,94] These nAChRs are generally ionotropic ligand-gated ion channels and are permeable to sodium and calcium ions.[95] The petameric nAChRs consist of five subunits in their structures to form a pore allowing ions influx.[95] Heterologous expression of these subunits comes from nine α (α2-α10) and three β (β2-β4) subunits. Heteromeric receptor type composed of α and β subunits such as α4β2 nAChR and has only two binding sites for the ligand. In contrast, homomeric receptor type composed of identical subunits such as α7 nAChR and has five binding sites for the ligand.[95] The α7 nAChRs are expressed by non-neuronal cells and produce regulatory function against immune response challenge in the CNS.[95,96] The α7 nAChRs are widely distributed in the different brain regions including prefrontal cortex, hippocampus, and other parts of the limbic system.[96] Previous studies reported that α7 nAChRs expressed by glial cells have important roles in regulating inflammatory processes in the CNS,[97–99] suggesting their critical involvement in MDD.

4.1 Alpha7 nicotinic receptor and neuroinflammation

The α7 nAChRs are homomeric type which consists of five α subunits and five binding sites between every two subunits.[95,96] The α7 nAChRs are classified as a neurotransmitter-gated ion channel and stimulation of these receptors by the ligand allows calcium influx but desensitizes within milliseconds.[96,100] As previously indicated that α7 nAChRs are expressed by brain microglia and show important modulatory influence on anti-inflammatory pathways and intracellular signaling.[101,102] For example, several reports suggested that the α7 nAChRs in the CNS regulate inflammatory processes caused by microglia[103–105] and the anti-inflammatory effects are likely mediated by microglial α7 nAChRs.[97] The underlying mechanisms of α7 nAChRs mediated effects are likely due to inhibition of microglial activation and decreased production proinflammatory cytokines.[104,106] It is noteworthy to mention that brain microglial α7 nAChRs exhibit dual ionotropic and metabotropic signaling system compared to neuronal α7 nAChRs.[97] In addition, these receptors are widely reported to modulate neuroinflammatory mechanisms associated with psychiatric and neurological disorders such as schizophrenia, Alzheimer's disease, and Parkinson's disease. Thus, the α7 nAChRs could be potential targets for treating neuropsychiatric disorders, including MDD associated with neuroinflammation.[107]

It should be noted here that cholinergic pathways in the brain are dysregulated by several neurodegenerative disorders and disturbances in cholinergic transmission may initiate the neuroinflammation by reducing the ACh availability on microglia.[108] Based on animal studies, the data indicate that α7 nAChRs could regulate the neuroinflammation[96] and therefore, the α7 nAChRs could be potential targets for treating MDD.[94,96] However, the activity of α7 nAChRs could be limited because of the desensitization during the receptor activation caused by full agonists.[100] Additional mechanistic understanding of microglial α7 nAChRs mediated effects were discussed to minimize this limitation.[97] For instance, compared to neuronal α7 nAChRs, microglial α7 nAChRs exhibit metabotropic function.[97] The downstream metabotropic signaling pathways of microglial α7 nAChRs are different from neuronal α7 nAChRs. Activation of microglial α7 nAChRs leads to stimulation of phospholipase C and enhancement of calcium release from intercellular stores which are sensitive to inositol trisphosphate.[105] This process results in the inhibition of NF-κB activation and IκB degradation.[79,101] As a result, the transcription signal produced by NF-κB will negatively influence the mRNAs for proinflammatory cytokines.[105,106] Therefore, pharmacological activation of α7 nAChRs by GTS-21, a full agonist, showed a decrease in LPS-induced activation of NF-κB and proinflammatory cytokine production.[109] Overall, this growing body of evidence indicates that targeting microglial α7 nAChRs could reduce neuroinflammation by decreasing proinflammatory cytokine production.

4.2 Alpha7 nicotinic receptor mediated modulators and major depressive disorder

Several studies have shown that α7 nAChR plays a role in depression-like behavior in animal models of MDD. For example, α7 nAChR knockout mice exhibit development of depression-like behavior compared to wild type mice.[110] The α7 nAChR agonist PNU 282987 shows antidepressant-like effects in rats.[111] GTS-21, α7 nAChR agonist, was found to reduce stress-induced depression-like behavior in mice.[112] However, as previously mentioned, full agonist was found to decrease α7 nAChR activity by inducing desensitization.[100] Thus, the α7 nAChR positive allosteric modulators were suggested to prevent desensitization and enhance the α7 nAChR activity. For instance, nAChRs have two major binding sites: orthosteric sites and allosteric sites.[113] Activation of the allosteric binding site allows the receptor to change in conformational states that facilitate for ACh to bind at the orthosteric site. Therefore, allosteric modulators work to stabilize the receptor for ACh binding based on modification of the configuration structure of the receptor.[113] In α7 nAChR, active open state of the receptor undergoes stabilization by allosteric substances. The physiological response produced by the receptor is increased by the effects of positive allosteric modulators even in the absence of tonic response. As a result, the receptor activity produced by full agonists is increased by positive allosteric modulator that does not show intrinsic activity.[114]

The classification of positive allosteric modulators is either type I, such as NS1738[115] or type II, such as *N*-(5-chloro-2,4-dimethoxyphenyl)-*N'*-(5-methyl-3-isoxazolyl)-urea (PNU 120596).[116] This classification was based on prevention of desensitization by positive allosteric modulators.[117] For instance, the positive allosteric modulators showing no or little effects on desensitization prevention belong to type I, whereas other positive allosteric modulator showing significant prevention on desensitization belong to type II.[118] Thus, type II positive allosteric modulators alter the receptor from desensitized state to active state via modifying the equilibrium between two states that leads to an increase in receptor-mediated response.[119] Furthermore, type II positive allosteric modulators have some ability to enhance the activity for already desensitized receptor.[119] Therefore, type II positive allosteric modulators is considered an alternative approach to avoid the desensitization and enhance the cholinergic signal in the brain. Recently, PNU 120596 was characterized as an α7 nAChR type II positive allosteric modulator.[120,121] For example, PNU 120596 was found to cross the blood-brain barrier[122,123]; PNU 120596 fails to exhibit intrinsic activity alone on α7 nAChR, whereas PNU 120596 significantly enhances ACh when they are combined.[119] In addition, PNU 120596 regulates the recovery from α7 nAChR desensitization induced by full agonist.[115]

The α7 nAChR positive allosteric modulators were studied for their potential effects in several CNS disorders due to their unique desensitization prevention property.[115] In addition, α7 nAChR positive allosteric modulators assist cholinergic neurotransmission and augment the action of full agonist without direct binding on the orthosteric site. Therefore, α7 nAChR positive allosteric modulators enhance and potentiate orthosteric site-mediated signaling of ACh and choline, an endogenous cholinergic ligand. Although these compounds do not have intrinsic activity but they are able to support endogenous cholinergic neurotransmission. The α7 nAChR positive allosteric modulators are found to elevate efficacy and potency of endogenous and exogenous α7 nAChR ligands. Several α7 nAChR positive allosteric modulators were investigated in animal models of depression. For example, cotinine, α7 nAChR positive allosteric modulator was found to decrease the cognitive deficit and depression-like behavior in a stress mouse model.[124] Additional α7 nAChR positive allosteric modulator, 3-furan-2-yl-N-p-tolyl-acrylamide exhibited antidepressant-like effects with chronic administration in mice.[125] The effects of α7 nAChR positive allosteric modulators on microglial activation and associated changes in the hippocampus and prefrontal cortex are also important with respect to MDD.

Consistent with this notion, recent evidence indicates that α7 nAChR positive allosteric modulator PNU 120596 reduces depressive-like behavior in the forced swim test, tail suspension test, and sucrose preference test following LPS administration, an animal model for MDD-like behavior measured by increased immobility time or decreased sucrose preference.[94] Pretreatment of methyllycaconitine (MLA), an α7 nAChR antagonist, significantly prevented the antidepressant-like effects of PNU 120596, suggesting the involvement of α7 nAChR. In addition, α7 nAChR positive allosteric modulator PNU 120596 reduces LPS-induced expression of ionized calcium binding adaptor molecule, a microglial marker, revealed in Western blot analysis and immunofluorescence staining in the hippocampus and prefrontal cortex, the brain regions implicated in MDD.[94] Clearly, these data suggest that inhibition of microglial activation via α7 nAChR is a possible underlying mechanism for reducing the MDD-related behaviors in preclinical models. It is noteworthy to mention that α7 nAChR positive allosteric modulator PNU 120596 likely increase ACh signaling to reduce inflammation caused by LPS associated with MDD. Several studies also reported that neuroinflammation can be modulated by stimulation of α7 nAChR likely expressed by brain microglia in the specific brain regions.[103,106]

Recently, the α7 nAChR positive allosteric modulator PNU 120596 was also found to reduce LPS-induced MDD-related behaviors such as anxiety and cognitive deficit.[99] Similarly, α7 nAChR antagonist MLA was reported to prevent the anxiolytic, and pro-cognitive-like effects of PNU 120596, suggesting the involvement of α7 nAChR.[99] Moreover, the α7 nAChR positive allosteric modulator PNU 120596 was found to decrease the expression of glial markers, such as ionized calcium binding adaptor molecule 1 and cluster of differentiation 11b, and astroglial marker, glial fibrillary acidic protein indicating the involvement of microglia and astrocytes in this preclinical MDD model.[99] Furthermore, the α7 nAChR positive allosteric modulator PNU 120596 was shown to reduce the expression of interleukin 1β and tumor necrosis factor-α, important inflammatory markers associated with MDD.[99] These data indicate that systemic LPS administration could cause anxiety, cognitive deficit, and depression-like behaviors in animal models as shown in elevated plus maze, Y-maze, and forced swim test after 24 h. Moreover, several studies suggest that systemic administration of LPS contributes to glial activation and upregulation of proinflammatory cytokines likely by increasing mRNA expression of these important glial markers such as CD11b, GFAP, IL-1β, and TNF-α in specific MDD-relevant brain regions, including hippocampus and prefrontal cortex.[36,46] The α7 nAChR positive allosteric modulator PNU 120596 was also shown to reduce the elevated CD11b, GFAP, IL-1β, and TNF-α expression in the hippocampus and prefrontal cortex,[99] suggesting anti-inflammatory effects of this positive allosteric modulator PNU 120596. Furthermore, antagonism of α7 nAChR with MLA was found to prevent the anti-inflammatory effects of PNU 120596 in these brain regions. Therefore, the preclinical data suggest that the anxiolytic, pro-cognitive, and antidepressant-like effects of PNU are likely mediated by attenuating the activation of glial cells and reducing the production of proinflammatory cytokines in the brain.

5. Conclusions

Converging lines of evidence indicates that glial inflammatory mechanisms in the CNS can play a major role in the pathophysiology of MDD. Activated microglia found in the hippocampus and prefrontal cortex is critical in MDD and related behaviors associated with several important inflammatory markers, transcription factors, and others. Brain nicotinic cholinergic pathway, specifically, microglial α7 nAChRs are believed to modulate microglial activation and expression of these inflammatory markers. Accumulating evidence suggests that the microglial α7 nAChRs are different from neuronal α7 nAChRs (i.e., inotropic only) by being dual inotropic and metabotropic in nature. We have demonstrated that microglial α7 nAChR positive allosteric modulator such as PNU120596 produces anti-inflammatory effects by deactivating microglial activation and reducing other inflammatory markers associated with MDD-related symptoms in animal models. Although targeting microglial α7 nAChR represents new therapeutic opportunity, further studies are required to provide a better understanding of the precise role of α7 nAChR positive allosteric modulator on glial transcription factors and brain derived neurotrophic factors and/or others in MDD-related behaviors. These studies will likely provide deeper understanding of the interplay between microglial activation, neuroinflammation and neuron-glia interactions in the MDD-relevant brain regions which may help to open a new road to treat MDD.

Acknowledgments

This work was supported in part by grant from South Dakota State University (SDSU) Research Foundation and SDSU COP&AHP (S.R.).

References

1. Kessler RC, Bromet EJ. The epidemiology of depression across cultures. *Annu Rev Public Health*. 2013;34:119–138.
2. World Health Organization (WHO). *Depression and Other Common Mental Disorders Global Health Estimates*. Geneva: World Health Organization; 2017.
3. Vos T, Abajobir AA, Abbafati C, et al. Global, regional, and national incidence, prevalence, and years lived with disability for 328 diseases and injuries for 195 countries, 1990-2016: a systematic analysis for the global burden of disease study 2016. *Lancet*. 2017;390:1211–1259.
4. Davis L, Uezato A, Newell JM, Frazier E. Major depression and comorbid substance use disorders. *Curr Opin Psychiatry*. 2008;21:14–18.
5. Rahman S. Targeting brain nicotinic acetylcholine receptors to treat major depression and co-morbid alcohol or nicotine addiction. *CNS Neurol Disord Drug Targets*. 2015;14:647–653.
6. Morozova M, Rabin RA, George TP. Co-morbid tobacco use disorder and depression: a re-evaluation of smoking cessation therapy in depressed smokers. *Am J Addict*. 2015;24:687–694.
7. Fassberg MM, Cheung G, Canetto SS, et al. A systematic review of physical illness, functional disability, and suicidal behaviour among older adults. *Aging Ment Health*. 2016;20:166–194.
8. Ferrari F, Villa RF. The neurobiology of depression: an integrated overview from biological theories to clinical evidence. *Mol Neurobiol*. 2017;54:4847–4865.
9. Bleakley S. Review of the choice and use of antidepressant drugs. *Prog Neurol Psychiatry*. 2013;17:18–26.
10. Krishnan V, Nestler EJ. The molecular neurobiology of depression. *Nature*. 2008;455(7215):894–902.
11. Nestler EJ. Role of the brain's reward circuitry in depression: transcriptional mechanisms. *Int Rev Neurobiol*. 2015;124:151–170.
12. Pandya M, Altinay M, Malone Jr DA, Anand A. Where in the brain is depression? *Curr Psychiatry Rep*. 2012;14:634–642.
13. Drevets WC, Price JL, Furey ML. Brain structural and functional abnormalities in mood disorders: implications for neurocircuitry models of depression. *Brain Struct Funct*. 2008;213:93–118.
14. Miller AH, Raison CL. The role of inflammation in depression: from evolutionary imperative to modern treatment target. *Nat Rev Immunol*. 2016;16:22–34.
15. Mechawar N, Savitz J. Neuropathology of mood disorders: do we see the stigmata of inflammation? *Transl Psychiatry*. 2016;6(11):e946.
16. Capuron L, Miller AH. Cytokines and psychopathology: lessons from interferon-alpha. *Biol Psychiatry*. 2004;56(11):819–824.
17. Dowlati Y, Herrmann N, Swardfager W, et al. A meta-analysis of cytokines in major depression. *Biol Psychiatry*. 2010;67:446–457.
18. Hannestad J, DellaGioia N, Bloch M. The effect of antidepressant medication treatment on serum levels of inflammatory cytokines: a meta-analysis. *Neuropsychopharmacology*. 2011;36:2452–2459.

19. Brydon L, Harrison NA, Walker C, Steptoe A, Critchley HD. Peripheral inflammation is associated with altered substantia nigra activity and psychomotor slowing in humans. *Biol Psychiatry*. 2008;63:1022–1029.
20. Eisenberger NI, Berkman ET, Inagaki TK, Rameson LT, Mashal NM, Irwin MR. Inflammation-induced anhedonia: endotoxin reduces ventral striatum responses to reward. *Biol Psychiatry*. 2010;68:748–754.
21. Harrison NA, Brydon L, Walker C, Gray MA, Steptoe A, Critchley HD. Inflammation causes mood changes through alterations in subgenual cingulate activity and mesolimbic connectivity. *Biol Psychiatry*. 2009;66:407–414.
22. Goshen I, Kreisel T, Ben-Menachem-Zidon O, et al. Brain interleukin-1 mediates chronic stress-induced depression in mice via adrenocortical activation and hippocampal neurogenesis suppression. *Mol Psychiatry*. 2008;13:717–728.
23. Kaster MP, Gadotti VM, Calixto JB, Santos AR, Rodrigues AL. Depressive-like behavior induced by tumor necrosis factor-alpha in mice. *Neuropharmacology*. 2012;62:419–426.
24. O'Connor JC, Andre C, Wang Y, et al. Interferon-gamma and tumor necrosis factor-alpha mediate the upregulation of indoleamine 2,3-dioxygenase and the induction of depressive-like behavior in mice in response to bacillus Calmette-Guerin. *J Neurosci*. 2009;29:4200–4209.
25. Koo JW, Duman RS. IL-1beta is an essential mediator of the antineurogenic and anhedonic effects of stress. *Proc Natl Acad Sci USA*. 2008;105:751–756.
26. Murray CL, Obiang P, Bannerman D, Cunningham C. Endogenous IL-1 in cognitive function and anxiety: a study in IL-1RI −/− mice. *PLoS One*. 2013;8(10):e78385.
27. Naude PJ, Dobos N, van der Meer D, et al. Analysis of cognition, motor performance and anxiety in young and aged tumor necrosis factor alpha receptor 1 and 2 deficient mice. *Behav Brain Res*. 2014;258:43–51.
28. Chaboub LS, Deneen B. Astrocyte form and function in the developing central nervous system. *Semin Pediatr Neurol*. 2013;20:230–235.
29. Lacagnina MJ, Rivera PD, Bilbo SD. Glial and neuroimmune mechanisms as critical modulators of drug use and abuse. *Neuropsychopharmacology*. 2017;42:156–177.
30. Linker KE, Cross SJ, Leslie FM. Glial mechanisms underlying substance use disorders. *Eur J Neurosci*. 2018;1–16. https://doi.org/10.1111/ejn.14163.
31. Colombo E, Farina C. Astrocytes: key regulators of neuroinflammation. *Trends Immunol*. 2016;37:608–620.
32. Dantzer R, O'Connor JC, Freund GG, Johnson RW, Kelley KW. From inflammation to sickness and depression: when the immune system subjugates the brain. *Nat Rev Neurosci*. 2008;9:46–56.
33. Liu T, Zhang L, Joo D, Sun SC. NF-kappaB signaling in inflammation. *Signal Transduct Target Ther*. 2017;2:17023.
34. O'Connor JC, Lawson MA, Andre C, et al. Lipopolysaccharide-induced depressive-like behavior is mediated by indoleamine 2,3-dioxygenase activation in mice. *Mol Psychiatry*. 2009;14:511–522.
35. Remus JL, Dantzer R. Inflammation models of depression in rodents: relevance to psychotropic drug discovery. *Int J Neuropsychopharmacol*. 2016;19(9). pyw028. https://doi.org/10.1093/ijnp/pyw028.
36. Parrott JM, Redus L, O'Connor JC. Kynurenine metabolic balance is disrupted in the hippocampus following peripheral lipopolysaccharide challenge. *J Neuroinflammation*. 2016;13(1):124.
37. Walker AK, Budac DP, Bisulco S, et al. NMDA receptor blockade by ketamine abrogates lipopolysaccharide-induced depressive-like behavior in C57BL/6J mice. *Neuropsychopharmacology*. 2013;38:1609–1616.

38. Rajkowska G, Stockmeier CA. Astrocyte pathology in major depressive disorder: insights from human postmortem brain tissue. *Curr Drug Targets*. 2013;14(11):1225–1236.
39. Torres-Platas SG, Hercher C, Davoli MA, et al. Astrocytic hypertrophy in anterior cingulate white matter of depressed suicides. *Neuropsychopharmacology*. 2011;36:2650–2658.
40. Maes M, Yirmyia R, Noraberg J, et al. The inflammatory & neurodegenerative (I&ND) hypothesis of depression: leads for future research and new drug developments in depression. *Metab Brain Dis*. 2009;24:27–53.
41. Middeldorp J, Hol EM. GFAP in health and disease. *Prog Neurobiol*. 2011;93:421–443.
42. Cullen DK, Simon CM, LaPlaca MC. Strain rate-dependent induction of reactive astrogliosis and cell death in three-dimensional neuronal-astrocytic co-cultures. *Brain Res*. 2007;1158:103–115.
43. Nedergaard M, Ransom B, Goldman SA. New roles for astrocytes: redefining the functional architecture of the brain. *Trends Neurosci*. 2003;26:523–530.
44. Miguel-Hidalgo JJ, Baucom C, Dilley G, et al. Glial fibrillary acidic protein immunoreactivity in the prefrontal cortex distinguishes younger from older adults in major depressive disorder. *Biol Psychiatry*. 2000;48:861–873.
45. Miguel-Hidalgo JJ, Waltzer R, Whittom AA, Austin MC, Rajkowska G, Stockmeier CA. Glial and glutamatergic markers in depression, alcoholism, and their comorbidity. *J Affect Disord*. 2010;127:230–240.
46. Laumet G, Edralin JD, Chiang AC, Dantzer R, Heijnen CJ, Kavelaars A. Resolution of inflammation-induced depression requires T lymphocytes and endogenous brain interleukin-10 signaling. *Neuropsychopharmacology*. 2018;43:2597–2605.
47. Goncalves CA, Leite MC, Nardin P. Biological and methodological features of the measurement of S100B, a putative marker of brain injury. *Clin Biochem*. 2008;41:755–763.
48. Rothermundt M, Arolt V, Wiesmann M, et al. S-100B is increased in melancholic but not in non-melancholic major depression. *J Affect Disord*. 2001;66:89–93.
49. Grabe HJ, Ahrens N, Rose HJ, Kessler C, Freyberger HJ. Neurotrophic factor S100 beta in major depression. *Neuropsychobiology*. 2001;44:88–90.
50. Paolicelli RC, Bolasco G, Pagani F, et al. Synaptic pruning by microglia is necessary for normal brain development. *Science*. 2011;333(6048):1456–1458.
51. Bachtell RK, Jones JD, Heinzerling KG, Beardsley PM, Comer SD. Glial and neuroinflammatory targets for treating substance use disorders. *Drug Alcohol Depend*. 2017;180:156–170.
52. Kierdorf K, Prinz M. Factors regulating microglia activation. *Front Cell Neurosci*. 2013;7:44.
53. Kim YS, Joh TH. Microglia, major player in the brain inflammation: their roles in the pathogenesis of Parkinson's disease. *Exp Mol Med*. 2006;38:333–347.
54. Schafer DP, Lehrman EK, Kautzman AG, et al. Microglia sculpt postnatal neural circuits in an activity and complement-dependent manner. *Neuron*. 2012;74:691–705.
55. Sierra A, Encinas JM, Deudero JJ, et al. Microglia shape adult hippocampal neurogenesis through apoptosis-coupled phagocytosis. *Cell Stem Cell*. 2010;7:483–495.
56. Duman RS, Aghajanian GK. Synaptic dysfunction in depression: potential therapeutic targets. *Science*. 2012;338(6103):68–72.
57. Prinz M, Priller J. Microglia and brain macrophages in the molecular age: from origin to neuropsychiatric disease. *Nat Rev Neurosci*. 2014;15:300–312.
58. Eyre H, Baune BT. Neuroplastic changes in depression: a role for the immune system. *Psychoneuroendocrinology*. 2012;37:1397–1416.
59. Yirmiya R, Rimmerman N, Reshef R. Depression as a microglial disease. *Trends Neurosci*. 2015;38:637–658.

60. Rock RB, Gekker G, Hu S, et al. Role of microglia in central nervous system infections. *Clin Microbiol Rev*. 2004;17:942–964.
61. Haapakoski R, Mathieu J, Ebmeier KP, Alenius H, Kivimaki M. Cumulative meta-analysis of interleukins 6 and 1beta, tumour necrosis factor alpha and C-reactive protein in patients with major depressive disorder. *Brain Behav Immun*. 2015;49:206–215.
62. Krabbe KS, Reichenberg A, Yirmiya R, Smed A, Pedersen BK, Bruunsgaard H. Low-dose endotoxemia and human neuropsychological functions. *Brain Behav Immun* 2005;19:453–460.
63. Bayer TA, Buslei R, Havas L, Falkai P. Evidence for activation of microglia in patients with psychiatric illnesses. *Neurosci Lett*. 1999;271:126–128.
64. Steiner J, Bielau H, Brisch R, et al. Immunological aspects in the neurobiology of suicide: elevated microglial density in schizophrenia and depression is associated with suicide. *J Psychiatr Res*. 2008;42:151–157.
65. Fenn AM, Gensel JC, Huang Y, Popovich PG, Lifshitz J, Godbout JP. Immune activation promotes depression 1 month after diffuse brain injury: a role for primed microglia. *Biol Psychiatry*. 2014;76:575–584.
66. Henry CJ, Huang Y, Wynne A, et al. Minocycline attenuates lipopolysaccharide (LPS)-induced neuroinflammation, sickness behavior, and anhedonia. *J Neuroinflammation*. 2008;5:15.
67. Bhattacharya A, Lord B, Grigoleit JS, et al. Neuropsychopharmacology of JNJ-55308942: evaluation of a clinical candidate targeting P2X7 ion channels in animal models of neuroinflammation and anhedonia. *Neuropsychopharmacology*. 2018;43:2586–2596.
68. Kettenmann H, Hanisch UK, Noda M, Verkhratsky A. Physiology of microglia. *Physiol Rev*. 2011;91:461–553.
69. Frick LR, Williams K, Pittenger C. Microglial dysregulation in psychiatric disease. *Clin Dev Immunol*. 2013;2013:608654.
70. Ohsawa K, Imai Y, Sasaki Y, Kohsaka S. Microglia/macrophage-specific protein Iba1 binds to fimbrin and enhances its actin-bundling activity. *J Neurochem*. 2004;88:844–856.
71. Torres-Platas SG, Cruceanu C, Chen GG, Turecki G, Mechawar N. Evidence for increased microglial priming and macrophage recruitment in the dorsal anterior cingulate white matter of depressed suicides. *Brain Behav Immun*. 2014;42:50–59.
72. Kreisel T, Frank MG, Licht T, et al. Dynamic microglial alterations underlie stress-induced depressive-like behavior and suppressed neurogenesis. *Mol Psychiatry*. 2014;19:699–709.
73. Reichenberg A, Yirmiya R, Schuld A, et al. Cytokine-associated emotional and cognitive disturbances in humans. *Arch Gen Psychiatry*. 2001;58:445–452.
74. Alcami A. Viral mimicry of cytokines, chemokines and their receptors. *Nat Rev Immunol*. 2003;3:36–50.
75. Godbout JP, Berg BM, Krzyszton C, Johnson RW. Alpha-tocopherol attenuates NFkappaB activation and pro-inflammatory cytokine production in brain and improves recovery from lipopolysaccharide-induced sickness behavior. *J Neuroimmunol*. 2005;169:97–105.
76. Bierhaus A, Wolf J, Andrassy M, et al. A mechanism converting psychosocial stress into mononuclear cell activation. *Proc Natl Acad Sci USA*. 2003;100:1920–1925.
77. Silverman HA, Dancho M, Regnier-Golanov A, et al. Brain region-specific alterations in the gene expression of cytokines, immune cell markers and cholinergic system components during peripheral endotoxin-induced inflammation. *Mol Med*. 2015;20:601–611.

78. Gao Y, Lecker S, Post MJ, et al. Inhibition of ubiquitin-proteasome pathway-mediated I kappa B alpha degradation by a naturally occurring antibacterial peptide. *J Clin Invest.* 2000;106:439–448.
79. Li Q, Verma IM. NF-kappaB regulation in the immune system. *Nat Rev Immunol.* 2002;2:725–734.
80. Pistis M, Melis M. From surface to nuclear receptors: the endocannabinoid family extends its assets. *Curr Med Chem.* 2010;17:1450–1467.
81. Genolet R, Wahli W, Michalik L. PPARs as drug targets to modulate inflammatory responses? *Curr Drug Targets Inflamm Allergy.* 2004;3:361–375.
82. Yang R, Wang P, Chen Z, et al. WY-14643, a selective agonist of peroxisome proliferator-activated receptor-alpha, ameliorates lipopolysaccharide-induced depressive-like behaviors by preventing neuroinflammation and oxido-nitrosative stress in mice. *Pharmacol Biochem Behav.* 2017;153:97–104.
83. Song L, Wang H, Wang Y-J, et al. Hippocampal PPARα is a novel therapeutic target for depression and mediates the antidepressant actions of fluoxetine in mice. *Br J Pharmacol.* 2018;175:2968–2987.
84. Dwivedi Y. Brain-derived neurotrophic factor: role in depression and suicide. *Neuropsychiatr Dis Treat.* 2009;5:433–449.
85. Lima Giacobbo B, Doorduin J, Klein HC, Dierckx R, Bromberg E, de Vries EFJ. Brain-derived neurotrophic factor in brain disorders: focus on neuroinflammation. *Mol Neurobiol.* 2019;56:3295–3312.
86. Marini AM, Jiang X, Wu X, et al. Role of brain-derived neurotrophic factor and NF-kappaB in neuronal plasticity and survival: from genes to phenotype. *Restor Neurol Neurosci.* 2004;22:121–130.
87. Tomaz VS, Cordeiro RC, Costa AM, et al. Antidepressant-like effect of nitric oxide synthase inhibitors and sildenafil against lipopolysaccharide-induced depressive-like behavior in mice. *Neuroscience.* 2014;268:236–246.
88. Zhang JC, Wu J, Fujita Y, et al. Antidepressant effects of TrkB ligands on depression-like behavior and dendritic changes in mice after inflammation. *Int J Neuropsychopharmacol.* 2015;18:pyu077. https://doi.org/10.1093/ijnp/pyu077.
89. Ferrini F, De Koninck Y. Microglia control neuronal network excitability via BDNF signalling. *Neural Plast.* 2013;2013:429815.
90. Luscher B, Shen Q, Sahir N. The GABAergic deficit hypothesis of major depressive disorder. *Mol Psychiatry.* 2011;16:383–406.
91. Schomberg SL, Bauer J, Kintner DB, et al. Cross talk between the GABA(A) receptor and the Na-K-Cl cotransporter is mediated by intracellular Cl. *J Neurophysiol.* 2003;89:159–167.
92. Goubert E, Altvater M, Rovira MN, et al. Bumetanide prevents brain trauma-induced depressive-like behavior. *Front Mol Neurosci.* 2019;12:12.
93. Dineley KT, Pandya AA, Yakel JL. Nicotinic ACh receptors as therapeutic targets in CNS disorders. *Trends Pharmacol Sci.* 2015;36:96–108.
94. Alzarea S, Rahman S. Effects of alpha-7 nicotinic allosteric modulator PNU 120596 on depressive-like behavior after lipopolysaccharide administration in mice. *Prog Neuropsychopharmacol Biol Psychiatry.* 2018;86:218–228.
95. Gotti C, Zoli M, Clementi F. Brain nicotinic acetylcholine receptors: native subtypes and their relevance. *Trends Pharmacol Sci.* 2006;27:482–491.
96. de Jonge WJ, Ulloa L. The alpha7 nicotinic acetylcholine receptor as a pharmacological target for inflammation. *Br J Pharmacol.* 2007;151:915–929.
97. King JR, Gillevet TC, Kabbani N. A G protein-coupled α7 nicotinic receptor regulates signaling and TNF-α release in microglia. *FEBS Open Bio.* 2017;7: 1350–1361.

98. Patel H, McIntire J, Ryan S, Dunah A, Loring R. Anti-inflammatory effects of astroglial α7 nicotinic acetylcholine receptors are mediated by inhibition of the NF-κB pathway and activation of the Nrf2 pathway. *J Neuroinflammation*. 2017;14:192.
99. Alzarea S, Rahman S. Alpha-7 nicotinic receptor allosteric modulator PNU120596 prevents lipopolysaccharide-induced anxiety, cognitive deficit and depression-like behaviors in mice. *Behav Brain Res*. 2019;366:19–28.
100. Papke RL, Kem WR, Soti F, Lopez-Hernandez GY, Horenstein NA. Activation and desensitization of nicotinic alpha7-type acetylcholine receptors by benzylidene anabaseines and nicotine. *J Pharmacol Exp Ther*. 2009;329:791–807.
101. Corradi J, Bouzat C. Understanding the bases of function and modulation of alpha7 nicotinic receptors: implications for drug discovery. *Mol Pharmacol*. 2016;90:288–299.
102. Wang H, Yu M, Ochani M, et al. Nicotinic acetylcholine receptor alpha7 subunit is an essential regulator of inflammation. *Nature*. 2003;421(6921):384–388.
103. Abbas M, Rahman S. Effects of alpha-7 nicotinic acetylcholine receptor positive allosteric modulator on lipopolysaccharide-induced neuroinflammatory pain in mice. *Eur J Pharmacol*. 2016;783:85–91.
104. Shytle RD, Mori T, Townsend K, et al. Cholinergic modulation of microglial activation by alpha 7 nicotinic receptors. *J Neurochem*. 2004;89:337–343.
105. Suzuki T, Hide I, Matsubara A, et al. Microglial alpha7 nicotinic acetylcholine receptors drive a phospholipase C/IP3 pathway and modulate the cell activation toward a neuroprotective role. *J Neurosci Res*. 2006;83:1461–1470.
106. Abbas M, Alzarea S, Papke RL, Rahman S. The alpha7 nicotinic acetylcholine receptor positive allosteric modulator attenuates lipopolysaccharide-induced activation of hippocampal IkappaB and CD11b gene expression in mice. *Drug Discov Ther*. 2017;11(4):206–211.
107. Kalkman HO, Feuerbach D. Modulatory effects of alpha7 nAChRs on the immune system and its relevance for CNS disorders. *Cell Mol Life Sci*. 2016;73(13):2511–2530.
108. Carnevale D, De Simone R, Minghetti L. Microglia-neuron interaction in inflammatory and degenerative diseases: role of cholinergic and noradrenergic systems. *CNS Neurol Disord Drug Targets*. 2007;6:388–397.
109. Yue Y, Liu R, Cheng W, et al. GTS-21 attenuates lipopolysaccharide-induced inflammatory cytokine production in vitro by modulating the Akt and NF-kappaB signaling pathway through the alpha7 nicotinic acetylcholine receptor. *Int Immunopharmacol*. 2015;29:504–512.
110. Zhang JC, Yao W, Ren Q, et al. Depression-like phenotype by deletion of alpha7 nicotinic acetylcholine receptor: role of BDNF-TrkB in nucleus accumbens. *Sci Rep*. 2016;6:36705.
111. Melis M, Scheggi S, Carta G, et al. PPARalpha regulates cholinergic-driven activity of midbrain dopamine neurons via a novel mechanism involving alpha7 nicotinic acetylcholine receptors. *J Neurosci*. 2013;33:6203–6211.
112. Zhao D, Xu X, Pan L, et al. Pharmacologic activation of cholinergic alpha7 nicotinic receptors mitigates depressive-like behavior in a mouse model of chronic stress. *J Neuroinflammation*. 2017;14:234.
113. Changeux JP, Edelstein SJ. Allosteric mechanisms of signal transduction. *Science*. 2005;308(5727):1424–1428.
114. Abdrakhmanova GR, Blough BE, Nesloney C, Navarro HA, Damaj MI, Carroll FI. In vitro and in vivo characterization of a novel negative allosteric modulator of neuronal nAChRs. *Neuropharmacology*. 2010;59:511–517.
115. Williams DK, Wang J, Papke RL. Positive allosteric modulators as an approach to nicotinic acetylcholine receptor-targeted therapeutics: advantages and limitations. *Biochem Pharmacol*. 2011;82:915–930.

116. Young GT, Zwart R, Walker AS, Sher E, Millar NS. Potentiation of alpha7 nicotinic acetylcholine receptors via an allosteric transmembrane site. *Proc Natl Acad Sci USA*. 2008;105:14686–14691.
117. Bertrand D, Gopalakrishnan M. Allosteric modulation of nicotinic acetylcholine receptors. *Biochem Pharmacol*. 2007;74:1155–1163.
118. Chatzidaki A, D'Oyley JM, Gill-Thind JK, Sheppard TD, Millar NS. The influence of allosteric modulators and transmembrane mutations on desensitisation and activation of alpha7 nicotinic acetylcholine receptors. *Neuropharmacology*. 2015;97:75–85.
119. Hurst RS, Hajos M, Raggenbass M, et al. A novel positive allosteric modulator of the alpha7 neuronal nicotinic acetylcholine receptor: in vitro and in vivo characterization. *J Neurosci*. 2005;25:4396–4405.
120. Arias HR. Positive and negative modulation of nicotinic receptors. *Adv Protein Chem Struct Biol*. 2010;80:153–203.
121. Gronlien JH, Hakerud M, Ween H, et al. Distinct profiles of alpha7 nAChR positive allosteric modulation revealed by structurally diverse chemotypes. *Mol Pharmacol*. 2007;72:715–724.
122. Freitas K, Carroll FI, Damaj MI. The antinociceptive effects of nicotinic receptors alpha7-positive allosteric modulators in murine acute and tonic pain models. *J Pharmacol Exp Ther*. 2013;344:264–275.
123. McLean SL, Idris NF, Grayson B, et al. PNU-120596, a positive allosteric modulator of alpha7 nicotinic acetylcholine receptors, reverses a sub-chronic phencyclidine-induced cognitive deficit in the attentional set-shifting task in female rats. *J Psychopharmacol*. 2012;26:1265–1270.
124. Grizzell JA, Iarkov A, Holmes R, Mori T, Echeverria V. Cotinine reduces depressive-like behavior, working memory deficits, and synaptic loss associated with chronic stress in mice. *Behav Brain Res*. 2014;268:55–65.
125. Targowska-Duda KM, Feuerbach D, Biala G, Jozwiak K, Arias HR. Antidepressant activity in mice elicited by 3-furan-2-yl-N-p-tolyl-acrylamide, a positive allosteric modulator of the alpha7 nicotinic acetylcholine receptor. *Neurosci Lett*. 2014;569:126–130.

Further reading

126. Camara ML, Corrigan F, Jaehne EJ, Jawahar MC, Anscomb H, Baune BT. Tumor necrosis factor alpha and its receptors in behaviour and neurobiology of adult mice, in the absence of an immune challenge. *Behav Brain Res*. 2015;290:51–60.

CHAPTER EIGHT

Its complicated: The relationship between alcohol and microglia in the search for novel pharmacotherapeutic targets for alcohol use disorders

Jennifer K. Melbourne[a], K. Ryan Thompson[a], Hui Peng[b], Kimberly Nixon[a,*]

[a]The University of Texas at Austin, College of Pharmacy, Division of Pharmacology & Toxicology, Austin, TX, United States
[b]University of Kentucky, College of Pharmacy, Department of Pharmaceutical Sciences, Lexington, KY, United States
*Corresponding author: e-mail address: kim.nixon@austin.utexas.edu

Contents

Progress in Molecular Biology and Translational Science, Volume 167
ISSN 1877-1173
https://doi.org/10.1016/bs.pmbts.2019.06.011

Abstract

Alcohol use disorder (AUD) is a chronic relapsing disorder with wide-ranging health consequences. Alcohol targets the central nervous system producing neurodegeneration and subsequent cognitive and behavioral deficits, but the mechanisms behind these effects remain unclear. Recently, evidence has been mounting for the role of neuroimmune activation in the pathogenesis of AUDs, but our nascent state of knowledge about the interaction of alcohol with the neuroimmune system supports that the relationship is complicated. As the resident macrophage of the central nervous system, microglia are a central focus. Human and animal research on the interplay between microglia and alcohol in AUDs has proven to be complex, and though early research focused on a pro-inflammatory phenotype of microglia, the anti-inflammatory and homeostatic roles of microglia must be considered. How these new roles for microglia should be incorporated into our thinking about the neuroimmune system in AUDs is discussed in the context of developing novel pharmacotherapies for AUDs.

1. Introduction

Excessive consumption of alcohol is associated with the development of dependence, wide-ranging health consequences and disability. These sequelae of alcohol abuse occur, in part, due to the effects of ethanol on the central nervous system (CNS). Clinical data and preclinical models demonstrate evidence of regional volume loss and neurodegeneration accompanied by glial activation and oxidative stress following acute and chronic exposure to alcohol. Microglia, the innate immune cells of the brain are hypothesized to mediate many of these neurotoxic effects; however, the role of these cells in alcohol use disorders (AUDs) is far from conclusive. Nonetheless, targeting the neuroimmune system with pharmacotherapeutics is a promising strategy in the treatment of alcoholism. In this review we begin by outlining the epidemiology and neuropathology of AUDs. Next, we focus on microglial biology and introduce new concepts in the field. The rich literature investigating neuroimmune activation in humans with AUDs and preclinical models is then discussed, with emphasis on the complexity of microglia-specific findings. Finally, existing treatments for AUDs, those currently in development at the clinical and preclinical stages, and future drug targets, are outlined and discussed in the context of the neuroimmune system.

2. Alcohol use disorders

Alcohol is among the most commonly used drugs in the United States.[1] In 2017, over 173 million, or 70.1%, of US adults aged 18 or older self-reported alcohol use in the past year, greater than reported tobacco or illicit drug use combined.[1] Of this population, a large portion consume alcohol in a risky pattern, with 26.4% reporting binge alcohol use (>4/5 drinks in a 2 h period for women/men, respectively) and 6.7% reporting heavy alcohol use (>7/14 drinks per week for women/men, respectively).[1] High risk drinking is just that, it puts one at risk for meeting diagnostic criteria for an AUD. Defined by the Diagnostic and Statistical Manual of Mental Disorders V,[2] a diagnosis of an AUD considers the severity of an individual's drinking behavior, past experiences of relapse, loss of control over drinking behavior, and drinking to the detriment of the individual's well-being. In the United States, 13.9% of adults 18 years or older, or >32 million people, are estimated to have an AUD,[3] a value that has increased substantially across the last several years.[4]

Alcohol has been linked to over 200 diseases and injury-related health issues, which results in a significant cost to society and burden to public health.[5] Excessive alcohol consumption costs the U.S. $459 billion annually, 71.9% of which is attributed to loss of productivity (impaired productivity, absenteeism, mortality, etc.), but also includes substantial costs attributable to alcohol-related healthcare and crime.[6] Indeed, alcohol is the most harmful drug of abuse.[7] Worldwide, 3 million people die per year from conditions related to alcohol use, which puts alcohol's mortality rate higher than other major health concerns like HIV/AIDS and diabetes.[8] In the U. S., although excessive alcohol consumption is the third leading cause of mortality, the greatest public health burden results from disability.[9,10] Neuropsychiatric and cognitive impairment are the largest cause of disability associated with AUDs.[11,12] The broad pharmacological actions and toxicity resulting from excessive alcohol consumption drive the mortality and disability associated with AUDs; however, there are currently no pharmacological treatments that successfully mitigate alcohol's toxic effects. Instead, drug discovery efforts have focused on the consummatory and rewarding effects of alcohol, but with only modest success.

Alcohol (ethanol) has many unique qualities compared to other drugs of abuse. Despite it being a relatively small, nonchiral molecule, most responses require ethanol doses in weights of grams. Acutely, alcohol exposure can act directly on a variety of ionotropic receptors and signaling molecules,[13] then with chronic alcohol exposure characteristic of an AUD, additional changes and neuroadaptations occur that result in cravings and relapse, hallmarks of alcohol dependence. As an AUD develops, alterations occur to many neurotransmitter systems, neuropeptides, signaling molecules, and as described more recently, activation of neuroimmune signaling.[13–16] Corrupted neuroplastic changes ranging to toxicities produce the neuropathological changes that contribute to the downward spiral into alcohol abuse and addiction.[17]

3. Alcohol-induced neuropathology

Excessive alcohol consumption is well-known to damage most organs of the body, including the brain which is particularly sensitive to the toxic effects of alcohol.[17] Chronic use is associated with multiple pathologies including: Wernicke's encephalopathy, the frequently co-diagnosed Korsakoff syndrome, hepatic encephalopathy, alcoholic cerebellar degeneration, Marchiafava-Bignami Disease, and alcohol related dementia.[18] Though ethanol is directly neurotoxic, indirect effects such as nutritional deficiencies and hepatic encephalopathy that occur in severe AUDs can exacerbate neuropathology. Individuals who lack these other alcohol-related pathologies are termed "uncomplicated AUD," which is the focus of the majority of studies in the AUD literature as well as this review. Research has shown that 50–80% of uncomplicated AUDs still show impairments in cognitive and motor functions.[12,19,20] Neurodegeneration and other neuropathological events are thought to underlie these impairments, but multiple layers of variability due to differences in amount, duration and patterns of intake, age of onset, as well as individual genetics and susceptibilities cloud our understanding.[17,21] Decades of research on the neuropathology of AUDs have demonstrated that the extent of neurodegeneration found in an individual varies depending on the individual's age at assessment, age of onset, sex, family history of alcoholism, and length of abstinence, among other variables (e.g., Refs. 22–26; see also Refs. 27–29 for more detailed reviews on this topic).

Despite these complexities, some patterns and consistencies have emerged. Brain imaging studies and post mortem analyses consistently have found neurodegeneration indicated by measures of smaller brain volumes, gray and white matter loss, enlarged ventricles and sulci, and higher volumes of cerebrospinal fluid (CSF) compared to healthy controls.[21,23,30–34] Evidence of volume loss has been found in most areas of the cortex, though the frontal lobes show the greatest vulnerabilities as both white and gray matter are damaged in this region.[24,35–39] Degeneration of the frontal cortex is consistent with the pattern of cognitive deficits such as compromised executive functions,[21,30,40] increased impulsivity[41,42] and impaired decision making.[43] Dysfunction of the mesolimbic reward system is often implicated in the development of AUDs as an essential site in motivation and reward seeking, but only recently have neurotoxic effects and structural deficits been reported. Reductions in the total volume of the reward system, including white and gray matter loss in the amygdala, prefrontal cortex, insula, and nucleus accumbens have been reported.[39,44–46] The hippocampus is another limbic region that has received much attention in the alcohol literature due to its roles in learning, memory, mood, and drug seeking/taking.[47,48] Impaired learning and memory are associated with cell and volume loss across the hippocampus.[33,34,49–53] Finally, hindbrain and cerebellar degeneration in gray and white matter has been consistently found to coincide with declines in functions such as gait and balance in AUD.[21,34,54–58] Thus, multiple sites of degeneration have been observed in AUDs, which has a broad impact on behavior and particularly the likelihood of recovery.[27,45]

While these neurodegenerative and associated cognitive deficits are well-established in AUDs, the specific cellular changes and mechanisms that underlie these pathologies are not well understood and remain an active area of inquiry and debate. Neuronal loss versus glial loss or shrinkage of neuronal soma and decreased dendritic arborization have all been attributed to the volume losses observed in AUDs, but the precise mechanisms of neurodegeneration remain unclear.[17,59–61] The observations that at least some volume and function recover in abstinence further complicates our understanding.[17,24,27] Neuronal loss has been reported in the cerebellum, frontal cortex, and hypothalamus[62–64] and more recently, perhaps thanks to technological advances, stronger evidence of hippocampal neuronal loss has emerged.[52,53] Process retraction especially in dendritic trees and the shrinkage of neuronal soma have been demonstrated.[29,65] Decreases in white matter volume in tandem with astrocyte and microglia loss, may all underlie volume loss as well.[66–68]

Animal models have been essential in establishing causality between excessive alcohol exposure and neurodegeneration. For example, the seminal work of Walker and colleagues demonstrated that chronic alcohol exposure in rats reduced the number of neurons in the hippocampus, which corresponded to learning and memory deficits on hippocampal-dependent tasks.[69,70] Studies in animal models have since shown that this neuronal loss is likely the result of cell death.[71–74] However the precise mechanism of cell death is still unknown and may depend on variables such as age and amount, duration and pattern of alcohol exposure.[17] In binge models of alcohol dependence, necrosis is observed,[72] while apoptosis evidenced by TUNEL staining, increases in caspase-3 and decreases in the anti-apoptotic protein

Bcl2, has been observed in other models *in vitro* and *in vivo*.[75–79] The mechanisms that lead to cell death are not known, but several groups provide evidence for oxidative stress playing a role.[79–82] The innate immune system of the CNS and microglia, in particular, have been implicated in producing reactive oxygen species in response to damage signals and may contribute to neurodegeneration.[17,83–85] Some hypothesize that alcohol acts directly on microglial receptors to activate pro-inflammatory pathways and drive alcohol-induced neurodegeneration.[83,85,86] Thus, microglia are an obvious suspect in mediating or contributing to alcohol-induced neurodegeneration. However, the role of the innate immune system of the CNS, following alcohol exposure is a hot topic of debate, and much remains to be discovered. In the next sections we will outline microglial biology and discuss studies that investigate neuroimmune function in humans and animal models of AUDs.

4. Microglia

Microglia are one of three kinds of glial cells in the CNS and comprise about 5–20% of glial cells in the brain.[87] Microglia are of myeloid origin,[88] derived from macrophage precursors in the yolk sac and migrate to the nervous system early on in embryogenesis.[89] After migration to the brain, microglia undergo stages of proliferation and migration during prenatal and postnatal development through adolescence.[90] After embryonic development, these brain parenchymal immune cells appear to be replenished locally with little to no contribution from the periphery under healthy conditions.[91] Microglia are also thought to be relatively long-lived, though their turnover rate is presently the subject of debate.[92] For example, one recent study estimated a high rate of microglia proliferation in both mouse and human brain using cell cycle markers,[93] while another estimated much lower proliferation rates in the human brain using carbon dating.[94]

Microglia, while best known for their role as immune cells in the CNS, have a much broader role in brain function that previously considered. Microglia contribute to synaptic plasticity through remodeling of synapses and influencing neurotransmission.[95] In the adult brain, "resting state" microglia exhibit a quiescent phenotype indicated by a ramified morphology with multiple thin branched processes extending from the soma and low expression of activation markers.[96] However, far from fulfilling the role implied by this label, microglia in the healthy brain are now understood to be highly motile and constantly surveying their parenchymal environment by extending and retracting processes and making contact with neurons and other glial cells.[97] This active sampling makes microglia exceptionally sensitive and quickly responsive to any perturbations in their local microenvironment.[98]

The function of microglia outside of the immune response has been particularly well studied in relation to CNS development, where they are thought to be critical for developmental synaptic pruning and the establishment of neural circuits.[97] Complement component 3 (C3) is thought to mark synapses for elimination by microglia, which express the C3 receptor (i.e., OX-42), in the developing brain in an activity-dependent manner.[99] A recent study using advanced imaging techniques directly demonstrated trogocytosis, but not phagocytosis, of synaptic elements by microglia as well as microglia-mediated induction of filopodia on established dendritic spines in developing hippocampal cultures.[100] In addition to their functions in development, microglia are now also known to be involved in adult CNS homeostatic processes.[101] Microglial depletion in adult mice led to deficits in hippocampal-mediated and motor learning as well as decreased spine dynamics in the motor cortex.[102] Microglia also modulate synaptic scaling via effects on astrocytes.[103] Additionally, microglia were shown to make contact with highly active neurons resulting in reduced neuronal activity.[104] This neuron-microglia communication is, at least in some instances, mediated by ATP release following NMDA receptor activation.[105] Microglia may also maintain the pool of oligodendrocyte progenitor cells as well as phagocytose myelin and dead or dying cells.[106] Thus, microglia play active homeostatic roles in CNS structure and function through their ability to shape synapses through remodeling, pruning and mediating neurotransmission.

4.1 Microglia and neuroinflammation

As the primary CNS resident (parenchymal) immune cells, microglia are equipped with a vast array of receptors to detect signaling molecules such as cytokines, chemokines and growth factors, as well as exogenous and endogenous molecular patterns that indicate damage or infection (termed the microglial "Sensome").[103,107] When microglia detect a pathogen or an indication of neuronal damage or systemic inflammation, they become "activated".[103] As part of activation, specific cell surface markers are upregulated, accompanied by morphological changes. Specifically, microglia undergo a graded decrease in branched processes, and enlargement of the soma, and, in the fully activated state, achieve an amoeboid-like appearance allowing for phagocytosis of pathogens or cellular debris.[108] Microglia may migrate to the site of injury and proliferate, and context-dependent secretion of effector molecules such as cytokines and trophic factors occurs. Understanding microglial activation as a temporally graded, but overall homogeneous, response to insult led to the establishment of a stage-based classification system of microglial activation.[109] Stage 0 represented resting microglia in the healthy brain. Early activation (stage 1) was associated with a state of alert, followed by homing (stage 2), phagocytosis (stage 3a) and progressing through to bystander activation (stage 3b). This initial response to clear the pathogen or damage is usually neuroprotective and restores homeostasis to the tissue, while prolonged activation is thought to be detrimental (see also, the discussion of phenotypes below).

Sustained microglial activation can cause damage to surrounding tissue due to unconstrained secretion of proinflammatory cytokines and reactive oxygen species. This chronic CNS inflammatory state, often termed "neuroinflammation," has been associated with a wide range of neurological and neuropsychiatric conditions.[110,111] The term "inflammation" was originally defined by Celsus with four cardinal signs: *tumor, rubor, calor et dolor*[112] and typically displays extravasation of blood cells. In established neuroinflammatory conditions, such as stroke and trauma, and viral, bacterial, and autoimmune diseases of the CNS, there is entry of circulating immune cells to the brain parenchyma. Outside of these conditions and in the healthy brain, the blood-brain barrier limits entry of these patrolling bone marrow-derived immune cells to the parenchyma. However, the concept of neuroinflammation has also been expanded to describe conditions that display none of Celsus's cardinal signs and don't attract inflammatory cells, including various types of neurodegeneration as classified by neuropathologists.[111] Therefore, the word "neuroinflammation" is used to identify a radically different set of conditions that are specific to the CNS. In neurological disorders characterized by neuroinflammation, changes are almost always associated with one cell type, microglia.

To further complicate this picture, while morphological evidence of microglial activation has been increasingly accepted as evidence of neuroinflammation, it is now apparent that the functional characteristics of activated microglia are context-dependent.[113] An understanding of the role of microglia in response to neuronal damage (e.g., brain injury, stroke) and in the context of neurodegenerative (e.g., Alzheimer's disease) and neuroinflammatory (e.g., multiple sclerosis) disorders historically focused on the inflammatory and neurotoxic properties of these cells.[114] However, microglial activation is now also understood to be critically involved in neuroprotection *and recovery*. For example, microglia are thought to be important for the clearance and encapsulation of amyloid beta in Alzheimer's disease.[115] Many of the protective functions of these cells are coming to light based on the outcomes of microglial depletion studies. For example, depletion of microglia following spinal cord injury was detrimental to neuronal and oligodendrocyte survival and motor recovery whereas boosting microglial proliferation was protective.[116]

4.2 Microglial phenotypes

This distinction between the damaging and protective functions of microglia has been described utilizing terminology from the peripheral macrophage literature, where macrophages have traditionally been categorized using "M1" or "classical activation" and "M2" or "alternative activation" based on *in vitro* responses to t-helper 1 and t-helper 2 polarizing stimuli (IFN-γ and IL-4, respectively).[117] As such, activated microglia were also divided into two subtypes: proinflammatory (M1) and anti-inflammatory (M2).[109,118,119] While the M2 phenotypes produce anti-inflammatory cytokines, promote tissue repair and phagocytose protein aggregates and cell debris, the M1 phenotypes are more liable to be detrimental to the brain by inducing neuronal toxicity through proinflammatory cytokine and chemokine secretion and reactive oxygen species production. These phenotypes combined with the phase specific nature of microglial activation led to a model whereby microglia initially respond to neural damage or dysfunction by acquiring a proinflammatory "M1-like" phenotype which may, depending on conditions, transition to an "M2-like" protective phenotype.[120] As it became clear that there are a wide variety of exogenous and endogenous signals that can impact both the macrophage and microglial phenotypes, spectrum models were adopted with M1 and M2 phenotypic extremes at either end.[117,121] However, the mechanisms responsible for microglia activation and/or polarization toward the various phenotypes are poorly understood.

The spectrum model of macrophage and microglial phenotypes is also now under question as data highlight the complex nature of myeloid phenotypes and associated functions *in vivo*.[122] Much of this advance in our understanding of macrophage/microglial phenotypes in health and disease has been propelled by recent advances in RNA sequencing, particularly single-cell, in conjunction with mass cytometry.[123] Such techniques have demonstrated that individual macrophages/microglia can display both M1 and M2 markers simultaneously in response to insult. Characterization of microglial signatures and surface markers using single cell approaches now indicates that, at rest, microglia exhibit a homeostatic signature that is shared across different brain regions in healthy mice and humans.[92] Functions associated with these homeostatic microglia include scanning, phagocytosis and neuronal support. Studies of microglia in disease states show that these homeostatic genes are downregulated and a subset of "disease- or degeneration-associated microglia (DAM)" are observed.[123] DAM express genes characteristic of both M1 and M2 microglial phenotypes as well as alterations to other functional signatures such as the interferon response, stress response, lysosomal function and lipid metabolism. This shared DAM signature is contrasted from microglial signatures in neuroinflammatory conditions, for example, in response to systemic LPS injection or the experimental autoimmune encephalitis model of multiple sclerosis, which demonstrate distinct transcriptional and cell surface changes. Ultimately, the beneficial versus protective nature of microglia appears to be context-dependent, and varies based on the disease or model investigated, temporal and spatial characteristics, and importantly which of the many microglial functions is affected.[92,123]

Similar to peripheral monocytes and macrophages, microglia also show evidence of immune memory. The concept "priming" when applied to these myeloid cells describes a phenomenon whereby an initial stimulus leads to a stronger immune response, particularly the release of proinflammatory cytokines, to subsequent challenges.[124] The application of this concept to microglia has been developed largely in relation to neurogenerative disease.[125,126] Priming has now been described in the brain in several disease models and physiological states including stress, normal aging, Parkinson's disease, and nerve crush-induced axonal degeneration.[127–129] In wild-type animals, while one peripheral LPS injection induced a mild inflammatory state in the CNS that resolved, a second hit induced a strong proinflammatory response, indicative of priming.[130] However, by the fourth LPS hit, there was no longer a cytokine response to challenge. This blocking of the immune response to challenge is called "tolerization" in peripheral monocyte and macrophage biology, and describes a state in which cells become desensitized to inflammatory stimuli and adopt a more anti-inflammatory phenotype.[124] Microglia-specific knockout of immune signaling pathways and epigenetic enzymes blocked the potentiation of the immune response following the second LPS stimulus, suggesting that these aspects of microglial biology are necessary for the described effects on immune sensitivity.[130] Interestingly, transcriptome profiling of primed microglia revealed a signature that overlaps with that of DAM, including the downregulation of homeostatic microglial genes.[124,131] The stimuli that lead to microglial priming, and the mechanisms, such as persistent chromatin modifications and/or increased receptor expression, that underly immune memory require further elucidation.

5. Neuroimmune activation in AUDs

The neuroimmune actions of alcohol have received significant attention as a potential driver of neuroplastic and neurodegenerative changes that occur in the pathogenesis of AUDs.[16,17,84] Alcohol-induced neuroinflammation, predominantly mediated by activated microglia in response to excessive ethanol consumption, is the hypothesized mechanism for how alcohol induces brain damage as well as how pro-inflammatory signaling hijacks behavioral control systems and mesolimbic reward systems to promote excessive alcohol consumption.[16,84,132] A strength of this hypothesis is that it unites what have been distinct lines of alcohol research, the damaging consequences of excessive alcohol intake on corticolimbic circuitry and neuroplastic changes in the mesolimbic reward system that are thought to underlie addiction. Compelling data from both lines of research implicate a role for neuroimmune signaling and its hallmark cell, microglia, in AUD pathogenesis. However, this hypothesis has really only promoted the pro-inflammatory effects of activated microglia and almost overlooked the fact that inhibiting or knocking-out microglia can have deleterious consequences for brain and behavior.[116] The neuroimmune hypothesis of addiction did not elaborate on the homeostatic role of microglia nor that microglia have various phenotypes and functions in disease, specifically that rather than driving neurotoxicity, microglia may promote repair or recovery mechanisms. Whether the activation of microglia is detrimental or protective in relation to alcohol induced neurodegeneration remains a subject of debate. Therefore, its complicated, when it comes to the role of the brain's neuroimmune effector in driving the development of AUDs.

As several recent, comprehensive reviews provide an eloquent and detailed synthesis of pro-inflammatory data in AUD models,[16,84,132,133] we discuss the more striking, contradictory and compelling findings from the preclinical and clinical literature below, beginning with non-cell specific neuroimmune measures, followed by data pertaining specifically to microglia. Finally, we discuss how these discoveries have translated to recent clinical trials of microglia-acting therapeutics.

5.1 Alcohol and neuroimmune effects in preclinical models

Preclinical models provide the clearest evidence for changes to neuroimmune measures following alcohol exposure. However, significant variability exists in these reports as alcohol's effects on the neuroimmune system obviously varies depending on the amount, duration and pattern of exposure. While it is not surprising that neuroimmune activation might occur in a severe models of AUD with significant cell death and degeneration, one of the earliest findings remains quite striking: even a single acute exposure to ethanol (5 g/kg i.g.) increased mRNA expression of TNF-α, CCL2 and COX-2 in whole brain in mice.[134] However, IL-1β and IL-10 mRNA expression were unchanged, and interestingly, no changes in TNF-α, CCL2 or IL-1β protein expression were observed. These findings are consistent with recent microdialysis data in which extracellular cytokine levels in the hippocampus did not change in response to acute ethanol (3 g/kg, i.p.)[135] even though region-dependent *decreases* in TNF-α and IL-1β, were found in the hippocampus, paraventricular nucleus and amygdala following 4 g/kg, i.p. ethanol exposure.[136,137] Thus, the acute effects of alcohol on neuroimmune gene expression depend on how, when and where one looks.

With longer exposures, reports are not particularly consistent between different AUD models, but there is some evidence of a species difference between rats and mice perhaps confounded by route of administration. For example, following 15 h of continuous ethanol infusion via catheter, TNF-α and IL-1β protein levels were increased in whole brain in mice.[138] With repeated exposure, using 5 or 6 g/kg, i.g. × 10 daily doses of ethanol, mRNA expression of CCL2 was increased across multiple studies (hippocampus, cerebellum, cortex, whole brain), whereas results for TNF-α were inconsistent and IL-6 was elevated only in the cerebellum.[134,139] Furthermore, IL1β, COX-2 and IL-10 mRNA expression were unchanged in whole brain.[134] A time course following a single binge ethanol treatment in mice demonstrated increased mRNA expression of the proinflammatory cytokines TNF-α and CCL2/MCP-1 at 18 h post-exposure which returned to baseline by 24 h, whereas the anti-inflammatory cytokine IL-4 remained elevated at 12, 18 and 24 h post-exposure[140] highlighting the transient nature of specific neuroimmune changes. Protein levels of TNF-α, CCL2/MCP-1 and IL-4 were also increased at the 18-h time point. The role of microglia in these increases is suggested by *in vitro* studies showing a direct effect of alcohol on microglia activation,[138] though astrocytes are activated as well.[73,141] In sum, inconsistent effects have been observed for TNF-α, the quintessential pro-inflammatory cytokine associated with M1-like, classical activation. As the TNF response is commonly used as a benchmark of a pro-inflammatory state, this mixed literature with effects predominantly in gene expression but not protein, results in a muddy picture of the role of neuroinflammation in models of AUDs.

In rat models of AUDs effects are even more subtle. When binge alcohol intake was modeled in adolescent rats in a chronic-intermittent model (2 days on, 2 days off of i.p. injections across 2 weeks), protein levels of iNOS and COX-2 were increased in the neocortex, hippocampus and cerebellum.[142] Using a 4-day binge model, which is considered to be a severe, acutely damaging model of AUD, alcohol treatment resulted in striking vimentin immunoreactivity in rat corticolimbic regions, indicating reactive astrogliosis in these areas.[73] However, little evidence of neuroinflammation has been found in 4-day binge models based on cytokine expression. For example, there was no change in TNF-α levels in the hippocampus following a 4-day binge in adolescent or adult rats.[143,144,145] Furthermore, anti-inflammatory cytokines IL-10 and TGF-β were *increased* in the hippocampus at day 7, while IL-6 was decreased in the entorhinal cortex in withdrawal.

In a 35-day vapor model with more moderate blood ethanol concentrations (BECs) in the 150–200 mg/dL range, there were morphological changes in microglia but few alterations in the protein expression of a panel of 18 cytokines, chemokines and other neuroimmune markers in the frontal cortex at either 1 or 28 days of withdrawal.[146] Finally, early studies of chronic ethanol consumption (5 months of ethanol liquid diet) resulted in increased mRNA and protein expression of iNOS, COX-2, TNF-α, IL-β and IL-6 in the cortex[86,141] but when the ethanol was only in the drinking water (20% *v*/v) for 6 months, there were no changes in mRNA expression of TNF-α or COX-2 in the hippocampus.[147] Thus, much more modest effects are observed in rats. As rats are thought to be superior to mice for studying inflammation and potentially more similar to the human condition,[148] we weight these results more strongly in our assessment of the neuroinflammatory state in models of AUDs.

Transcriptome profiling using microarrays and RNA sequencing and proteomic analyses in rodent models similarly highlight the complexity of immune changes following alcohol exposure. In an acute intoxication models, genes enriched for the immune response were upregulated while genes enriched for regulation of the JAK/STAT cascade were downregulated in the cerebellum which strikingly contrasts with a lack of pro-inflammatory proteomic signature in acute alcohol-exposed microglia cell lines.[149,150] Differentially expressed genes in the mouse prefrontal cortex following inflammatory challenge (two LPS injections) compared with three models of chronic ethanol consumption demonstrated significant overlap for a chronic intermittent two-bottle choice paradigm, but not chronic two-bottle choice or drinking in the dark paradigms.[151] A subsequent study by the same group investigated the temporal and brain region-dependent characteristics of gene expression alterations in a chronic intermittent ethanol vapor paradigm.[152] Here, immune-related genes were enriched in all regions examined (prefrontal cortex, amygdala and nucleus accumbens) at the 8-h post-exposure time point, but not at 0 or 120 h. Additionally, cell specific signatures indicated enrichment of microglial and astrocytic differentially expressed genes across all time points and regions. Finally, pathway analysis following RNA sequencing in adolescent alcohol-preferring rats that were given intermittent access to alcohol for 3 weeks, indicated enrichment of immune pathway genes in the prefrontal cortex (both increased and decreased expression) 3 h post ethanol access, whereas these pathways were not enriched in the hippocampus.[153] While genome-wide expression profiling has the advantage of allowing for analyses that identify dysregulated gene networks rather than relying on a small subset of markers, altered gene expression does not necessarily reflect changes at the protein level. Thus, the nature of changes to the immune signature at the genome and proteome level following alcohol exposure appears to depend greatly on factors including chronicity and route of alcohol treatment, the time points examined post ethanol, the brain regions selected for analyses and the species used[149]. Discrepancies between studies may reflect differences in the methodologies and markers used for quantification, which only adds to the complexity in understanding the role that the innate immune system plays in the pathogenesis of AUDs.

The signaling pathways that respond to and mediate the secretion of immune effectors are activated in response to pathogen- or damage-associated molecular signal (PAMP and DAMP) detection by pattern recognition receptors (PRRs).[154] Both DAMPs and PRRs are implicated in driving alcohol-induced neuroimmune activation.[84,86,155] Studies demonstrating that manipulation of these signaling systems can block alcohol induced damage in the CNS provide the most compelling evidence for the role of immune activation in the pathophysiology of AUDs.[16,84,86] In the first study to support a role of PRRs, alcohol-induced upregulation of multiple protein markers of inflammation and glial activation, decreases in neuronal number, and increased FluoroJade B staining, a marker of neuronal degeneration, was seen in the cortex of wild-type mice following 5 months of ethanol treatment but not in those with knockout of the PRR, toll like receptor 4 (TLR4).[86,156] However, a comprehensive series of studies ruled out a role of TLR4 as a driver of ethanol consumption,[157] but this merely rules out *one* of the receptors for DAMP signaling. It is of note that the DAMP, HMGB1, induces inflammatory cascades through a variety of receptors including TLR4, RAGE, Integrin/Mac1 and in complex with other factors through TLR 2, 3, 7, 9, IL-1R1 and CXCR4 (for review, see Ref. 158). A role for HMGB1 has been established in alcohol-neurotoxicity[159–162] and is leading to the investigation of other PRRs in AUDs (e.g., see Ref.163). Importantly, HMGB1 can elicit pro-inflammatory or anti-inflammatory effects and is implicated in microglia priming,[154,158,164] which may play a role in the pathogenesis of AUDs.[165]

5.2 Alcohol and neuroimmune effects in human AUDs

Studies from humans, though quite limited in number, provide the most compelling evidence for a role of neuroimmune activation in AUDs. Changes in proinflammatory cytokines have been observed in post-mortem brain from alcoholics. For example, elevated CCL2 (MCP-1) was found in the ventral tegmental area, substantia nigra, hippocampus and amygdala.[166] Another study demonstrated increased levels of IL-6 in the cortex, but there was no difference in expression of IL-4, IL-10 or IFN-γ.[167] Genome-wide expression profiling in AUD post-mortem brain has revealed both up- and down-regulation of genes involved in immunity and inflammation in the hippocampus and frontal cortex.[168,169] Pathway analysis following RNA sequencing of prefrontal cortex from post mortem brain, using the largest study sample to date (63 alcoholics and 73 controls), showed that interferon signaling had the strongest overall enrichment in alcoholics.[170] In addition, there was strong enrichment of the "inflammatory response" biological process among differentially expressed genes in subjects with AUD.

In individuals with AUDs there are obvious methodological limitations in assessing the presence of immune activation in the brain. Early studies measured circulating proinflammatory cytokine levels in AUDs and highlighted increases related primarily to the presence of liver disease, with minimal changes in those with uncomplicated alcoholism.[171] Subsequently, increased levels of IL-6 and TNF-α were found on the first day of withdrawal in two separate studies,[172] one of which confirmed that participants were non-cirrhotic.[173] Interestingly, measures of intestinal permeability and circulating LPS were elevated at this time point, returning to control levels following 18 days of abstinence, whereas cytokines remained somewhat elevated.[173] However, the literature is riddled with inconsistencies: another study showed increased IFN-γ and decreased IL-4 in both uncomplicated alcoholics and those with liver disease relative to controls, whereas TNF-α was decreased in uncomplicated alcohols compared to controls and there were no differences in IL-6 levels.[174,175] Finally, it was recently demonstrated that participants with AUDs who were currently abstinent had elevated TNF-α and IFN-γ relative to controls, but this elevation was only present in those with concurrent hepatitis C infection (n=16 of 81 AUDs vs 54 controls).[176] CSF measurements more closely reflect the inflammatory milieu of the CNS. CSF levels of CCL2/MCP-1 were increased 4 and 25 days following alcohol withdrawal, though TNF-α was decreased at day 25 and there were no changes in IL-1β compared to controls.[177] CCL2/MCP-1 in the CSF was further correlated with measures of liver function early in withdrawal. Proinflammatory cytokines in circulation are proposed to contribute to neuroimmune activation in AUD,[134,178] and multiple brain-immune communication routes have been described.[84,179,180] Immune activation in the brain likely still occurs independent of peripheral immune signals; however, as there is significant evidence that endogenous ligands released following neuronal damage activate neuroimmune signaling.[84,158,161,162,181]

6. The role of microglia in neuroimmune effects of alcohol

Microglia, the resident macrophages in the brain, can be directly or indirectly activated by alcohol, cytokines, and likely neuronal damage after excessive alcohol exposure.[83,84,141,143,144,182–185] Both clinical and animal model studies suggest that the pathogenesis of alcohol-induced changes in the brain is mediated by an innate immune response.[83,86] The findings discussed thus far demonstrate changes in the expression of specific immune genes and effectors in the brain following alcohol use in preclinical models and in the post-mortem brain and circulation of humans with alcoholism. However, the presence of immune markers in brain tissue does not necessarily reflect neuroinflammatory processes mediated by activated microglia. Several cell types, including neurons, astrocytes, oligodendrocytes and endothelial cells, can secrete these effectors, the function of which varies depending on cell type and context. Thus, investigations that provide improved cellular specificity are required to understand the role of microglia and other cell types in alcohol use disorder.

6.1 Preclinical studies on microglia and AUDs

Several studies have attempted to determine the role of microglia, specifically, in neuroimmune activation in AUDs. High doses of ethanol (4.5 and 6 g/kg, i.g.) were required to induce alterations to mRNA expression of microglia-specific markers Iba1 (decreased) and CD68 (increased) in whole brain at 18 h post binge exposure.[140] To determine which alcohol-induced transcriptional changes were mediated by microglia, these cells were depleted using the CSF1R inhibitor PLX5622 and mRNA expression of a panel of 43 genes were measured in response to ethanol compared to control mice. Microglial depletion decreased ethanol induced increases in TNF-α and CCL2/MCP-1 expression at 18 h post-binge, and resulted in increased expression of iNOS, IL-4 and IL-10. With regards to basal expression, microglial depletion decreased levels of TNF-α and TGF-β but the expression of many other neuroimmune markers assayed were unchanged, highlighting the role of other brain cell types in expressing these genes.[140] These techniques are some of the first that assess the causal role of microglia in models of AUD, suggesting that microglia contribute to ethanol induced increases in certain proinflammatory markers.

A number of studies have directly assessed alterations to microglia based on morphology and the expression of cell surface markers in models of alcohol use disorder. Immunohistochemistry for microglia-specific markers helps to inform the interpretations of these studies; for example, immunoreactivity of OX42 (complement receptor 3) is increased immediately following 4-day binge alcohol exposure through at least 28 days later in rats.[82,143] Stereological estimates showed that IBA1-positive cells were increased in the hippocampus starting at 1 week of abstinence through 28 days post 4-day binge in rats.[165,183] Even in rats exposed to moderate levels of alcohol (e.g., BEC = 162.6 ± 8.3 mg/dL) increases in microglial activation measured by IBA1 immunoreactivity was observed in all brain regions assessed (frontal cortex, amygdala, hippocampus, substantia nigra, cerebellum) immediately following 35 days of chronic intermittent ethanol vapor and at 1 day and 5 weeks post-ethanol, which remained elevated in the frontal cortex, amygdala and hippocampus 28 days post-ethanol.[146] Investigation of microglial morphology in preclinical models of AUDs demonstrated alterations to cell shape and size.[144,183] In separate studies of adult and adolescent rats, hippocampal microglia in control rats had small soma and highly ramified processes associated with a resting phenotype, whereas microglia in 4-day binge ethanol exposed rats had large soma and processes that were shorter and thicker indicative of a partially activated or "primed" phenotype.[144,183] A twofold increase in the number of these partially activated microglia was seen in the hippocampus in a 6-month chronic ethanol model as well as following 2 months of withdrawal.[147] Again, no microglia with an ameboid phenotype, indicative of full activation, were observed. Only in one study with a valid model of an AUD did microglia display an activated ameboid phenotype, in the motor cortex following 5 months of chronic ethanol treatment.[186]

The 18-kDa translocator protein (TSPO) has been used as a biomarker of neuroinflammation as it is upregulated predominantly following microglial activation though it can also bind to astrocytes and endothelial cells.[187] Radiolabeling of TSPO using [^{3}H]PK-11195 demonstrated increased binding in the hippocampus and entorhinal cortex immediately following a 4-day binge ethanol exposure in rats,[143] which was still detectable in the piriform cortex months later.[188] Replications in a rat vapor model of an AUD confirmed these findings for [^{3}H]PK-11195 in hippocampus and thalamus as well as for [^{3}H]PBR28.[189] *In vivo* PET imaging of the TSPO ligand ^{18}F-DPA-714 in an adolescent non-human primate model carried out before, during and after ethanol binge revealed increased binding during the binge across several brain regions, which decreased but remained elevated relative to pre-exposure levels 7–12 months later.[190] However, no difference between ethanol and control animals was observed with [^{11}C]PBR28 *in vivo* in the same rat vapor model that showed differences with [^{3}H]PK-11195 and [^{3}H]PBR28 autoradiography.[189]

Increases in both TSPO ligand binding and IBA1 immunoreactivity may represent an increase in microglia numbers and/or an upregulation of TSPO in individual cells. The increase in microglia, is supported by studies that demonstrate microglial proliferation in response to ethanol as well as an estimated increase in IBA1+ cells in female rats though not males 7 days post binge.[144,183,191] New IBA1 expressing cells were present in male rat hippocampus 2 days after a 4-day binge, and IBA1+ cells are increased following a double 4-day binge in males.[165,191] Later studies demonstrated that these cells were still present at day 7, and returned to controls levels by day 28 in CA1 and dentate gyrus but remained elevated in CA2/3.[143] However, contrary to the microglial proliferation shown in the hippocampus in the

4-day binge model, there is initially microglia loss in male rats after 4-day binge exposure and in a chronic 6-month ethanol model there were no changes in the overall number of microglia in the hippocampus immediately after treatment or after 2 months of withdrawal.[147,183] This discrepancy may be due to differences in BECs or the time points examined.

6.2 Human studies on microglia and AUDs

Human post-mortem histological studies predominantly agree with these reports from preclinical models of AUDs. Increased IBA1 immunoreactivity has been observed in the prefrontal cortex[192] and cingulate cortex of male alcoholics, though not ventral tegmental area, midbrain or amygdala.[166] Both of these studies used immunoreactivity which can be difficult to interpret as to whether IBA1 is upregulated in cells, the cell morphology changes or there are more microglia. Another study demonstrated altered microglial morphology, consisting of hypertrophied cell bodies and shorter thicker processes in proliferating cells in the post-mortem cortex from alcoholics with hepatic encephalopathy.[167] Indeed, the observation of proliferating microglia in hepatic encephalopathy corresponds to what has been observed in the 4-day binge model in rats.[144,191]

TSPO has been assessed in PET imaging studies in humans with AUDs. Increased TSPO may reflect changes in TSPO expression on individual cells and/or a change in the *number* of TSPO-expressing cells. Using the radioligand [^{11}C]PBR28 across multiple brain regions showed decreased binding in the hippocampus of participants with AUDs.[193] Importantly, these participants had undergone medically assisted withdrawal from alcohol dependence within the last 2–4 weeks, and the effects of withdrawal medications (benzodiazepines) on microglial activation using a ligand to what was originally called the peripheral benzodiazepine receptor is unknown. A different study using [^{11}C]PBR28 in participants who had undergone withdrawal without benzodiazepines and had their last drink within the previous 4 days, also found overall decreased binding in participants with AUDs.[67] Finally, another group of participants who had their last drink within the last week, 5 of 19 of whom had been treated with benzodiazepines, found a decrease in binding only in a subgroup of medium-affinity, but not high-affinity, [^{11}C]PBR28 binders.[194] Given the discrepancy between results of radiolabeling studies in pre-clinical models and radioligand *in vivo* imaging in clinical populations with AUDs both techniques were compared in a rat model.[189] As mentioned above, increased binding was demonstrated using the TSPO ligands [^{3}H]PK11195 and [^{3}H]PBR28 in the hippocampus and thalamus *in vitro*, but there were no differences in [^{11}C]PBR28 binding *in vivo*. These data highlight the need for further work to understand this lack of congruence between methodologies. Ligands for alternative targets to TSPO are being developed, such as [^{11}C]CPPC which binds to CSF1R and show promise for future use in human PET imaging.[195]

6.3 Preclinical studies on microglial phenotype in AUDs

While the data discussed thus far provide evidence of microglial activation in animal models and potentially in humans with AUD, the markers used and morphological assessment do not differentiate between the range of microglial functional phenotypes that exist. In addition to establishing the presence of microglial activation, microglial functional capacity and associated phenotype can be determined using immunohistochemistry to detect the presence and levels of relevant proteins.[196] In adult rats, OX-42, which binds to CD11b and is one of the first markers upregulated in activated microglia, was increased immediately following the last binge in all regions examined, and remained elevated in CA1, CA2/3 and the entorhinal cortex, but not the dentate gyrus at post-binge day 28.[143] On the other hand, ED-1 (CD68), a marker of phagocytic microglia, did not show any immunoreactivity in the hippocampus of control or 4-day binge adolescent or adult rats, nor were there changes in numbers of MHCII positive (OX-6 antibody stained) microglia, which indicates a lack of an M1 proinflammatory microglial phenotype.[143,144] In contrast, in a chronic ethanol mouse model there were increased numbers of MHCII-positive cells and increased CD68 staining in the cortex compared to controls.[186] Again, mice are showing the greater pro-inflammatory response compared to rats though route of administration differs as well.

While examining these various markers and their expression patterns in brain tissue sections provides improved phenotypic specificity, the approach is somewhat laborious. More recently, microglial isolation techniques have been applied that allow for downstream molecular analyses that are cell-type specific.[197] Using isolated microglia, RNA sequencing highlighted enrichment of TGF-β signaling and the chronic inflammatory response among genes that were up- and down-regulated in the prefrontal cortex of mice following chronic intermittent (30 drinking days every other day) voluntary ethanol consumption.[198] In another study, microglia were isolated from the hippocampus and entorhinal cortices after 4-day binge exposure in rats and examined for established M1 (MHCII, CD68, CD32) and M2 (CD206) cell surface markers using flow cytometry.[199] The percentage of MHCII, CD68, CD32 and CD206 positive cells was increased in both the hippocampus and entorhinal cortex at day 2, but not days 0, 4 or 7 following the binge. CD206, the M2 marker, was strikingly elevated compared to changes in M1 markers supporting other evidence of a more beneficial activation state.[143,144,147] This increase in both M1 and M2 markers demonstrates microglial activation, and further indicates microglial tissue heterogeneity or a mixed microglial phenotype in response to binge ethanol. These findings highlight the need for continued microglia-specific investigations in AUD as well as focus on the functional phenotype of these cells beyond the imposed M1 versus M2 dichotomy.[199]

While the findings outlined thus far do appear to show microglial reactivity following alcohol exposure, there are few studies that indicate the presence of a "fully activated" proinflammatory phenotype, determined by morphology or cell markers, following exposure to alcohol. As the field of microglial biology has developed, it becomes clear that the activity of these cells in response to insult is contextually and temporally dynamic, and the concept of microglial sensitization, or priming, has developed. Here, an initial activating stimulus potentiates the subsequent microglial response to further challenge. There is a strong possibility that the differences in results between short term and chronic exposure models of AUD may reflect priming of microglia over time. Indeed, 4-day binge exposure resulted in increased OX-42 staining in all hippocampal regions and the entorhinal cortex, and repeated binge increased OX-42 staining further in CA1, CA2/3 and the entorhinal cortex, but not the dentate gyrus.[165] Interestingly, ED-1 and OX-6 staining, markers of classical activation, remained unchanged even after two 4-day binge cycles.[143,165] These data indicate that microglia are not classically activated immediately following the second binge, but that multiple alcohol hits do exacerbate microglial reactivity.

As such, the role of the immune system in contributing to neurodegeneration versus neuroprotection following alcohol exposure remains to be determined. Though any changes in immune activity, and the presence of non-specific markers of microglial activation (regardless of phenotype) historically have been presumed to signify neuroinflammation and to be deleterious, recent findings highlight the functional complexity of these cells under a wide range of physiological and disease-related conditions. Going forward, investigations will benefit from continued focus on cell-specificity, temporal characteristics of microglial activation, and the cumulative effects of multiple alcohol hits, combined with other environmental and/or genetic factors, on cell function. Acknowledgement of this complexity is necessary given the overlap of alcohol abuse with stress, anxiety and depression, all of which may be impacted by neuroimmune functioning.[200–202]

7. Anti-inflammatory approaches as potential drug targets for AUDs

The ultimate goal of this body of work on the role of neuroimmune activation in AUDs is to drive the translation of basic neuroscience discoveries to help identify a novel, druggable target for the treatment of AUDs. To date, there are only four FDA-approved medications for the treatment of AUDs: Disulfiram (Antabuse®), Acamprosate (Campral®) and two delivery formulations of Naltrexone, a 50 mg tablet (Revia®) commonly taken once a day and an extended-release depot injection (Vivitrol®; for more extensive reviews, see Refs.203–206). Disulfiram, first approved by the FDA in 1951, was for many years the only approved treatment for AUDs. Disulfiram inhibits aldehyde dehydrogenase, part of the ethanol metabolic pathway, which results in the accumulation of acetaldehyde and therefore adverse symptoms such as flushing, sweating, tachycardia, headache and nausea/vomiting. This state, in theory, should discourage the abuse of alcohol, though in practice, the efficacy of this approach is modest and rare adverse drug reactions remain a concern.[203,204,207] While the drug's biggest drawback is that the patient can decide not to take it, Disulfiram appears effective in heavily supervised or highly motivated, treatment-seeking individuals.[208–210] Naltrexone was approved for use in AUDs in 1994, but renewed interest in opioid receptor antagonists has emerged recently due to their action on neuroimmune signaling. Naltrexone antagonizes μ, κ and to a much lesser extent δ receptors, which corresponds to a reduction in alcohol consumption and craving, though its efficacy has also been modest at best.[206] While early studies were promising, its blunting of alcohol-induced euphoria[211] extends to the common complaint of anhedonia, which impacted compliance and therefore success of this therapeutic approach.[212] However, the realization that naltrexone enantiomers were anti-inflammatory coupled with the discoveries on the role of neuroimmune activation in the development of AUDs has resulted in a renewed interest in this pharmacotherapeutic approach, if anhedonia side effects can be avoided. Acamprosate gained FDA approval in 2004, though it had been used in Europe since 1989. While its mechanism of action for reducing drinking is not fully understood, its glutamatergic (metabotropic and NMDA) actions are most often implicated in its modest effects though GABAergic or calcium may mimic what appear as glutamatergic effects.[213,214] Acamprosate is most effective in those who exhibit alcohol dependence as it appears to aid in the transition from withdrawal to abstinence.[203,206,215]

There are a host of medications being prescribed off label for the treatment of AUDs, most of which target behaviors/conditions that drive consumption such as antidepressants or anxiolytics (drink to self-medicate; e.g., Duloxetine; Fluoxetine), anticonvulsants (regulate maladaptive plasticity in alcohol dependence, e.g., Gabapentin, Topiramate), and others that impact general motivation/reward system signaling (e.g., Baclofen, a $GABA_B$ receptor agonist, or Ondansetron, the commonly prescribed antiemetic and selective $5HT_3$ receptor antagonist.[16,203,216] Discouragingly, all of the drugs described above only work modestly, perhaps due, in part, to the fact that people use and abuse alcohol for a variety of reasons and study groups have not been stratified to match type of addict or pharmacogenetics with biology until only recently.[217,218] Thus, there is a great need for novel pharmacotherapeutic approaches to treating AUDs, especially those outside of the traditional consummatory and motivational circuits of the mesolimbic dopamine system.[205]

The convergence of novel discoveries in the genomic arena that the neuroimmune system is dysregulated in human AUDs[168,219,220] coupled with basic science discoveries on the mechanisms of alcohol-induced neuropathology[17,166,188] creates a strong rationale to probe the role of the immune system in the development and treatment of AUDs.[216] This new focus on neuroimmune signaling also provides a wealth of novel potential drug targets. However, targeting this system is complicated by our lack of knowledge regarding the effect of alcohol on the neuroimmune system and how that impacts the structural and functional consequences of excessive alcohol consumption. Below, we review recent drug discovery and development efforts where there are current human clinical trials for microglia-relevant, neuroimmune-modulating compounds for the treatment of AUDs. For broader reviews of other novel neuroimmune acting agents being pursued preclinically for the treatment of AUDs see also Refs. 84, 216, or 221.

7.1 Avermectins

While the potential use of the avermectins did not emerge directly out of its effects on microglia, drug discovery work in this space coincided with the emergence of the neuroimmune hypothesis of addiction.[16,84] Avermectins are derived from the natural product *Streptomyces avermitilis* and have been used traditionally as anthelmintics in agriculture, as reviewed.[222] Thus far, the greatest work has been conducted on Ivermectin, a semi-synthetic macrocyclic lactone that is an FDA-approved, broad-spectrum anti-parasitic.[223] Although its anti-parasitic actions are through the activation of a non-mammalian glutamate-gated Cl^- channel, its actions in mammals appear to be as positive allosteric modulators of cys-loop superfamily receptors including $GABA_A$, glycine and nicotinic acetylcholinergic receptors as well as the ligand-gated ion channels of the purinergic (P_2X) superfamily.[221–224] It is this latter pharmacological action that has been focus of recent development efforts. Several studies across the last 10 years show a role for P_2X_4 receptors in the regulation of alcohol consumption in animal models.[225] These receptors are notably found on neurons and glia, including microglia, in the mesolimbic and mesocortical reward circuitry supporting their roles in regulating dopaminergic and glutamatergic neurotransmission.[225] Indeed, P_2X_4 receptors mediate the effect of alcohol on critical cell signaling pathways in alcohol-induced microglia functions.[226] Ivermectin dose-dependently reduces alcohol drinking and preference, and operant self-administration.[227,228] Moxidectin, an ivermectin analogue with a better safety profile, dose-dependently reduces ethanol intake in two-bottle choice and drinking in the dark models.[229] Despite the strong effects in preclinical studies for ivermectin, a human clinical trial of ivermectin was more modest; however, safety of its use was established and additional doses can be tested going forward (see ClinicalTrials.gov: NCT02046200).[230]

7.2 Minocycline

Minocycline is a broad-spectrum antibiotic originally synthesized from the natural product antibiotic, tetracycline. While minocycline has been used traditionally for the treatment of bacterial infections, its non-antibiotic effects as an anti-inflammatory and subsequent successful treatment of acne and rheumatoid arthritis have led to its investigation in a multitude of inflammatory conditions.[231,232] Minocycline is the most lipid soluble of the tetracycline antibiotics (tetracycline, doxycycline, minocycline), allowing for greater brain bioavailability and potentially anti-neuroinflammatory activity.[232] As such, minocycline has been investigated in animal models as well as human clinical trials for conditions with a neuroimmune component such as Alzheimer's disease, schizophrenia, stroke, multiple sclerosis, Parkinson's disease, pain and spinal cord and traumatic brain injuries.[232,233] Its anti-inflammatory effects in both peripheral organ systems and the CNS are well-established,[231] but the mechanism by which it exerts its anti-inflammatory effects remains unknown. It is often described as a highly selective microglial *inhibitor*, but its effects are not specific to microglia nor does it necessarily inhibit microglia functions.[231] Minocycline has direct effects on microglia, astrocytes, oligodendrocytes and even neurons.[231,234] Interestingly, multiple potential mechanisms may contribute to its net anti-inflammatory and/or neuroprotective effects that overlap with known contributors to alcohol toxicity and alcohol dependence.[221] Of note, its actions as an anti-convulsant, anti-apoptotic, free radicals scavenger and that it inhibits glutamate excitotoxicity, suggest multiple means of neuroprotection.[221,231,232,234] Despite the lack of knowledge over its exact anti-inflammatory or neuroprotective mechanisms of action, several preclinical studies supported that minocycline reduces alcohol drinking. For example, minocycline treatment reduced alcohol consumption under two-bottle choice in C57Bl/6J male and female mice[235] and reduced microglia activation in 10-day exposure models in adult, male mice.[139,236,237] The reduction in alcohol drinking may be due to reduced alcohol preference as minocycline reduced alcohol conditioned place preference similar to naltrexone in mice but also increased alcohol-induced sedation,[238] both of which regulate alcohol consumption. Another tetracycline, doxycycline, also increased the sedative and motor impairing effects of alcohol, an effect that did not quite reach significance ($P = 0.08$) for similarly increasing BECs despite reduced consumption in a drinking in the dark model in mice.[239] Thus, while there is strong rationale for minocycline to reduce alcohol consumption in animal models, there will be critical caveats for drug discovery and there is at least one clinical trial recruiting at this time (ClinicalTrials.gov: NCT02187211). A drug that is too sedating will cause compliance issues.[240] Furthermore, the timing of when minocycline is known to be most effective is not congruent with the therapeutic needs in treating AUDs. Minocycline has been a highly effective neuroprotectant/anti-inflammatory when administered *before* the event, with few protective effects after[241,242] though others have seen some success with acute administration.[243] On the other hand, minocycline administration has also been associated with a worsening of damage/degeneration[244–246] as was the case with a clinical trial in ALS, which was then mimicked in an animal model.[231,247,248] One might postulate that minocycline either inhibits beneficial activation, or it primes microglia similar to other "microglia inhibitors."[249]

7.3 Phosphodiesterase inhibitors

Cyclic nucleotide phosphodiesterases (PDEs) regulate cyclic adenosine monophosphate and cyclic guanosine monophosphate, second messenger systems essential in immune function and plasticity, including learning, memory, cell survival and repair mechanisms.[250] As all of these are involved in or result from the development of AUDs, there are multiple lines of evidence supporting the use of PDE inhibitors for the treatment of AUDs (for review, see Ref. 251). PDEs have over 100 isoforms classified into 11 different families of which $PDE_{1,\ 3,\ 4,\ 5}$ and $_{10}$ have been examined for roles in alcohol drinking (see table 3 of Ref. 16 or table 1 of Ref. 251 for review). Specific inhibition of PDE_4 or PDE_{10A} reduces proinflammatory neuroimmune activation while also reducing alcohol drinking, and 10A inhibition also impacts saccharin self-administration.[252] However, the strongest evidence for a role of PDEs in AUDs is PDE_4 which regulates alcohol intake, preference and alcohol-induced neuroimmune activation. PDE_4 has well known anti-inflammatory actions[253] and recent work has shown that PDE_{4b} gene expression is increased by alcohol.[254] PDE4 inhibitors decrease alcohol intake in both rats[255,256] and mice,[257,258] as well as operant self-administration of ethanol in a strain of alcohol preferring rats. The most specific PDE_4 inhibitors were more efficacious in reducing alcohol drinking though tolerance to its effects was observed for some compounds.[259] While inhibiting PDE4 has been a successful approach in reducing inflammation/inflammatory conditions and alcohol drinking in rodent models, adverse gastrointestinal drug reactions including nausea and vomiting have hindered the development and success of PDE4 inhibitors in humans.[260]

Ibudilast is a pyrazolo-pyridine (small molecule 230 MW) that non-selectively inhibits PDE_{3A} and PDE_4 as its primary sites of action but also PDE_{10} and PDE_{11}.[221,261] Ibudilast was originally approved for use in asthma, but is now in phase II clinical trials for neuroinflammatory conditions such as multiple sclerosis, neuropathic pain and alcohol and drug use disorders.[261] Its ability to dose-dependently inhibit microglial activation while promoting GDNF expression[260] if not phenotype switch M1-like to M2-like microglia makes it an ideal candidate for use in AUDs. Preclinical studies in alcohol preferring P rats, HAD1 rats, and c57Bl/6J mice all showed reduced alcohol drinking.[262] Ibudilast is also being investigated in clinical trials for reductions in the use of opiates, cocaine, or methamphetamine.[221,261] Importantly, Ibudilast has been shown to be effective at reducing neuroinflammatory effects at doses where gastro-intestinal side effects are tolerable.[260] However, despite all of these promising basic science studies that provided strong rationale to develop a PDE4 specific inhibitor for human AUDs, the results, though positive, were not particularly compelling.[263] The safety of Ibudilast was established and additional doses can be examined as well as in different aspects of alcohol use, abuse and dependence.[221,263] Currently, two clinical trials exploring Ibudilast for the treatment of alcohol use disorder and alcohol withdrawal-related dysphoria are ongoing (ClinicalTrials.gov: NCT03594435; NCT03489850).

7.4 Peroxisome proliferator-activated receptor (PPAR) agonists

Peroxisome proliferator-activated receptor (PPAR) agonists have emerged as promising repurposed drugs due to their anti-inflammatory activities. PPARs are Type II nuclear receptors, which function in complexes as ligand-activated transcription factors and are best known for their roles in development and metabolism.[264] PPARs regulate neuroinflammation through a receptor-dependent transrepression. Namely, upon ligand binding (e.g., free fatty acids or eicosanoids), PPAR forms a receptor-transcription factor complex with NFκB or AP1. A corepressor is then recruited to prevent NFκB from binding to the promotor region of their pro-inflammatory gene targets.[264] Indeed, PPAR inhibition of inflammatory signaling through NFκB, which is notably activated in models of AUD and implicated in the human condition,[82,265,266] provides strong rationale for this pharmacological approach. Of the three isoforms of PPARs, PPARα and PPARγ have existing FDA-approved drugs that are being considered for AUDs.

PPARα agonists, fibrates such as fenofibrate or gemfibrozil, were originally approved for the treatment of dyslipidemia and hypercholesterolemia. The discoveries that PPARα are found in adult brain and that they may modulate dopamine neurotransmission suggests that they may play a role in the reinforcing properties of alcohol consumption.[267] Indeed, fenofibrate dose-dependently reduced voluntary alcohol drinking in two bottle choice studies in mice[268,269] and rats,[270] as well as dose-dependently in contingency-dependent EtOH reinforcement[271] while the PPARα partial agonist, gemfibrozil, also decreased voluntary alcohol drinking in rats.[272] While these studies are in contrast with an old paper,[273] significant methodological differences likely explain the discrepancy. Fenofibrate also affects addiction relevant behaviors such as increased conditioned taste aversion and increased withdrawal severity, both of which when coupled with its effect of increasing ethanol metabolism, should be highly protective against developing an AUD.[274] Questions about the mechanism of action of Fenofibrate in AUD models led to the elegant study that Fenofibrate works through PPARα, as it does not inhibit alcohol drinking in PPARα knock out mice.[274] However, PPARα agonists impact ethanol pharmacokinetics. They increase Adh1 and decrease Adh2 mRNA in the liver, which results in an accumulation of acetaldehyde levels, an effect that would be aversive similar to Disulfiram, as discussed above.[275] Intriguingly, genomic studies supported that fenofibrate more likely works through dopamine neurotransmission in neurons than through its anti-inflammatory effects.[269] Thus, preclinical studies suggest that PPARα agonists are promising from multiple pharmacological perspectives. A phase II clinical trial (ClinicalTrials.gov: NCT02158273) of Fenofibrate although not yet published, lists modest effects where the one dose tried (145 mg/day for 9 days) did not appear to affect the number of drinks per week, but improvements in craving measures were more promising.

PPARγ agonists such as pioglitazone are used more commonly for the treatment of Type II diabetes; however, there is compelling data for their anti-inflammatory activity.[276] PPARγ is the primary isoform expressed in microglia.[277,278] As such, agonists at the γ isoform are better known for their roles in neuroprotection and are under investigation for efficacy in neurological conditions where inflammation plays a role such as Alzheimer's disease, Parkinson's disease, amyotrophic lateral sclerosis, multiple sclerosis, brain and spinal cord injuries and stroke (for review, see Refs.279–281). PPARγ agonists such as pioglitazone are highly effective at inhibiting pro-inflammatory microglia activation[280] and reduced microglia activation in a mouse model of fetal alcohol spectrum disorders.[282] Pioglitazone inhibited alcohol drinking and preference in multiple mouse[283] and rat models of AUDs, including stress induced relapse.[284,285] Based on these promising preclinical data, three clinical trials are listed though one has terminated for feasibility issues (ClinicalTrials.gov: NCT01631630). Importantly, a variety of different agents act as agonists at this receptor including NSAIDs and fatty acids such as the endocannabinoids OEA and PEA (which are also affected in models of AUDs[286–289]) and DHA,[290] which opens up a wealth of drug discovery possibilities for anti-inflammatory actions at this drug target.

8. Conclusion

We have described a rich body of literature pertaining to microglia as primary neuroimmune effectors in the pathophysiology and treatment of AUDs. Clinical studies, while methodologically challenging, provide evidence to support an alteration to the neuroimmune system in AUD. These human studies, by nature, cannot demonstrate causality, and pre-clinical models are therefore necessary to investigate whether immune activation in the CNS is a cause or consequence of alcohol addiction. These pre-clinical studies, while methodologically more straightforward, highlight the complexity of the relationship of alcohol with microglia and the neuroimmune system. The effects of ethanol on the neuroimmune system depend on experimental variables such as the route of drug administration, species used, sex, dosage and chronicity of treatment. Additionally, the outcomes are region- and cell type-dependent and temporally dynamic. As major advances in the understanding of microglial biology in health and disease have taken place, hypotheses regarding the potential function of these cells in AUDs have also evolved. Indeed, whether microglial activation in AUD is harmful or protective may not be as clear-cut as previously assumed, rather, dependent on which specific function is activated or disrupted under any given context. As such, the success of pharmacologically targeting microglia and other components of the neuroimmune system in AUD will be bolstered by an improved understanding of these contextually sensitive alcohol-neuroimmune interactions.

References

1. SAMHSA. *National Survey on Drug Use and Health*. Substance Abuse and Mental Health Services Administration; 2017; 2018.
2. American Psychiatric Association. *Diagnostic and Statistical Manual of Mental Disorders*. 5th ed. Arlington, VA: American Psychiatric Publishing; 2013.
3. Grant BF, Goldstein RB, Saha TD, et al. Epidemiology of DSM-5 alcohol use disorder. *JAMA Psychiat*. 2015;72(8):757–766.
4. Grant BF, Chou SP, Saha TD, et al. Prevalence of 12-month alcohol use, high-risk drinking, and DSM-IV alcohol use disorder in the United States, 2001-2002 to 2012-2013: results from the national epidemiologic survey on alcohol and related conditions. *JAMA Psychiat*. 2017;74(9):911–923.
5. Frank MG, Fonken LK, Dolzani SD, et al. Immunization with *Mycobacterium vaccae* induces an anti-inflammatory milieu in the CNS: attenuation of stress-induced microglial priming, alarmins and anxiety-like behavior. *Brain Behav Immun*. 2018;73:352–363.
6. Sacks JJ, Gonzales KR, Bouchery EE, Tomedi LE, Brewer RD. 2010 National and state costs of excessive alcohol consumption. *Am J Prev Med*. 2015;49(5):e73–e79.
7. Nutt DJ, King LA, Phillips LD. Drug harms in the UK: a multicriteria decision analysis. *Lancet*. 2010;376(9752):1558–1565.
8. WHO. *Global Status Report on Alcohol and Health*. World Health Organization Management of Substance Abuse Team; 2018. 2018 2018. 978-92-4-156563-9.
9. Mokdad AH, Marks JS, Stroup DF, Gerberding JL. Actual causes of death in the United States, 2000. *JAMA*. 2004;291(10):1238–1245.
10. Rehm J, Dawson D, Frick U, et al. Burden of disease associated with alcohol use disorders in the United States. *Alcohol Clin Exp Res*. 2014;38(4):1068–1077.

11. Rehm J. The risks associated with alcohol use and alcoholism. *Alcohol Res Health.* 2011;34(2):135–143.
12. Parsons OA, Nixon SJ. Neurobehavioral sequelae of alcoholism. *Neurol Clin.* 1993;11(1):205–218.
13. Vengeliene V, Bilbao A, Molander A, Spanagel R. Neuropharmacology of alcohol addiction. *Br J Pharmacol.* 2008;154(2):299–315.
14. Fadda F, Rossetti ZL. Chronic ethanol consumption: from neuroadaptation to neurodegeneration. *Prog Neurobiol.* 1998;56(4):385–431.
15. Tabakoff B, Hoffman PL. The neurobiology of alcohol consumption and alcoholism: an integrative history. *Pharmacol Biochem Behav.* 2013;113(0):20–37.
16. Erickson EK, Grantham EK, Warden AS, Harris RA. Neuroimmune signaling in alcohol use disorder. *Pharmacol Biochem Behav.* 2019;177:34–60.
17. Crews FT, Nixon K. Mechanisms of neurodegeneration and regeneration in alcoholism. *Alcohol Alcohol.* 2009;44(2):115–127.
18. Zahr NM, Kaufman KL, Harper CG. Clinical and pathological features of alcohol-related brain damage. *Nat Rev Neurol.* 2011;7(5):284–294.
19. Eckardt MJ, Martin PR. Clinical assessment of cognition in alcoholism. *Alcohol Clin Exp Res.* 1986;10(2):123–127.
20. Bernardin F, Maheut-Bosser A, Paille F. Cognitive impairments in alcohol-dependent subjects. *Front Psych.* 2014;5:78.
21. Chanraud S, Martelli C, Delain F, et al. Brain morphometry and cognitive performance in detoxified alcohol-dependents with preserved psychosocial functioning. *Neuropsychopharmacology.* 2007;32(2):429–438.
22. De Bellis MD, Narasimhan A, Thatcher DL, Keshavan MS, Soloff P, Clark DB. Prefrontal cortex, thalamus, and cerebellar volumes in adolescents and young adults with adolescent-onset alcohol use disorders and comorbid mental disorders. *Alcohol Clin Exp Res.* 2005;29(9):1590–1600.
23. Cardenas V, Studholme C, Meyerhoff DJ, Song E, Weiner MW. Chronic active heavy drinking and family history of problem drinking modulate regional brain tissue volumes. *Psychiatry Res Neuroimaging.* 2005;138(2):115–130.
24. Cardenas V, Studholme C, Gazdzinski S, Durazzo TC, Meyerhoff DJ. Deformation based morphometry of brain changes in alcohol dependence and abstinence. *Neuroimage.* 2007;34(3):879–887.
25. Sullivan EV, Marsh L, Mathalon DH, Lim KO, Pfefferbaum A. Age-related decline in MRI volumes of temporal lobe gray matter but not hippocampus. *Neurobiol Aging* 1995;16(4):591–606.
26. Pfefferbaum A, Rosenbloom M, Deshmukh A, Sullivan E. Sex differences in the effects of alcohol on brain structure. *Am J Psychiatry.* 2001;158(2):188–197.
27. Sullivan EV, Pfefferbaum A. Neurocircuitry in alcoholism: a substrate of disruption and repair. *Psychopharmacology (Berl).* 2005;180:583–594.
28. Zahr NM, Pfefferbaum A. Alcohol's effects on the brain: neuroimaging results in humans and animal models. *Alcohol Res.* 2017;38(2):183–206.
29. de la Monte SM, Kril JJ. Human alcohol-related neuropathology. *Acta Neuropathol.* 2014;127(1):71–90.
30. Jang D-P, Namkoong K, Kim J-J, et al. The relationship between brain morphometry and neuropsychological performance in alcohol dependence. *Neurosci Lett.* 2007;428(1):21–26.
31. Pfefferbaum A, Lim KO, Zipursky RB, et al. Brain gray and white matter volume loss accelerates with aging in chronic alcoholics: a quantitative MRI study. *Alcohol Clin Exp Res.* 1992;16(6):1078–1089.
32. Jernigan TL, Butters N, DiTriaglia G, et al. Reduced cerebral grey matter observed in alcoholics using magnetic resonance imaging. *Alcohol Clin Exp Res.* 1991;15(3):418–427.

33. Sullivan EV, Marsh L, Mathalon DH, Lim KO, Pfefferbaum A. Anterior hippocampal volume deficits in nonamnesic, aging chronic alcoholics. *Alcohol Clin Exp Res*. 1995;19(1):110–122.
34. Mechtcheriakov S, Brenneis C, Egger K, Koppelstaetter F, Schocke M, Marksteiner J. A widespread distinct pattern of cerebral atrophy in patients with alcohol addiction revealed by voxel-based morphometry. *J Neurol Neurosurg Psychiatry*. 2007;78(6):610–614.
35. Chanraud S, Martelli C, Delain F, et al. Brain morphometry and cognitive performance in detoxified alcohol-dependents with preserved psychosocial functioning. *Neuropsychopharmacology*. 2007;32(2):429–438.
36. Pitel A-L, Chételat G, Le Berre A-P, Desgranges B, Eustache F, Beaunieux H. Macrostructural abnormalities in Korsakoff syndrome compared with uncomplicated alcoholism. *Neurology*. 2012;78(17):1330–1333.
37. Jernigan TL, Butters N, DiTraglia G, et al. Reduced cerebral grey matter observed in alcoholics using magnetic resonance imaging. *Alcohol Clin Exp Res*. 1991;15(3):418–427.
38. Pfefferbaum A, Sullivan EV, Mathalon DH, Lim KO. Frontal lobe volume loss observed with magnetic resonance imaging in older chronic alcoholics. *Alcohol Clin Exp Res*. 1997;21(3):521–529.
39. Yang X, Tian F, Zhang H, et al. Cortical and subcortical gray matter shrinkage in alcohol-use disorders: a voxel-based meta-analysis. *Neurosci Biobehav Rev*. 2016;66:92–103.
40. Oscar-Berman M, Evert DL. Impairments of brain and behavior. *Alcohol Health Res World*. 1997;21(1):11.
41. Bechara A. Decision making, impulse control and loss of willpower to resist drugs: a neurocognitive perspective. *Nat Neurosci*. 2005;8(11):1458–1463.
42. Crews FT, Boettiger CA. Impulsivity, frontal lobes and risk for addiction. *Pharmacol Biochem Behav*. 2009;93(3):237–247.
43. Galandra C, Basso G, Cappa S, Canessa N. The alcoholic brain: neural bases of impaired reward-based decision-making in alcohol use disorders. *Neurol Sci*. 2018;39(3):423–435.
44. Fein G, Landman B, Tran H, et al. Brain atrophy in long-term abstinent alcoholics who demonstrate impairment on a simulated gambling task. *Neuroimage*. 2006;32(3):1465–1471.
45. Cardenas V, Durazzo T, Gazdzinski S, Mon A, Studholme C, Meyerhoff D. Brain morphology at entry into treatment for alcohol dependence is related to relapse propensity. *Biol Psychiatry*. 2011;70(6):561–567.
46. Makris N, Oscar-Berman M, Kim S, et al. Decreased volume of the brain reward system in alcoholism. *Biol Psychiatry*. 2008;64(3):192–202.
47. Nixon K, Pauly JR, Hayes DM. Alcohol, nicotine, and adult neurogenesis. In: Olive MF, ed. *Drug Addiction and Adult Neurogenesis*. Kerala, India: Research Signpost; 2011:95–119. vol 1.
48. Fuchs RA, Evans KA, Ledford CC, et al. The role of the dorsomedial prefrontal cortex, basolateral amygdala, and dorsal hippocampus in contextual reinstatement of cocaine seeking in rats. *Neuropsychopharmacology*. 2005;30(2):296–309.
49. Parsons OA. Impaired neuropsychological cognitive functioning in sober alcoholics. In: Hunt WA, Nixon SJ, eds. *Alcohol-Induced Brain Damage (NIAAA Research Monograph No. 22)*. Rockville, MD: National Institutes of Health; 1993:173–194.
50. Laakso MP, Vaurio O, Savolainen L, et al. A volumetric MRI study of the hippocampus in type 1 and 2 alcoholism. *Behav Brain Res*. 2000;109(2):177–186.
51. Gazdzinski S, Durazzo TC, Yeh PH, Hardin D, Banys P, Meyerhoff DJ. Chronic cigarette smoking modulates injury and short-term recovery of the medial temporal lobe in alcoholics. *Psychiatry Res*. 2008;162(2):133–145.

52. Wilson S, Bair JL, Thomas KM, Iacono WG. Problematic alcohol use and reduced hippocampal volume: a meta-analytic review. *Psychol Med*. 2017;47(13):2288–2301.
53. Dhanabalan G, Le Maître TW, Bogdanovic N, Alkass K, Druid H. Hippocampal granule cell loss in human chronic alcohol abusers. *Neurobiol Dis*. 2018;120:63–75.
54. Sullivan EV, Rosenbloom MJ, Lim KO, Pfefferbaum A. Longitudinal changes in cognition, gait, and balance in abstinent and relapsed alcoholic men: relationships to changes in brain structure. *Neuropsychology*. 2000;14(2):178–188.
55. Sullivan EV, Rosenbloom MJ, Pfefferbaum A. Pattern of motor and cognitive deficits in detoxified alcoholic men. *Alcohol Clin Exp Res*. 2000;24(5):611–621.
56. Sullivan EV, Deshmukh A, Desmond JE, Lim KO, Pfefferbaum A. Cerebellar volume decline in normal aging, alcoholism, and Korsakoff's syndrome: relation to ataxia. *Neuropsychology*. 2000;14(3):341–352.
57. Fitzpatrick LE, Jackson M, Crowe SF. Characterization of cerebellar ataxia in chronic alcoholics using the International Cooperative Ataxia Rating Scale (ICARS). *Alcohol Clin Exp Res*. 2012;36(11):1942–1951.
58. Shear PK, Sullivan EV, Lane B, Pfefferbaum A. Mammillary body and cerebellar shrinkage in chronic alcoholics with and without amnesia. *Alcohol Clin Exp Res*. 1996;20(8):1489–1495.
59. Harper C, Kril J. Patterns of neuronal loss in the cerebral cortex in chronic alcoholic patients. *J Neurol Sci*. 1989;92(1):81–89.
60. Harper C, Kril J, Daly J. Are we drinking our neurones away? *BMJ*. 1987;294(6571):534–536.
61. Harper C, Corbett D. Changes in the basal dendrites of cortical pyramidal cells from alcoholic patients—a quantitative Golgi study. *J Neurol Neurosurg Psychiatry*. 1990;53(10):856–861.
62. Harding AJ, Halliday GM, Ng JL, Harper CG, Kril JJ. Loss of vasopressin-immunoreactive neurons in alcoholics is dose-related and time-dependent. *Neuroscience*. 1996;72(3):699–708.
63. Baker KG. Neuronal loss in functional zones of the cerebellum of chronic alcoholics with and without Wernicke's encephalopathy. *Neuroscience*. 1999;91(2):429–438.
64. Harper C. Are we drinking our neurones away? *Br Med J (Clin Res Ed)*. 1987;294(6571):534–536.
65. Sutherland GT, Sheedy D, Kril JJ. Neuropathology of alcoholism. *Handb Clin Neurol*. 2014;125:603–615.
66. Korbo L. Glial cell loss in the hippocampus of alcoholics. *Alcohol Clin Exp Res*. 1999;23(1):164–168.
67. Hillmer AT, Sandiego CM, Hannestad J, et al. In vivo imaging of translocator protein, a marker of activated microglia, in alcohol dependence. *Mol Psychiatry*. 2017;22(12):1759–1766.
68. Sullivan EV, Marsh L, Mathalon DH, Lim KO, Pfefferbaum A. Relationship between alcohol withdrawal seizures and temporal lobe white matter volume deficits. *Alcohol Clin Exp Res*. 1996;20(2):348–354.
69. Walker DW. Short-term memory impairment following chronic alcohol consumption in rats. *Neuropsychologia*. 1978;16(5):545–553.
70. Walker DW, Barnes DE, Zornetzer SF, Hunter BE, Kubanis P. Neuronal loss in hippocampus induced by prolonged ethanol consumption in rats. *Science*. 1980;209:711–713.
71. Collins MA, Corse TD, Neafsey EJ. Neuronal degeneration in rat cerebrocortical and olfactory regions during subchronic "binge" intoxication with ethanol: possible explanation for olfactory deficits in alcoholics. *Alcohol Clin Exp Res*. 1996;20(2):284–292.

72. Obernier JA, Bouldin TW, Crews FT. Binge ethanol exposure in adult rats causes necrotic cell death. *Alcohol Clin Exp Res*. 2002;26(4):547–557.
73. Kelso ML, Liput DJ, Eaves DW, Nixon K. Upregulated vimentin suggests new areas of neurodegeneration in a model of an alcohol use disorder. *Neuroscience*. 2011;197:381–393.
74. Crews FT, Broau CJ, Hoplight B, et al. Binge ethanol consumption causes differential brain damage in young adolescent rats compared with adult rats. *Alcohol Clin Exp Res*. 2000;24(11):1712–1723.
75. Alfonso-Loeches S, Pascual M, Guerri C. Gender differences in alcohol-induced neurotoxicity and brain damage. *Toxicology*. 2013;311(1):27–34.
76. Ou X-M, Johnson C, Lu D, et al. Ethanol increases TIEG2-MAO B cell death cascade in the prefrontal cortex of ethanol-preferring rats. *Neurotox Res*. 2011;19(4):511–518.
77. Haorah J, Ramirez SH, Floreani N, Gorantla S, Morsey B, Persidsky Y. Mechanism of alcohol-induced oxidative stress and neuronal injury. *Free Radic Biol Med*. 2008;45(11):1542–1550.
78. Wu D. Alcohol, oxidative stress, and free radical damage. *Alcohol Res Health*. 2003;27(4):277–284.
79. Nordmann R. Oxidative stress from alcohol in the brain. *Alcohol Alcohol Suppl*. 1987;1:75–82.
80. Johnsen-Soriano S, Bosch-Morell F, Miranda M, et al. Ebselen prevents chronic alcohol-induced rat hippocampal stress and functional impairment. *Alcohol Clin Exp Res*. 2007;31(3):486–492.
81. Collins MA, Zou JY, Neafsey EJ. Brain damage due to episodic alcohol exposure in vivo and in vitro: furosemide neuroprotection implicates edema-based mechanism. *FASEB J*. 1998;12:221–230.
82. Crews F, Nixon K, Kim D, et al. BHT blocks NF-kappaB activation and ethanol-induced brain damage. *Alcohol Clin Exp Res*. 2006;30(11):1938–1949.
83. Qin L, Crews FT. NADPH oxidase and reactive oxygen species contribute to alcohol-induced microglial activation and neurodegeneration. *J Neuroinflammation*. 2012;9:5.
84. Crews FT, Lawrimore CJ, Walter TJ, Coleman Jr LG. The role of neuroimmune signaling in alcoholism. *Neuropharmacology*. 2017;122:56–73.
85. Block ML, Hong JS. Microglia and inflammation-mediated neurodegeneration: multiple triggers with a common mechanism. *Prog Neurobiol*. 2005;76(2):77–98.
86. Alfonso-Loeches S, Pascual-Lucas M, Blanco AM, Sanchez-Vera I, Guerri C. Pivotal role of TLR4 receptors in alcohol-induced neuroinflammation and brain damage. *J Neurosci*. 2010;30(24):8285–8295.
87. Gosselin D, Skola D, Coufal NG, et al. An environment-dependent transcriptional network specifies human microglia identity. *Science*. 2017;356(6344):eaal3222.
88. Ginhoux F, Greter M, Leboeuf M, et al. Fate mapping analysis reveals that adult microglia derive from primitive macrophages. *Science*. 2010;330(6005):841–845.
89. Ginhoux F, Lim S, Hoeffel G, Low D, Huber T. Origin and differentiation of microglia. *Front Cell Neurosci*. 2013;7:45.
90. Harry GJ. Microglia during development and aging. *Pharmacol Ther*. 2013;139(3):313–326.
91. Goldmann T, Blank T, Prinz M. Fine-tuning of type I IFN-signaling in microglia—implications for homeostasis, CNS autoimmunity and interferonopathies. *Curr Opin Neurobiol*. 2016;36:38–42.
92. Dubbelaar ML, Kracht L, Eggen BJL, Boddeke E. The kaleidoscope of microglial phenotypes. *Front Immunol*. 2018;9:1753.
93. Askew K, Li K, Olmos-Alonso A, et al. Coupled proliferation and apoptosis maintain the rapid turnover of microglia in the adult brain. *Cell Rep*. 2017;18(2):391–405.

94. Reu P, Khosravi A, Bernard S, et al. The lifespan and turnover of microglia in the human brain. *Cell Rep*. 2017;20(4):779–784.
95. Tremblay ME, Stevens B, Sierra A, Wake H, Bessis A, Nimmerjahn A. The role of microglia in the healthy brain. *J Neurosci*. 2011;31(45):16064–16069.
96. Ransohoff RM, Perry VH. Microglial physiology: unique stimuli, specialized responses. *Annu Rev Immunol*. 2009;27:119–145.
97. Li Q, Cheng Z, Zhou L, et al. Developmental heterogeneity of microglia and brain myeloid cells revealed by deep single-cell RNA sequencing. *Neuron*. 2019;101(2): 207–223.e210.
98. Nimmerjahn A, Kirchhoff F, Helmchen F. Resting microglial cells are highly dynamic surveillants of brain parenchyma in vivo. *Science*. 2005;308(5726):1314–1318.
99. Schafer DP, Lehrman EK, Kautzman AG, et al. Microglia sculpt postnatal neural circuits in an activity and complement-dependent manner. *Neuron*. 2012;74(4): 691–705.
100. Weinhard L, di Bartolomei G, Bolasco G, et al. Microglia remodel synapses by presynaptic trogocytosis and spine head filopodia induction. *Nat Commun*. 2018;9(1):1228.
101. Salter MW, Beggs S. Sublime microglia: expanding roles for the guardians of the CNS. *Cell*. 2014;158(1):15–24.
102. Parkhurst CN, Yang G, Ninan I, et al. Microglia promote learning-dependent synapse formation through brain-derived neurotrophic factor. *Cell*. 2013;155(7):1596–1609.
103. Colonna M, Butovsky O. Microglia function in the central nervous system during health and neurodegeneration. *Annu Rev Immunol*. 2017;35:441–468.
104. Li Y, Du XF, Liu CS, Wen ZL, Du JL. Reciprocal regulation between resting microglial dynamics and neuronal activity in vivo. *Dev Cell*. 2012;23(6):1189–1202.
105. Dissing-Olesen L, LeDue JM, Rungta RL, Hefendehl JK, Choi HB, MacVicar BA. Activation of neuronal NMDA receptors triggers transient ATP-mediated microglial process outgrowth. *J Neurosci*. 2014;34(32):10511–10527.
106. Li Q, Barres BA. Microglia and macrophages in brain homeostasis and disease. *Nat Rev Immunol*. 2018;18(4):225–242.
107. Hickman SE, Kingery ND, Ohsumi TK, et al. The microglial sensome revealed by direct RNA sequencing. *Nat Neurosci*. 2013;16(12):1896–1905.
108. Kreutzberg GW. Microglia: a sensor for pathological events in the CNS. *Trends Neurosci* 1996;19(8):312–318.
109. Raivich G, Bohatschek M, Kloss CU, Werner A, Jones LL, Kreutzberg GW. Neuroglial activation repertoire in the injured brain: graded response, molecular mechanisms and cues to physiological function. *Brain Res Brain Res Rev*. 1999;30(1):77–105.
110. Lull ME, Block ML. Microglial activation and chronic neurodegeneration. *Neurotherapeutics*. 2010;7(4):354–365.
111. Filiou MD, Arefin AS, Moscato P, Graeber MB. 'Neuroinflammation' differs categorically from inflammation: transcriptomes of Alzheimer's disease, Parkinson's disease, schizophrenia and inflammatory diseases compared. *Neurogenetics*. 2014;15(3):201–212.
112. Rocha e Silva M. A brief survey of the history of inflammation. *Agents Actions*. 1978;8(1–2):45–49.
113. Colton C, Wilcock DM. Assessing activation states in microglia. *CNS Neurol Disord Drug Targets*. 2010;9(2):174–191.
114. Block ML, Zecca L, Hong JS. Microglia-mediated neurotoxicity: uncovering the molecular mechanisms. *Nat Rev Neurosci*. 2007;8(1):57–69.
115. Condello C, Yuan P, Grutzendler J. Microglia-mediated neuroprotection, TREM2, and Alzheimer's disease: evidence from optical imaging. *Biol Psychiatry*. 2018;83(4):377–387.
116. Bellver-Landete V, Bretheau F, Mailhot B, et al. Microglia are an essential component of the neuroprotective scar that forms after spinal cord injury. *Nat Commun*. 2019;10(1):518.

117. Murray PJ, Allen JE, Biswas SK, et al. Macrophage activation and polarization: nomenclature and experimental guidelines. *Immunity*. 2014;41(1):14–20.
118. Beynon SB, Walker FR. Microglial activation in the injured and healthy brain: what are we really talking about? Practical and theoretical issues associated with the measurement of changes in microglial morphology. *Neuroscience*. 2012;225:162–171.
119. Benarroch EE. Microglia: multiple roles in surveillance, circuit shaping, and response to injury. *Neurology*. 2013;81(12):1079–1088.
120. Prinz M, Priller J. Microglia and brain macrophages in the molecular age: from origin to neuropsychiatric disease. *Nat Rev Neurosci*. 2014;15(5):300–312.
121. Cherry JD, Olschowka JA, O'Banion MK. Neuroinflammation and M2 microglia: the good, the bad, and the inflamed. *J Neuroinflammation*. 2014;11:98.
122. Ransohoff RM. A polarizing question: do M1 and M2 microglia exist? *Nat Neurosci*. 2016;19(8):987–991.
123. Song WM, Colonna M. The identity and function of microglia in neurodegeneration. *Nat Immunol*. 2018;19(10):1048–1058.
124. Neher JJ, Cunningham C. Priming microglia for innate immune memory in the brain. *Trends Immunol*. 2019;40(4):358–374.
125. Perry VH, Cunningham C, Holmes C. Systemic infections and inflammation affect chronic neurodegeneration. *Nat Rev Immunol*. 2007;7(2):161–167.
126. Perry VH, Holmes C. Microglial priming in neurodegenerative disease. *Nat Rev Neurol*. 2014;10(4):217–224.
127. Fonken LK, Frank MG, Gaudet AD, et al. Neuroinflammatory priming to stress is differentially regulated in male and female rats. *Brain Behav Immun*. 2018;70:257–267.
128. Pott Godoy MC, Tarelli R, Ferrari CC, Sarchi MI, Pitossi FJ. Central and systemic IL-1 exacerbates neurodegeneration and motor symptoms in a model of Parkinson's disease. *Brain*. 2008;131(pt 7):1880–1894.
129. Palin K, Cunningham C, Forse P, Perry VH, Platt N. Systemic inflammation switches the inflammatory cytokine profile in CNS Wallerian degeneration. *Neurobiol Dis*. 2008;30(1):19–29.
130. Wendeln AC, Degenhardt K, Kaurani L, et al. Innate immune memory in the brain shapes neurological disease hallmarks. *Nature*. 2018;556(7701):332–338.
131. Holtman IR, Raj DD, Miller JA, et al. Induction of a common microglia gene expression signature by aging and neurodegenerative conditions: a co-expression meta-analysis. *Acta Neuropathol Commun*. 2015;3:31.
132. Guerri C, Pascual M. Impact of neuroimmune activation induced by alcohol or drug abuse on adolescent brain development. *Int J Dev Neurosci*. 2018; in press. https://doi.org/10.1016/j.ijdevneu.2018.11.006.
133. Roberto M, Patel RR, Bajo M. Ethanol and cytokines in the central nervous system. *Handb Exp Pharmacol*. 2018;248:397–431.
134. Qin L, He J, Hanes RN, Pluzarev O, Hong JS, Crews FT. Increased systemic and brain cytokine production and neuroinflammation by endotoxin following ethanol treatment. *J Neuroinflammation*. 2008;5:10.
135. Gano A, Vore AS, Sammakia MN, Deak T. Assessment of extracellular cytokines in the hippocampus of the awake behaving rat using large-molecule microdialysis combined with multiplex arrays after acute and chronic ethanol exposure. *Alcohol Clin Exp Res*. 2019;43(4):640–654.
136. Doremus-Fitzwater TL, Buck HM, Bordner K, Richey L, Jones ME, Deak T. Intoxication- and withdrawal-dependent expression of central and peripheral cytokines following initial ethanol exposure. *Alcohol Clin Exp Res*. 2014;38(8):2186–2198.
137. Doremus-Fitzwater TL, Gano A, Paniccia JE, Deak T. Male adolescent rats display blunted cytokine responses in the CNS after acute ethanol or lipopolysaccharide exposure. *Physiol Behav*. 2015;148:131–144.

138. Rajayer SR, Jacob A, Yang WL, Zhou M, Chaung W, Wang P. Cold-inducible RNA-binding protein is an important mediator of alcohol-induced brain inflammation. *PLoS One*. 2013;8(11):e79430.
139. Kane CJ, Phelan KD, Douglas JC, et al. Effects of ethanol on immune response in the brain: region-specific changes in adolescent versus adult mice. *Alcohol Clin Exp Res*. 2014;38(2):384–391.
140. Walter TJ, Crews FT. Microglial depletion alters the brain neuroimmune response to acute binge ethanol withdrawal. *J Neuroinflammation*. 2017;14(1):86.
141. Valles SL, Blanco AM, Pascual M, Guerri C. Chronic ethanol treatment enhances inflammatory mediators and cell death in the brain and in astrocytes. *Brain Pathol*. 2004;14(4):365–371.
142. Pascual M, Blanco AM, Cauli O, Minarro J, Guerri C. Intermittent ethanol exposure induces inflammatory brain damage and causes long-term behavioural alterations in adolescent rats. *Eur J Neurosci*. 2007;25(2):541–550.
143. Marshall SA, McClain JA, Kelso ML, Hopkins DM, Pauly JR, Nixon K. Microglial activation is not equivalent to neuroinflammation in alcohol-induced neurodegeneration: the importance of microglia phenotype. *Neurobiol Dis*. 2013;54:239–251.
144. McClain JA, Morris SA, Deeny MA, et al. Adolescent binge alcohol exposure induces long-lasting partial activation of microglia. *Brain Behav Immun*. 2011;25(suppl 1): S120–S128.
145. Zahr NM, Luong R, Sullivan EV, Pfefferbaum A. Measurement of serum, liver, and brain cytokine induction, thiamine levels, and hepatopathology in rats exposed to a 4-day alcohol binge protocol. *Alcohol Clin Exp Res*. 2010;34(11):1858–1870.
146. Sanchez-Alavez M, Nguyen W, Mori S, et al. Time course of microglia activation and brain and blood cytokine/chemokine levels following chronic ethanol exposure and protracted withdrawal in rats. *Alcohol*. 2019;76:37–45.
147. Cruz C, Meireles M, Silva SM. Chronic ethanol intake induces partial microglial activation that is not reversed by long-term ethanol withdrawal in the rat hippocampal formation. *Neurotoxicology*. 2017;60:107–115.
148. Seok J, Warren HS, Cuenca AG, et al. Genomic responses in mouse models poorly mimic human inflammatory diseases. *Proc Natl Acad Sci USA*. 2013;110(9):3507–3512.
149. Bell-Temin H, Zhang P, Chaput D, et al. Quantitative proteomic characterization of ethanol-responsive pathways in rat microglial cells. *J Proteome Res*. 2013;12(5):2067–2077.
150. Zhang Y, Wei G, Wang Y, Jing L, Zhao Q. Gene expression profile analysis of rat cerebellum under acute alcohol intoxication. *Gene*. 2015;557(2):188–194.
151. Osterndorff-Kahanek E, Ponomarev I, Blednov YA, Harris RA. Gene expression in brain and liver produced by three different regimens of alcohol consumption in mice: comparison with immune activation. *PLoS One*. 2013;8(3):e59870.
152. Osterndorff-Kahanek EA, Becker HC, Lopez MF, et al. Chronic ethanol exposure produces time- and brain region-dependent changes in gene coexpression networks. *PLoS One*. 2015;10(3):e0121522.
153. McClintick JN, McBride WJ, Bell RL, et al. Gene expression changes in the ventral hippocampus and medial prefrontal cortex of adolescent alcohol-preferring (P) rats following binge-like alcohol drinking. *Alcohol*. 2018;68:37–47.
154. Scaffidi P, Misteli T, Bianchi ME. Release of chromatin protein HMGB1 by necrotic cells triggers inflammation. *Nature*. 2002;418(6894):191–195.
155. Pascual M, Montesinos J, Guerri C. Role of the innate immune system in the neuropathological consequences induced by adolescent binge drinking. *J Neurosci Res*. 2018;96(5):765–780.
156. Pascual M, Do Couto BR, Alfonso-Loeches S, Aguilar MA, Rodriguez-Arias M, Guerri C. Changes in histone acetylation in the prefrontal cortex of ethanol-exposed adolescent rats are associated with ethanol-induced place conditioning. *Neuropharmacology*. 2012;62(7):2309–2319.

157. Harris RA, Bajo M, Bell RL, et al. Genetic and pharmacologic manipulation of TLR4 has minimal impact on ethanol consumption in rodents. *J Neurosci*. 2017;37(5):1139–1155.
158. Yang H, Wang H, Chavan SS, Andersson U. High mobility group box protein 1 (HMGB1): the prototypical endogenous danger molecule. *Mol Med*. 2015;21(suppl 1): S6–S12.
159. Vetreno RP, Crews FT. Adolescent binge drinking increases expression of the danger signal receptor agonist HMGB1 and toll-like receptors in the adult prefrontal cortex. *Neuroscience*. 2012;226:475–488.
160. Coleman Jr LG, Zou J, Crews FT. Microglial-derived miRNA let-7 and HMGB1 contribute to ethanol-induced neurotoxicity via TLR7. *J Neuroinflammation*. 2017;14(1):22.
161. Coleman Jr LG, Zou J, Qin L, Crews FT. HMGB1/IL-1beta complexes regulate neuroimmune responses in alcoholism. *Brain Behav Immun*. 2018;72:61–77.
162. Crews FT, Qin L, Sheedy D, Vetreno RP, Zou J. High mobility group box 1/Toll-like receptor danger signaling increases brain neuroimmune activation in alcohol dependence. *Biol Psychiatry*. 2013;73(7):602–612.
163. Warden AS, Azzam M, DaCosta A, et al. Toll-like receptor 3 dynamics in female C57BL/6J mice: regulation of alcohol intake. *Brain Behav Immun*. 2019;77:66–76.
164. Weber MD, Frank MG, Tracey KJ, Watkins LR, Maier SF. Stress induces the danger-associated molecular pattern HMGB-1 in the hippocampus of male Sprague Dawley rats: a priming stimulus of microglia and the NLRP3 inflammasome. *J Neurosci*. 2015;35(1):316–324.
165. Marshall SA, Geil CR, Nixon K. Prior binge ethanol exposure potentiates the microglial response in a model of alcohol-induced neurodegeneration. *Brain Sci*. 2016;6(2):16.
166. He J, Crews FT. Increased MCP-1 and microglia in various regions of the human alcoholic brain. *Exp Neurol*. 2008;210(2):349–358.
167. Dennis CV, Sheahan PJ, Graeber MB, Sheedy DL, Kril JJ, Sutherland GT. Microglial proliferation in the brain of chronic alcoholics with hepatic encephalopathy. *Metab Brain Dis*. 2014;29(4):1027–1039.
168. Liu J, Lewohl JM, Harris RA, et al. Patterns of gene expression in the frontal cortex discriminate alcoholic from nonalcoholic individuals. *Neuropsychopharmacology*. 2006;31(7):1574–1582.
169. McClintick JN, Xuei X, Tischfield JA, et al. Stress-response pathways are altered in the hippocampus of chronic alcoholics. *Alcohol*. 2013;47(7):505–515.
170. Kapoor M, Wang JC, Farris SP, et al. Analysis of whole genome-transcriptomic organization in brain to identify genes associated with alcoholism. *Transl Psychiatry*. 2019;9(1):89.
171. Achur RN, Freeman WM, Vrana KE. Circulating cytokines as biomarkers of alcohol abuse and alcoholism. *J Neuroimmune Pharmacol*. 2010;5(1):83–91.
172. Heberlein A, Kaser M, Lichtinghagen R, et al. TNF-alpha and IL-6 serum levels: neurobiological markers of alcohol consumption in alcohol-dependent patients? *Alcohol*. 2014;48(7):671–676.
173. Leclercq S, Cani PD, Neyrinck AM, et al. Role of intestinal permeability and inflammation in the biological and behavioral control of alcohol-dependent subjects. *Brain Behav Immun*. 2012;26(6):911–918.
174. Gonzalez-Reimers E, Sanchez-Perez MJ, Santolaria-Fernandez F, et al. Changes in cytokine levels during admission and mortality in acute alcoholic hepatitis. *Alcohol*. 2012;46(5):433–440.
175. Gonzalez-Reimers E, Santolaria-Fernandez F, Medina-Garcia JA, et al. TH-1 and TH-2 cytokines in stable chronic alcoholics. *Alcohol Alcohol*. 2012;47(4):390–396.

176. Zahr NM. Peripheral TNFalpha elevations in abstinent alcoholics are associated with hepatitis C infection. *PLoS One*. 2018;13(2):e0191586.
177. Umhau JC, Schwandt M, Solomon MG, et al. Cerebrospinal fluid monocyte chemoattractant protein-1 in alcoholics: support for a neuroinflammatory model of chronic alcoholism. *Alcohol Clin Exp Res*. 2014;38(5):1301–1306.
178. Crews FT, Bechara R, Brown LA, et al. Cytokines and alcohol. *Alcohol Clin Exp Res*. 2006;30(4):720–730.
179. de Timary P, Starkel P, Delzenne NM, Leclercq S. A role for the peripheral immune system in the development of alcohol use disorders? *Neuropharmacology*. 2017;122:148–160.
180. Qin L, Wu X, Block ML, et al. Systemic LPS causes chronic neuroinflammation and progressive neurodegeneration. *Glia*. 2007;55(5):453–462.
181. Wang X, Chu G, Yang Z, et al. Ethanol directly induced HMGB1 release through NOX2/NLRP1 inflammasome in neuronal cells. *Toxicology*. 2015;334:104–110.
182. Boyadjieva NI, Sarkar DK. Role of microglia in ethanol's apoptotic action on hypothalamic neuronal cells in primary cultures. *Alcohol Clin Exp Res*. 2010;34(11):1835–1842.
183. Barton EA, Baker C, Leasure JL. Investigation of sex differences in the microglial response to binge ethanol and exercise. *Brain Sci*. 2017;7(10):E139.
184. Boyadjieva NI, Sarkar DK. Cyclic adenosine monophosphate and brain-derived neurotrophic factor decreased oxidative stress and apoptosis in developing hypothalamic neuronal cells: role of microglia. *Alcohol Clin Exp Res*. 2013;37(8):1370–1379.
185. Chastain LG, Sarkar DK. Role of microglia in regulation of ethanol neurotoxic action. *Int Rev Neurobiol*. 2014;118:81–103.
186. Alfonso-Loeches S, Urena-Peralta J, Morillo-Bargues MJ, Gomez-Pinedo U, Guerri C. Ethanol-induced TLR4/NLRP3 neuroinflammatory response in microglial cells promotes leukocyte infiltration across the BBB. *Neurochem Res*. 2015;41:193–209.
187. Chen MK, Guilarte TR. Translocator protein 18 kDa (TSPO): molecular sensor of brain injury and repair. *Pharmacol Ther*. 2008;118(1):1–17.
188. Obernier JA, White AM, Swartzwelder HS, Crews FT. Cognitive deficits and CNS damage after a 4-day binge ethanol exposure in rats. *Pharmacol Biochem Behav*. 2002;72(3):521–532.
189. Tyler RE, Kim SW, Guo M, et al. Detecting neuroinflammation in the brain following chronic alcohol exposure in rats: a comparison between in vivo and in vitro TSPO radioligand binding. *Eur J Neurosci*. 2019;1–12.
190. Saba W, Goutal S, Auvity S, et al. Imaging the neuroimmune response to alcohol exposure in adolescent baboons: a TSPO PET study using (18) F-DPA-714. *Addict Biol*. 2018;23(5):1000–1009.
191. Nixon K, Kim DH, Potts EN, He J, Crews FT. Distinct cell proliferation events during abstinence after alcohol dependence: microglia proliferation precedes neurogenesis. *Neurobiol Dis*. 2008;31(2):218–229.
192. Rubio-Araiz A, Porcu F, Perez-Hernandez M, et al. Disruption of blood-brain barrier integrity in postmortem alcoholic brain: preclinical evidence of TLR4 involvement from a binge-like drinking model. *Addict Biol*. 2017;22(4):1103–1116.
193. Kalk NJ, Guo Q, Owen D, et al. Decreased hippocampal translocator protein (18 kDa) expression in alcohol dependence: a [(11)C]PBR28 PET study. *Transl Psychiatry*. 2017;7(1):e996.
194. Kim SW, Wiers CE, Tyler R, et al. Influence of alcoholism and cholesterol on TSPO binding in brain: PET [(11)C]PBR28 studies in humans and rodents. *Neuropsychopharmacology*. 2018;43(9):1832–1839.

195. Horti AG, Naik R, Foss CA, et al. PET imaging of microglia by targeting macrophage colony-stimulating factor 1 receptor (CSF1R). *Proc Natl Acad Sci USA*. 2019;116(5):1686–1691.
196. Guillemin GJ, Brew BJ. Microglia, macrophages, perivascular macrophages, and pericytes: a review of function and identification. *J Leukoc Biol*. 2004;75(3):388–397.
197. Frank MG, Wieseler-Frank JL, Watkins LR, Maier SF. Rapid isolation of highly enriched and quiescent microglia from adult rat hippocampus: immunophenotypic and functional characteristics. *J Neurosci Methods*. 2006;151(2):121–130.
198. McCarthy GM, Warden AS, Bridges CR, Blednov YA, Harris RA. Chronic ethanol consumption: role of TLR3/TRIF-dependent signaling. *Addict Biol*. 2018;23(3):889–903.
199. Peng H, Nickell CRG, Chen KY, Nixon K. Increased expression of M1 and M2 phenotypic markers in isolated microglia after four-day binge alcohol exposure in male rats. *Alcohol*. 2017;62:29–40.
200. Stephen Rich J, Martin PR. Co-occurring psychiatric disorders and alcoholism. *Handb Clin Neurol*. 2014;125:573–588.
201. Wohleb ES. Neuron-microglia interactions in mental health disorders: "for better, and for worse". *Front Immunol*. 2016;7:544.
202. Wohleb ES, Franklin T, Iwata M, Duman RS. Integrating neuroimmune systems in the neurobiology of depression. *Nat Rev Neurosci*. 2016;17(8):497–511.
203. Akbar M, Egli M, Cho YE, Song BJ, Noronha A. Medications for alcohol use disorders: an overview. *Pharmacol Ther*. 2018;185:64–85.
204. Franck J, Jayaram-Lindstrom N. Pharmacotherapy for alcohol dependence: status of current treatments. *Curr Opin Neurobiol*. 2013;23(4):692–699.
205. Van Skike CE, Maggio SE, Reynolds AR, et al. Critical needs in drug discovery for cessation of alcohol and nicotine polysubstance abuse. *Prog Neuropsychopharmacol Biol Psychiatry*. 2016;65:269–287.
206. Kranzler HR, Soyka M. Diagnosis and pharmacotherapy of alcohol use disorder: a review. *JAMA*. 2018;320(8):815–824.
207. Mohapatra S, Rath NR. Disulfiram induced psychosis. *Clin Psychopharmacol Neurosci*. 2017;15(1):68–69.
208. Skinner MD, Lahmek P, Pham H, Aubin HJ. Disulfiram efficacy in the treatment of alcohol dependence: a meta-analysis. *PLoS One*. 2014;9(2):e87366.
209. Fuller RK, Branchey L, Brightwell DR, et al. Disulfiram treatment of alcoholism. A veterans administration cooperative study. *JAMA*. 1986;256(11):1449–1455.
210. Krampe H, Spies CD, Ehrenreich H. Supervised disulfiram in the treatment of alcohol use disorder: a commentary. *Alcohol Clin Exp Res*. 2011;35(10):1732–1736.
211. O'Brien CP, Volpicelli LA, Volpicelli JR. Naltrexone in the treatment of alcoholism: a clinical review. *Alcohol*. 1996;13(1):35–39.
212. Garbutt JC. Efficacy and tolerability of naltrexone in the management of alcohol dependence. *Curr Pharm Des*. 2010;16(19):2091–2097.
213. Seneviratne C, Johnson BA. Advances in medications and tailoring treatment for alcohol use disorder. *Alcohol Res*. 2015;37(1):15–28.
214. Spanagel R, Vengeliene V, Jandeleit B, et al. Acamprosate produces its anti-relapse effects via calcium. *Neuropsychopharmacology*. 2014;39(4):783–791.
215. Jonas DE, Amick HR, Feltner C, et al. Pharmacotherapy for adults with alcohol use disorders in outpatient settings: a systematic review and meta-analysis. *JAMA*. 2014;311(18):1889–1900.
216. Warden A, Erickson E, Robinson G, Harris RA, Mayfield RD. The neuroimmune transcriptome and alcohol dependence: potential for targeted therapies. *Pharmacogenomics*. 2016;17(18):2081–2096.

217. Litten RZ, Ryan ML, Falk DE, Reilly M, Fertig JB, Koob GF. Heterogeneity of alcohol use disorder: understanding mechanisms to advance personalized treatment. *Alcohol Clin Exp Res*. 2015;39(4):579–584.
218. Heilig M, Goldman D, Berrettini W, O'Brien CP. Pharmacogenetic approaches to the treatment of alcohol addiction. *Nat Rev Neurosci*. 2011;12(11):670–684.
219. Mayfield RD, Lewohl JM, Dodd PR, Herlihy A, Liu J, Harris RA. Patterns of gene expression are altered in the frontal and motor cortices of human alcoholics. *J Neurochem*. 2002;81(4):802–813.
220. Mulligan MK, Ponomarev I, Hitzemann RJ, et al. Toward understanding the genetics of alcohol drinking through transcriptome meta-analysis. *Proc Natl Acad Sci USA*. 2006;103(16):6368–6373.
221. Ray LA, Roche DJ, Heinzerling K, Shoptaw S. Opportunities for the development of neuroimmune therapies in addiction. *Int Rev Neurobiol*. 2014;118:381–401.
222. Khoja S, Huynh N, Warnecke AMP, Asatryan L, Jakowec MW, Davies DL. Preclinical evaluation of avermectins as novel therapeutic agents for alcohol use disorders. *Psychopharmacology (Berl)*. 2018;235(6):1697–1709.
223. Bortolato M, Yardley MM, Khoja S, et al. Pharmacological insights into the role of P2X4 receptors in behavioural regulation: lessons from ivermectin. *Int J Neuropsychopharmacol*. 2013;16(5):1059–1070.
224. Asatryan L, Popova M, Perkins D, Trudell JR, Alkana RL, Davies DL. Ivermectin antagonizes ethanol inhibition in purinergic P2X4 receptors. *J Pharmacol Exp Ther*. 2010;334(3):720–728.
225. Franklin KM, Asatryan L, Jakowec MW, Trudell JR, Bell RL, Davies DL. P2X4 receptors (P2X4Rs) represent a novel target for the development of drugs to prevent and/or treat alcohol use disorders. *Front Neurosci*. 2014;8:176.
226. Gofman L, Fernandes NC, Potula R. Relative role of Akt, ERK and CREB alcohol-induced microglia P2X4R receptor expression. *Alcohol Alcohol*. 2016;51(6): 647–654.
227. Yardley MM, Neely M, Huynh N, et al. Multiday administration of ivermectin is effective in reducing alcohol intake in mice at doses shown to be safe in humans. *Neuroreport*. 2014;25(13):1018–1023.
228. Yardley MM, Wyatt L, Khoja S, et al. Ivermectin reduces alcohol intake and preference in mice. *Neuropharmacology*. 2012;63(2):190–201.
229. Huynh N, Arabian N, Naito A, et al. Preclinical development of moxidectin as a novel therapeutic for alcohol use disorder. *Neuropharmacology*. 2017;113(pt A):60–70.
230. Roche DJ, Yardley MM, Lunny KF, et al. A pilot study of the safety and initial efficacy of ivermectin for the treatment of alcohol use disorder. *Alcohol Clin Exp Res*. 2016;40(6):1312–1320.
231. Moller T, Bard F, Bhattacharya A, et al. Critical data-based re-evaluation of minocycline as a putative specific microglia inhibitor. *Glia*. 2016;64(10):1788–1794.
232. Garrido-Mesa N, Zarzuelo A, Galvez J. Minocycline: far beyond an antibiotic. *Br J Pharmacol*. 2013;169(2):337–352.
233. Kim HS, Suh YH. Minocycline and neurodegenerative diseases. *Behav Brain Res*. 2009;196(2):168–179.
234. Yrjanheikki J, Tikka T, Keinanen R, Goldsteins G, Chan PH, Koistinaho J. A tetracycline derivative, minocycline, reduces inflammation and protects against focal cerebral ischemia with a wide therapeutic window. *Proc Natl Acad Sci USA*. 1999;96(23):13496–13500.
235. Agrawal RG, Hewetson A, George CM, Syapin PJ, Bergeson SE. Minocycline reduces ethanol drinking. *Brain Behav Immun*. 2011;25(suppl 1):S165–S169.
236. Qin L, Crews FT. Chronic ethanol increases systemic TLR3 agonist-induced neuroinflammation and neurodegeneration. *J Neuroinflammation*. 2012;9(1):130.

237. Kane CJ, Phelan KD, Douglas JC, et al. Effects of ethanol on immune response in the brain: region-specific changes in aged mice. *J Neuroinflammation*. 2013;10:66.
238. Wu Y, Lousberg EL, Moldenhauer LM, et al. Attenuation of microglial and IL-1 signaling protects mice from acute alcohol-induced sedation and/or motor impairment. *Brain Behav Immun*. 2011;25(suppl 1):S155–S164.
239. McIver SR, Muccigrosso MM, Haydon PG. The effect of doxycycline on alcohol consumption and sensitivity: consideration for inducible transgenic mouse models. *Exp Biol Med (Maywood)*. 2012;237(10):1129–1133.
240. Swift RM, Aston ER. Pharmacotherapy for alcohol use disorder: current and emerging therapies. *Harv Rev Psychiatry*. 2015;23(2):122–133.
241. Kelso ML, Scheff NN, Scheff SW, Pauly JR. Melatonin and minocycline for combinatorial therapy to improve functional and histopathological deficits following traumatic brain injury. *Neurosci Lett*. 2011;488(1):60–64.
242. Harper KM, Knapp DJ, Park MA, Breese GR. Differential effects of single versus repeated minocycline administration-lack of significant interaction with chronic alcohol history. *Pharmacol Biochem Behav*. 2018;168:33–42.
243. Gajbhiye SV, Tripathi RK, Petare A, Potey AV, Shankar A. Minocycline in alcohol withdrawal induced anxiety and alcohol relapse in rats. *Curr Clin Pharmacol*. 2018;13(1):65–72.
244. Scott G, Zetterberg H, Jolly A, et al. Minocycline reduces chronic microglial activation after brain trauma but increases neurodegeneration. *Brain*. 2018;141(2):459–471.
245. Diguet E, Fernagut PO, Wei X, et al. Deleterious effects of minocycline in animal models of Parkinson's disease and Huntington's disease. *Eur J Neurosci*. 2004;19(12):3266–3276.
246. Diguet E, Gross CE, Tison F, Bezard E. Rise and fall of minocycline in neuroprotection: need to promote publication of negative results. *Exp Neurol*. 2004;189(1):1–4.
247. Gordon PH, Moore DH, Miller RG, et al. Efficacy of minocycline in patients with amyotrophic lateral sclerosis: a phase III randomised trial. *Lancet Neurol*. 2007;6(12):1045–1053.
248. Keller AF, Gravel M, Kriz J. Treatment with minocycline after disease onset alters astrocyte reactivity and increases microgliosis in SOD1 mutant mice. *Exp Neurol*. 2011;228(1):69–79.
249. Choi J, Ifuku M, Noda M, Guilarte TR. Translocator protein (18 kDa)/peripheral benzodiazepine receptor specific ligands induce microglia functions consistent with an activated state. *Glia*. 2011;59(2):219–230.
250. Knott EP, Assi M, Rao SN, Ghosh M, Pearse DD. Phosphodiesterase inhibitors as a therapeutic approach to neuroprotection and repair. *Int J Mol Sci*. 2017;18(4):696.
251. Logrip ML. Phosphodiesterase regulation of alcohol drinking in rodents. *Alcohol*. 2015;49(8):795–802.
252. Logrip ML, Vendruscolo LF, Schlosburg JE, Koob GF, Zorrilla EP. Phosphodiesterase 10A regulates alcohol and saccharin self-administration in rats. *Neuropsychopharmacology*. 2014;39(7):1722–1731.
253. Pearse DD, Hughes ZA. PDE4B as a microglia target to reduce neuroinflammation. *Glia*. 2016;64(10):1698–1709.
254. Avila DV, Myers SA, Zhang J, et al. Phosphodiesterase 4b expression plays a major role in alcohol-induced neuro-inflammation. *Neuropharmacology*. 2017;125:376–385.
255. Franklin KM, Hauser SR, Lasek AW, et al. Reduction of alcohol drinking of alcohol-preferring (P) and high-alcohol drinking (HAD1) rats by targeting phosphodiesterase-4 (PDE4). *Psychopharmacology (Berl)*. 2015;232(13):2251–2262.

256. Wen RT, Zhang M, Qin WJ, et al. The phosphodiesterase-4 (PDE4) inhibitor rolipram decreases ethanol seeking and consumption in alcohol-preferring fawn-hooded rats. *Alcohol Clin Exp Res*. 2012;36(12):2157–2167.
257. Hu W, Lu T, Chen A, et al. Inhibition of phosphodiesterase-4 decreases ethanol intake in mice. *Psychopharmacology (Berl)*. 2011;218(2):331–339.
258. Liu X, Hao PD, Yang MF, et al. The phosphodiesterase-4 inhibitor roflumilast decreases ethanol consumption in C57BL/6J mice. *Psychopharmacology (Berl)*. 2017;234(16):2409–2419.
259. Blednov YA, Benavidez JM, Black M, Harris RA. Inhibition of phosphodiesterase 4 reduces ethanol intake and preference in C57BL/6J mice. *Front Neurosci*. 2014;8:129.
260. Mizuno T, Kurotani T, Komatsu Y, et al. Neuroprotective role of phosphodiesterase inhibitor ibudilast on neuronal cell death induced by activated microglia. *Neuropharmacology*. 2004;46(3):404–411.
261. Rolan P, Hutchinson M, Johnson K. Ibudilast: a review of its pharmacology, efficacy and safety in respiratory and neurological disease. *Expert Opin Pharmacother*. 2009;10(17):2897–2904.
262. Bell RL, Lopez MF, Cui C, et al. Ibudilast reduces alcohol drinking in multiple animal models of alcohol dependence. *Addict Biol*. 2015;20(1):38–42.
263. Ray LA, Bujarski S, Shoptaw S, Roche DJ, Heinzerling K, Miotto K. Development of the neuroimmune modulator ibudilast for the treatment of alcoholism: a randomized, placebo-controlled, human laboratory trial. *Neuropsychopharmacology*. 2017;42(9):1776–1788.
264. Glass CK, Saijo K. Nuclear receptor transrepression pathways that regulate inflammation in macrophages and T cells. *Nat Rev Immunol*. 2010;10(5):365–376.
265. Nennig SE, Schank JR. The role of NFkB in drug addiction: beyond inflammation. *Alcohol Alcohol*. 2017;52(2):172–179.
266. Okvist A, Johansson S, Kuzmin A, et al. Neuroadaptations in human chronic alcoholics: dysregulation of the NF-kappaB system. *PLoS One*. 2007;2(9):e930.
267. Le Foll B, Di Ciano P, Panlilio LV, Goldberg SR, Ciccocioppo R. Peroxisome proliferator-activated receptor (PPAR) agonists as promising new medications for drug addiction: preclinical evidence. *Curr Drug Targets*. 2013;14(7):768–776.
268. Blednov YA, Black M, Benavidez JM, Stamatakis EE, Harris RA. PPAR agonists: I. Role of receptor subunits in alcohol consumption in male and female mice. *Alcohol Clin Exp Res*. 2016;40(3):553–562.
269. Ferguson LB, Most D, Blednov YA, Harris RA. PPAR agonists regulate brain gene expression: relationship to their effects on ethanol consumption. *Neuropharmacology*. 2014;86:397–407.
270. Karahanian E, Quintanilla ME, Fernandez K, Israel Y. Fenofibrate—a lipid-lowering drug—reduces voluntary alcohol drinking in rats. *Alcohol*. 2014;48(7):665–670.
271. Haile CN, Kosten TA. The peroxisome proliferator-activated receptor alpha agonist fenofibrate attenuates alcohol self-administration in rats. *Neuropharmacology*. 2017;116:364–370.
272. Barson JR, Karatayev O, Chang GQ, et al. Positive relationship between dietary fat, ethanol intake, triglycerides, and hypothalamic peptides: counteraction by lipid-lowering drugs. *Alcohol*. 2009;43(6):433–441.
273. Schlicht I. Enhancement of voluntary alcohol consumption in rats by clofibrate feeding. *Alcohol*. 1987;4(3):199–206.
274. Blednov YA, Black M, Benavidez JM, Stamatakis EE, Harris RA. PPAR agonists: II. Fenofibrate and tesaglitazar alter behaviors related to voluntary alcohol consumption. *Alcohol Clin Exp Res*. 2016;40(3):563–571.

275. Karahanian E, Rivera-Meza M, Quintanilla ME, Munoz D, Fernandez K, Israel Y. PPARalpha agonists reduce alcohol drinking: do they act in the brain or in the liver? *Alcohol Alcohol.* 2015;50(6):717–718.
276. Jiang C, Ting AT, Seed B. PPAR-gamma agonists inhibit production of monocyte inflammatory cytokines. *Nature.* 1998;391(6662):82–86.
277. Heneka MT, Landreth GE, Hull M. Drug insight: effects mediated by peroxisome proliferator-activated receptor-gamma in CNS disorders. *Nat Clin Pract Neurol.* 2007;3(9):496–504.
278. Heneka MT, Landreth GE. PPARs in the brain. *Biochim Biophys Acta.* 2007;1771(8):1031–1045.
279. Zolezzi JM, Santos MJ, Bastias-Candia S, Pinto C, Godoy JA, Inestrosa NC. PPARs in the central nervous system: roles in neurodegeneration and neuroinflammation. *Biol Rev Camb Philos Soc.* 2017;92(4):2046–2069.
280. Storer PD, Xu J, Chavis J, Drew PD. Peroxisome proliferator-activated receptor-gamma agonists inhibit the activation of microglia and astrocytes: implications for multiple sclerosis. *J Neuroimmunol.* 2005;161(1–2):113–122.
281. Kane CJ, Drew PD. Inflammatory responses to alcohol in the CNS: nuclear receptors as potential therapeutics for alcohol-induced neuropathologies. *J Leukoc Biol.* 2016;100(5):951–959.
282. Drew PD, Johnson JW, Douglas JC, Phelan KD, Kane CJ. Pioglitazone blocks ethanol induction of microglial activation and immune responses in the hippocampus, cerebellum, and cerebral cortex in a mouse model of fetal alcohol spectrum disorders. *Alcohol Clin Exp Res.* 2015;39(3):445–454.
283. Blednov YA, Benavidez JM, Black M, et al. Peroxisome proliferator-activated receptors alpha and gamma are linked with alcohol consumption in mice and withdrawal and dependence in humans. *Alcohol Clin Exp Res.* 2015;39(1):136–145.
284. Stopponi S, de Guglielmo G, Somaini L, et al. Activation of PPARgamma by pioglitazone potentiates the effects of naltrexone on alcohol drinking and relapse in msP rats. *Alcohol Clin Exp Res.* 2013;37(8):1351–1360.
285. Stopponi S, Somaini L, Cippitelli A, et al. Activation of nuclear PPARgamma receptors by the antidiabetic agent pioglitazone suppresses alcohol drinking and relapse to alcohol seeking. *Biol Psychiatry.* 2011;69(7):642–649.
286. Liput DJ, Prendergast MA, Nixon K. Endocannabinoids mediate neurotoxicity and neuroprotection in an in vitro model of ethanol-induced neurotoxicity. *Alcohol Clin Exp Res.* 2012;36(suppl 1):20A (abstract; Gordis Research Award winner, biomedical predoc).
287. Liput DJ, Tsakalozou E, Hammell DC, Paudel KS, Nixon K, Stinchcomb AL. Quantification of anandamide, oleoylethanolamide and palmitoylethanolamide in rodent brain tissue using high performance liquid chromatography-electrospray mass spectroscopy. *J Pharm Anal.* 2014;4(4):234–241.
288. Pava MJ, Woodward JJ. A review of the interactions between alcohol and the endocannabinoid system: implications for alcohol dependence and future directions for research. *Alcohol.* 2012;46(3):185–204.
289. Pava MJ, Woodward JJ. Chronic ethanol alters network activity and endocannabinoid signaling in the prefrontal cortex. *Front Integr Neurosci.* 2014;8:58.
290. Collins MA, Tajuddin N, Moon KH, Kim HY, Nixon K, Neafsey EJ. Alcohol, phospholipase A2-associated neuroinflammation, and omega3 docosahexaenoic acid protection. *Mol Neurobiol.* 2014;50(1):239–245.

CHAPTER NINE

Microglia as possible therapeutic targets for autism spectrum disorders

Megumi Andoh, Yuji Ikegaya, Ryuta Koyama*
Laboratory of Chemical Pharmacology, Graduate School of Pharmaceutical Sciences, The University of Tokyo, Tokyo, Japan
*Corresponding author: e-mail address: rkoyama@mol.f.u-tokyo.ac.jp

Contents

Abstract

Malfunctions of the nervous and immune systems are now recognized to be fundamental causes of autism spectrum disorders (ASDs). Studies have suggested that the brain's resident immune cells, microglia are possible key players in ASDs. Specifically, deficits in synaptic pruning by microglia may underlie the pathogenesis of ASDs, in which excess synapses are occasionally reported. This idea has driven researchers to investigate causal links between microglial dysfunction and ASDs. In this review, we first introduce the characteristics of microglia in ASD brains and discuss their possible roles in the pathogenesis of ASDs. We also refer to immunomodulatory agents that could be potentially used as symptomatic therapies for ASDs in light of their ability to modify microglial functions. Finally, we will mention a possible strategy to radically cure some of the symptoms reported in ASDs through reorganizing neural circuits *via* microglia-dependent synaptic pruning.

Progress in Molecular Biology and Translational Science, Volume 167
ISSN 1877-1173
https://doi.org/10.1016/bs.pmbts.2019.06.012

1. Introduction

Some decades ago, it was revealed that the synaptic density in the human brain is drastically increased during childhood and decreased during adolescence to the adult level.[1] The decrease in synaptic density has been called synaptic pruning or synapse elimination and has been suggested to be crucial for the normal development of brain functions.[2,3] Accumulating studies have shown that microglia, the brain-resident immune cells, are the executors of synaptic pruning.[2,4] Deficits in synaptic pruning are suggested to perturb refinement of neural circuits, leading to abnormal development of higher-order brain functions such as cognition and social communication, which are the main symptoms of autism spectrum disorders (ASDs), one of the neurodevelopmental diseases.[3,5] Therefore, many researchers have focused on the relationship between ASDs and microglia.

Neuropathological analysis using postmortem brains of ASD patients shows that microglial number is increased in that population and that microglia exhibit the activated form (enlarged cell body and thick process) in multiple regions, including the frontal lobes.[6,7] PET/CT scan analysis also shows that microglia are activated in ASD frontal lobes and cerebellum.[8] Moreover, in the ASD prefrontal cortex (PFC), the microglia-neuron distance is reduced, and microglial processes enclose the neuronal cell body, which suggests that the interaction between microglia and neurons is facilitated in ASD brains.[9] It should be noted, however, that the likely activated status of microglia in ASD brains is reported from studies that analyzed postmortem specimens or individuals who had been already diagnosed with ASDs; whether microglia are activated or even inactivated during the development of ASDs remains to be elucidated.

Recently, the complement molecules C1q and C3, which stimulate the microglial complement receptor CR to prune synapses,[4] have been suggested to be related to ASDs. In the serum of ASD patients, the levels of C1q and C-reactive protein, which promote the inflammatory response through binding to C1q, are elevated.[10,11] An increased concentration of complement molecules in ASDs may enhance microglia-dependent synaptic pruning. It is also possible that excess complements mask the differences in complement deposition levels on each synapse, which may result in the disruption of complement-targeted specific synaptic pruning.

Almost all of the reports from human studies have been based on samples from patients already diagnosed with ASDs or on postmortem brain specimens. To investigate microglial contributions to the development of ASDs in detail, animal models of ASDs have been widely used. In maternal immune activation (MIA) models, which mimic maternal infection during pregnancy as a risk factor for ASDs, microglial density and proinflammatory cytokine levels are increased in the cerebral cortex and the hippocampus.[12,13] Neurons also exhibit unusual morphologies in these models. For example, hippocampal granule cells have more spines.[14] Furthermore, the expression levels of CX3CR1, a molecule suggested in neuron-microglia interactions, were decreased in the hippocampus, suggesting that MIA perturbs microglia-dependent synaptic pruning.[14] Fmr1 KO mice, in which the risk gene for fragile X syndrome is depleted, were also used as a model of ASD and exhibited impaired microglial phagocytosis of postsynaptic protein PSD95.[15] In addition to these primary ASD models, the BTBR mouse strain, which is used as an idiopathic ASD model, has more activated microglia with higher levels of MHC-II compared to the C57BL/6J mouse strain.[16] It has also been revealed that hippocampal pyramidal cells show increased spontaneous excitatory postsynaptic currents (EPSCs) in BTBR mice, suggesting deficits in synaptic pruning.[17]

Some reports have shown that deficits in microglia-dependent synaptic pruning cause ASD-like features. Inhibition of microglial autophagy decreases microglial synaptic phagocytosis, leading to increased spine density and impaired social interaction.[18,19] Mice lacking CX3CR1, which is expressed by microglia and regulates synaptic pruning, exhibited decreased functional connectivity in ASD-related brain regions, impaired social interaction, and increased repetitive behaviors.[20] More recently, TREM2, which is a risk gene of Alzheimer's disease, has been suggested to be necessary for developmental synaptic pruning, and TREM2 KO mice exhibit reduced functional connectivity in the brain, impaired social interaction and increased repetitive behavior.[21]

The male-to-female ratio of ASD morbidity is approximately 4 to 1 in humans; thus, it is believed that there are relationships between sexes and ASDs. Microglia have attracted attention as a factor that causes sexual differences in ASDs. A genome-wide association study (GWAS) has revealed that genes expressing at higher levels in prenatal human males are also upregulated in postmortem brains of ASD patients and, importantly, these genes include microglial markers.[22] Sex differences in microglial properties are now being investigated energetically. One such study has shown that expression levels of the lysosomal protein CD68 are higher in female mice during development when surplus synapses are eliminated, suggesting that phagocytic capacity differs between sexes.[23] This finding may suggest an excess of synapses underlying the elevated ASD morbidity rate in males.

Astrocytes are the most abundant glia cells in the brain. They contact adjacent synapses and form the tripartite synapse, modulating synaptogenesis and synaptic plasticity *via* providing and receiving ions and cytokines. These astrocytes' functions suggest their involvement in the pathophysiology of ASDs, which is also referred to as a synaptopathy.[24,25] Indeed, astrocytes are persistently activated in the cerebellum of a mouse model of Fragile X syndrome.[26] However, a report shows that morphology and number of astrocytes do not change in postmortem brains of ASD patients.[27] Furthermore, few studies have investigated astrocyte-related genes and molecules in ASD animal models. Thus, further investigation is necessary to clarify the role of astrocytes in the pathogenesis of ASDs.

2. Therapeutic targets for ASD treatment and possible involvement of microglia

To date, two drugs, risperidone and aripiprazole, are approved by the US FDA to be used for the treatment of ASD symptoms.[28] Some reports have suggested that these drugs modulate microglial functions. For example, risperidone inhibits the production of inflammatory cytokines in microglia *in vitro,* while aripiprazole reduces microglial activation in an MIA model mouse.[29,30] In the serum and cerebrospinal fluid (CSF) of ASD patients, inflammatory cytokines such as IL-1β and IL-6 are expressed at higher levels, which implies immunological dysfunction in ASDs.[31] This chapter discusses the use of immunomodulatory agents as a potential therapeutic strategy for ASDs (listed in Table 1). We also introduce how these agents possibly modify microglial functions. In the latter part, we discuss cell therapy and its future prospects.

2.1 Drug therapy

2.1.1 Luteolin

Luteolin is a naturally occurring flavonoid. Luteolin decreases the serum level of tumor necrosis factor (TNF) and IL-6 and has been shown to improve the social interaction of ASD patients.[32,33] In an MIA model mouse, luteolin also ameliorates elevated anxiety and impaired sociability.[34] Additionally, BV-2 microglial activation and subsequent neurotoxicity induced by lipopolysaccharide (LPS) are inhibited by luteolin application *in vitro*.[59]

Table 1 Drug therapies for ASDs.

Drug	Subject species	Administration route Dosage	Finding	Reference
Luteolin	Human ASD patients	Orally 100 mg	Decreased serum levels of IL-6 and TNF Improved communication and social skills	32
	Human ASD patients	Orally 200 mg/10 kg	Improved GI and allergy symptoms Increased social interaction	33
	IL-6-induced ASD model mice	Orally 10 mg/kg/day	Decreased TNF-α and IL-1β in brains Decreased repetitive behavior Improved social interaction	34
	BV-2 microglia *in vitro*	~50 μM	Inhibited proinflammatory pathway Enhanced anti-inflammatory pathway	35
Minocycline	Human ASD patients	50 mg twice/day	Decreased hyperactivity and Irritability	36
	VPA-induced ASD model rats	Orally 25 and 50 mg/kg	Decreased BBB permeability Improved social interaction Decreased repetitive behavior	37
	Microglia prepared from Poly(I:C)-induced ASD model mice *in vitro*	Orally 3 mg/kg/day	Increased phagocytic capacity	38
Suramin	Human ASD patients	Intravenously 20 mg/kg	Improved social interaction Decreased repetitive behavior	39
	Poly(I:C)-induced ASD model mice	Intraperitoneally 20 mg/kg	Improved social interaction Improved hippocampus-dependent memory	40
Vitamin D	Human ASD patients	Orally 300 IU/kg/day	Improved social interaction Decreased irritability	41
	Mouse microglia *in vitro*	10 nM	Decreased proinflammatory cytokines Increased anti-inflammatory cytokines	42

Continued

Table 1 Drug therapies for ASDs.—cont'd

Drug	Subject species	Administration route Dosage	Finding	Reference
Gut microbiota	Poly(I:C)-induced ASD model mice	Orally ND	Corrected GI deficits Decreased repetitive and anxiety behavior Improved communicative behavior	43
Immunoglobulin	Human ASD patients	Intravenously 0.4 g/kg	Reduced hyperactivity and irritability Improved communication ability	44
	Mouse microglia *in vitro*	20 μM	Increased Aβ clearance by microglia	45
Pioglitazone	Human ASD patients	Orally 30 mg/day	Decreased hyperactivity and irritability	46
	Human ASD patients	Orally 30 or 60 mg/day	Decreased hyperactivity, irritability and stereotypy	47
Pioglitazone	APP/PS1	Orally ~150 mg/kg/day	Increased Aβ phagocytosis by microglia	48
	HAIP microglia	~20 μM	Inhibited proinflammatory cytokine expression induced by LPS	49
Sulforaphane	Human ASD patients	Orally 50–150 μmol	Reduced hyperactivity, irritability and repetitive behavior Improved communication ability	50
	BV2 microglia Primary mouse microglia	2.5 μM	Reduced proinflammatory cytokine expression induced by LPS	51
	N9 microglia	~10 μM	Reduced proinflammatory cytokine expression and ROS production induced by LPS	52

Celecoxib	Human ASD patients	Orally 100–300 mg/day	Improved social interaction Decreased irritability and repetitive behavior	53
	Rats	Intraperitoneally 20 mg/kg	Decreased the number of activated microglia in substantia nigra and striatum	54
Lenalidomide	Human ASD patients	Orally 2.5 mg/day	Decreased serum levels of TNF-α Improved social interaction Decreased irritability and repetitive behavior	55
	Parkinson's disease model mice BV-2 microglia	NS 100 mg/kg 250 μg/mL	Decreased microgliosis in striatum Reduced proinflammatory cytokine expression induced by α-synuclein	56
Spironolactone	Human ASD patients	Orally 2 mg/kg/day	Decreased irritability, hyperactivity and repetitive behavior	57
	Rats	Intrathecally ND	Suppressed microglial secretion of inflammatory cytokines induced by chronic compression of the dorsal root ganglion	58

ND, not determined; *NS*, not shown.

2.1.2 Minocycline

Minocycline is a tetracycline analog and is widely used to suppress microglial activation, although its mechanism of action remains to be revealed. Minocycline has been suggested as an effective adjuvant of risperidone because patients cotreated with minocycline and risperidone show greater improvements in hyperactivity and irritability than patients treated with risperidone alone.[60] In a valproic acid (VPA)-induced rat model of ASDs, which mimics a drug-induced ASD, minocycline ameliorates ASD-like behaviors and disruption of the blood-brain-barrier (BBB).[37] Loss of BBB integrity is one of the pathological features of ASDs, and it is shown that BBB disruption causes microglial activation *in vivo*.[61,62] Minocycline has also been shown to rescue the decreased phagocytic ability of microglia prepared from MIA model mice *in vitro*.[38] Thus, minocycline possibly normalizes synaptic density in the ASD brain *via* inducing engulfment of surplus synapses by microglia.

2.1.3 Suramin

Suramin is a polyanionic compound with an unknown mechanism of action. Intravenous administration of suramin improves social interaction and repetitive behavior in ASD patients.[39] MIA model mice also show improved social interaction and memory after suramin treatment.[40] Suramin inhibits microglial activation in the spinal cord *in vivo*.[63] Additionally, suramin inhibits P2X7-dependent secretion of IL-6, CCL2 and TNF-α from mouse primary microglia.[64] A recent report has shown that P2X7 drives ASD behaviors in MIA model mice.[35] Thus, P2X7 can be a possible therapeutic target for ASDs.

2.1.4 Vitamin D

In ASD patients, the serum levels of vitamin D and its receptor are reduced.[36] Vitamin D supplementation ameliorates irritability and impaired social interaction in ASD patients.[41] Vitamin D decreases the expression of inflammatory cytokines such as IL-6 and TNF-α, while it increases the expression of antiinflammatory cytokine IL-10 from mouse primary microglia stimulated by LPS or IFN-γ.[42]

2.1.5 Gut microbiota

Because many ASD patients suffer from gastrointestinal disturbances, the gut microbiota has been considered.[65] Microbiota transfer therapy improves irritability, communication skills and sociability in ASD patients.[43] The gut microbiota regulates microglial development and homeostasis.[66] For example, microglia in mice without gut microbiota were hyper-ramified and expressed higher levels of maturation and activation marker, while their immune response to LPS infection was diminished.[66] In a mouse model of MIA-induced ASDs, the composition of microbiota is disrupted, and tight junctions of intestinal mucosa become fragile.[67] Transplantation of human commensal *Bacteroides fragilis* into a mouse model of MIA-induced ASDs improves repetitive and anxiety behaviors.[67]

2.1.6 Immunoglobulin

Intravenous administration of immunoglobulin reduces hyperactivity and irritability and elevates communication ability in ASD patients.[44] Furthermore, IgG has been shown to promote Aβ phagocytosis by murine microglia *in vitro*.[45] Thus, it is possible that IgG induces engulfment of increased synapses by microglia in the ASD brain.

2.1.7 Pioglitazone

Pioglitazone is a thiazolidinedione with antidiabetic properties and potential antineoplastic activity. Oral administration of pioglitazone reduces hyperactivity, irritability and repetitive behaviors in ASD patients.[46,47] Pioglitazone enhances Aβ phagocytosis by microglia in Alzheimer's disease model mice, while it suppresses secretion of IL-6, TNF-α, and iNOS from microglia stimulated by LPS *in vitro*.[48,49] Therefore, pioglitazone could promote engulfment of excess synapses in the ASD brain by microglia without inflammation and neurodegeneration.

2.1.8 Sulforaphane

Sulforaphane is the most characterized isothiocyanate. Oral administration of sulforaphane reduces hyperactivity, irritability and repetitive behaviors in ASD patients in addition to improving their communication ability.[50] Sulforaphane inhibits LPS-induced microglial activation processes, such as proinflammatory cytokines (IL-1β, IL-6 and TNF-α) expression and reactive oxygen species (iNOS) release, in murine microglial cultures.[51,52]

2.1.9 Celecoxib

Celecoxib is a pyrazole derivative and a selective cyclooxygenase 2 inhibitor. Celecoxib, in combination with risperidone, ameliorates irritability, repetitive behavior and impaired social interaction in ASD patients.[53] The LPS-induced increase in the number of activated microglia is inhibited by celecoxib in neonatal rats.[54]

2.1.10 Lenalidomide

Lenalidomide is a phthalimide and piperidone derivative that has immunomodulatory and antiangiogenic properties. Lenalidomide reduces the serum levels of TNF-α in ASD patients.[55] In addition, ASD patients have shown improved irritability, repetitive behavior and social interaction after lenalidomide administration.[55] Furthermore, lenalidomide inhibits the expression of inflammatory cytokines such as IL-1β, IL-6 and TNF-α, while it increases the expression of antiinflammatory cytokine IL-10 in Parkinson's disease model mice and α-synuclein-stimulated activation of microglia *in vitro*.[56]

2.1.11 Spironolactone

Spironolactone is a potassium-sparing diuretic that acts by antagonism of aldosterone in the distal renal tubules. Oral administration of spironolactone reduced irritability, hyperactivity and repetitive behavior in ASD patients.[57] Spironolactone also suppresses microglial secretion of inflammatory cytokines such as IL-1β and TNF-α induced by chronic compression of the dorsal root ganglion in rats.[58]

2.2 Cell-transplantation-based therapy

Cell-transplantation-based therapies may provide us with some clues regarding the future application of microglial transplantation as a therapeutic strategy for ASDs. Injection of bone marrow-derived mononuclear cell (BMMNC) intrathecally or subcuntaneously[68,69] and embryonic stem (ES) cells intravenously[70,71] to ASD patients rescues ASD behaviors, including hyperactivity, repetitive behavior and impaired social communication. Additionally, intravenous injection of umbilical cord blood cells improved social communication in ASD patients.[72] Transplantation of cord blood mononuclear cells to veins in combination with umbilical cord-derived mesenchymal stem cells reduces repetitive behavior and improved sociability.[73]

It has also been suggested that BMMNC injection increases glucose metabolism at the prefrontal cortex and the hippocampus in ASD patients.[68,69] Glucose metabolism regulates activity-dependent synaptic vesicle transmission, suggesting the possible contribution of BMMNCs to the normalization of activity-dependent signal transmission.[74] This idea is supported by the transplantation of bone marrow mesenchymal stem cells into the cerebellum of a mouse model of Niemann-Pick disease: Purkinje cells increase spontaneous activity, promoting the formation of functional neural circuits after transplantation.[75] Recently, it has been reported that microglia-like cells can be generated from iPSCs, ES, and blood-derived monocytes.[76–79]

These microglia-like cells can be engrafted into the brain parenchyma, and they exhibit phagocytic properties.[76] Thus, it might be possible that microglia-like cells eliminate surplus synapses in the ASD brain. Moreover, microglia-like cells are suggested to be useful for drug screening.[78,79]

3. Synaptic deficits in ASDs

As mentioned above, immunomodulatory agents that suppress immune responses, such as luteolin and sulforaphane, are expected to be options for the treatment of ASDs; however, these strategies are symptomatic therapies. Furthermore, patients are usually treated with these agents chronically, which could cause multiple side effects. Another issue in ASD treatment is that it is difficult to determine the criteria for ASD severity because ASDs are spectrum disorders. In general, medication is prescribed if patients exhibit symptoms interfering with daily life, such as hyperactivity, aggression and self-injurious behavior. However, patients without those problems but with ASD-related problems, including deficits in social communication and their family, would also wish treatment for higher quality of life. These limits of medication led us to discover another therapeutic strategy.

In developing brains where synaptic pruning is robustly carried out, neural-activity-dependent signal transmissions are promoted, while they are suppressed after the critical period of developmental plasticity.[80,81] Deficits in developmental synapse elimination have been suggested to cause ASDs, but the mechanisms remain elusive. ASD patients usually suffer impaired processing of higher-order cognitive functions, but many reports imply abnormalities in primary sensory processing.[82] In sensory cortexes, including the visual and auditory cortices, neural activity induced by sensory stimuli causes neural circuits to form.[83] This circuit formation is regulated by modulation of synaptic maturation and function. Thus, it is possible that activity-dependent synaptic maturation and further circuit formation are not normally conducted in ASD brains.[84] Previous reports have demonstrated that microglia likely phagocytose less active synapses, indicating that microglia are able to decipher the strength of neuronal activity.[85] These findings have motivated us to examine whether microglia-dependent reorganization of synaptic connections could be a therapeutic strategy for ASDs; we hypothesize that microglia are unable to select synapses to phagocytose due to weak contrast between activities of each synapse in ASD brains.

Most animal models of ASDs are based on mutations of synapse-related genes, especially those that impair activity-dependent signal transmission (listed in Table 2).[95] For example, knockout of a member of the neurexin

Table 2 Animal models for ASDs.

Gene	Associated disease Function	Mutation Animal species	Behavioral abnormality	Neuroanatomical abnormality	Reference
NRX1	ASDs, schizophrenia, epilepsy, Pitt-Hopkins syndrome Presynaptic adhesion molecule	Nrxn1α knockout C57BL6/SV129 mice	Decreased social interaction	NA	86
SHANK	ASDs Postsynaptic scaffolding protein	Shank3 knockout C57 mice	Decreased social interaction, increased repetitive behavior	Decreased cortico-striatal synaptic transmission, striatal hypertrophy	87
		Shank2 knockout C57BL/6J mice	Decreased social interaction, impaired communication, enhanced anxiety	Impaired NMDAR function	88
FMR1	Fragile X syndrome mRNA trafficking, dendrite maturation, synaptic plasticity	Fmr1 knockout C57BL/6J mice	NA	Increased spines in hippocampal CA1, decreased microglia-dependent synaptic engulfment	15
		Fmr1 knockout C57BL/6 mice	NA	Dysregulation of glutamatergic signaling maturation	89
		Fmr1 knockout C57BL/6 mice	Impaired social interaction, increased repetitive behavior	Enhanced LTD and increased spines in hippocampal CA1	90

MECP2	Rett syndrome Transcriptional regulation	MeCP2 S421A knock-in C57BL/6 mice	Impaired social interaction	Increased dendritic complexity and increased mIPSC amplitude in visual cortex	91
PTEN	ASDs Inhibition of mTOR activation	Nse-cre; $Pten^{loxP/loxP}$ C57/BL6 mice	Decreased social interaction, decreased pre-pulse inhibition, decreased anxiety	Hippocampus hypertrophy, increased spines and ectopic dendrites in dentate gyrus	92
TSC1/2	Tuberous sclerosis Inhibition of mTOR activation	TSC1/2 hetero knockout C57BL/6J mice	Impaired context discrimination	Decreased LTD in hippocampal CA1	93
UBE3A	ASDs, Angelman syndrome	Ube3A knockout	NA	Increased Arc expression and decreased number of AMPAR at synapses in hippocampus	94

NA, not analyzed.

family, NRX1α, leads to decreased social interaction,[86] which suggests the possible involvement of deficits in neural activity and synapse formation in ASDs because the structural binding of neurexin and neuroligin, which are located at the presynaptic and postsynaptic membranes, respectively, promotes synaptic maturation associated with neural activity.[96,97] SHANK2, a scaffold protein found in the postsynaptic densities of excitatory synapses, is crucial for normal long-term potentiation (LTP) and long-term depression (LTD).[87] SHANK2 knockout mice exhibit impaired social interaction, but the abnormal behavior is improved by restoring synaptic plasticity *via* NMDA receptor stimulation. Mutations in another SHANK-family protein, SHANK3, weaken excitatory neuronal transmission in corticostriatal synapses, dysfunctions of which are implicated in ASDs.[88] Risk genes for symptomatic ASDs, including fragile X syndrome and Rett syndrome, are also related to activity-dependent synaptic modulation. In fmr1 knockout mice, a model of fragile X syndrome, activity-dependent synaptic maturation and pruning are impaired.[89,90] Recently, fmr1 knockout has been revealed to reduce microglia-dependent synaptic pruning,[15] indicating the possible involvement of microglia in ASD-related symptoms. The Rett syndrome risk gene MECP2 is necessary for synaptic maturation.[91] Knockout of PTEN, an ASD risk factor gene, reduced synaptic pruning in the hippocampus.[92] Mutations of the tuberous sclerosis risk gene TSC1/2, which encodes a downstream protein of PTEN, hinder activity-dependent gene translation and synaptic plasticity.[93] In an Angelman model, mice with mutations in UBE3A, which is also an ASD risk gene, exhibit reduced activity-dependent gene expression and excitatory neural transmission.[94]

4. Physical exercise, possibly a potent therapeutic strategy for ASDs

Mild therapeutic strategies for ASDs, as opposed to drug or cell-transplantation therapies, are ideal. Recently, exercise has attracted attention as a potential therapeutic strategy for abnormal behaviors in neurodevelopmental disorders, including ASDs, but the cellular and molecular mechanisms by which exercise ameliorates the symptoms remain unrevealed.[98–100] In the following paragraphs, we discuss the possibility that physical exercise is a viable treatment to induce synaptic pruning by microglia, resulting in improved ASD-related behaviors.

Previous studies have reported that exercise may help improve cognitive performance in humans and rodents, probably through enhancing the production of neurotrophic factors and increasing hippocampal neurogenesis.[101–104] The hippocampus and prefrontal cortex regulate cognition and social behavior, deficits of which are the main symptoms of ASD, highlighting these regions as possible target brain regions for ASD treatment.[105] Consistent with this possibility, exercise increases neural activity in multiple regions including the hippocampus and prefrontal cortex in humans and rodents.[106,107] These regions show increased expression of brain-derived neurotrophic factor (BDNF) and insulin-like growth factor-1 (IGF-1) in response to exercise.[108,109] BDNF is secreted depending on neural activity, promoting synaptic maturation.[110] IGF-1 also stabilizes synapses by increasing the expression of synapse-related molecules such as Synapsin I and PSD95.[111,112] These molecules may accelerate activity-dependent synaptic maturation after exercise.

Another major effect of exercise is the enhancement of adult neurogenesis in the hippocampus.[104] Adult-born hippocampal granule cells exhibit higher excitability than preexisting mature granule cells, which may generate synaptic competition between newborn and existing cells.[113] Thus, it is possible that microglia detect differences in neural activity therein. It was recently revealed that in MIA-induced ASD model mice, hippocampal neurogenesis is inhibited during both the immediate postnatal period and adulthood, leading to the impaired maturation of newborn granule cells.[114] Thus, inadequate neurogenesis may underlie ASD development. The activity of neural precursor cells is also enhanced by exercise, and this process is modulated by the CX3CL1-CX3CR1 axis, one of the important signaling pathways regulating microglia-neuron interaction, suggesting an active role of microglia in the exercise-induced activity of neural precursors.[115]

In order to generate contrast in synaptic activity, which is necessary for microglia-dependent synaptic pruning, it will be also useful to focus on inhibitory neurotransmission, as GABAergic signaling has been shown to regulate synaptic pruning in the cerebellum.[116] In addition, impaired inhibitory neurotransmission has been reported in some animal models of ASDs.[117–119] Furthermore, ASD patients express reduced levels of glutamic acid decarboxylase (GAD) and GABA receptor subunits, which implies a causal relationship between impaired inhibitory neural transmission and ASDs.[120–122] Exercise increases the numbers of interneurons and inhibitory synapses, enhancing the secretion of GABA.[123,124] Therefore, exercise is expected to induce synaptic competition by modulating both excitatory and inhibitory neural transmission.

Neural activity promotes the dynamics of microglial processes *in vivo*.[125] Therefore, elevation of neural activity by exercise might induce microglial contact with synapses, resulting in the elimination of less active synapses by microglia. It has been reported that exercise decreases the expression of MHC-II and CD68 in microglia, both of which are regarded as markers of activated microglia.[126,127] These results indicate that exercise induces synaptic pruning without overactivating microglia. We are now testing the hypothesis that exercise induces microglia-dependent synaptic pruning through synaptic competition. Our study will link exercise to microglia-dependent changes in the status of the brain.

5. Concluding remarks

We discussed the involvement of microglia in ASDs and the possibility of microglia as the therapeutic targets for ASDs. As introduced in the first section, several researches have suggested that microglia are likely involved in the pathogenesis and development of ASDs. In the second section, we introduced immunomodulatory agents that could attenuate ASD-related behaviors probably *via* regulation of microglial functions. Some of them are under clinical trial, but the side effects should be carefully addressed before the actual clinical use. We also mentioned the cell-transplantation-based therapy, but the method itself is invasive and could result in undesired activation of innate microglia. In the third and fourth sections, we further focused on synaptic remodeling by microglia in the ASD brain. Specifically, we mentioned the possibility that exercise cures ASD behaviors *via* inducing synaptic pruning by microglia. Though it should be noted that some ASD patients have difficulties in exercise or may feel stress by compulsive exercise, exercise, if it is effective, is useful in that it can be basically a noninvasive treatment and the exercise intensity can be changed according to the severity of ASD symptoms. A few clinical studies have shown that various types of exercise actually ameliorate ASD-related behaviors. We are currently examining the cellular and molecular mechanisms underlying exercise-induced modulation of ASD behaviors.

Acknowledgment

This work was supported in part by a Grant-in-Aid for Scientific Research (B) (17H03988 to R.K.) from JSPS, by JST PRESTO (JPMJPR18H4 to R.K.), and by JST ERATO (JPMJER1801 to Y.I.).

References

1. Huttenlocher PR. Morphometric study of human cerebral cortex development. *Neuropsychologia*. 1990;28(6):517–527.
2. Paolicelli RC, Bolasco G, Pagani F, et al. Synaptic pruning by microglia is necessary for normal brain development. *Science*. 2011;333(6048):1456–1458.
3. Penzes P, Cahill ME, Jones KA, VanLeeuwen JE, Woolfrey KM. Dendritic spine pathology in neuropsychiatric disorders. *Nat Neurosci*. 2011;14(3):285–293.
4. Stevens B, Allen NJ, Vazquez LE, et al. The classical complement cascade mediates CNS synapse elimination. *Cell*. 2007;131(6):1164–1178.
5. Koyama R, Ikegaya Y. Microglia in the pathogenesis of autism spectrum disorders. *Neurosci Res*. 2015;100:1–5.
6. Morgan JT, Chana G, Pardo CA, et al. Microglial activation and increased microglial density observed in the dorsolateral prefrontal cortex in autism. *Biol Psychiatry*. 2010;68(4):368–376.
7. Tetreault NA, Hakeem AY, Jiang S, et al. Microglia in the cerebral cortex in autism. *J Autism Dev Disord*. 2012;42(12):2569–2584.
8. Suzuki K, Sugihara G, Ouchi Y, et al. Microglial activation in young adults with autism spectrum disorder. *JAMA Psychiatry*. 2013;70(1):49–58.
9. Morgan JT, Chana G, Abramson I, Semendeferi K, Courchesne E, Everall IP. Abnormal microglial-neuronal spatial organization in the dorsolateral prefrontal cortex in autism. *Brain Res*. 2012;1456:72–81.
10. Corbett BA, Kantor AB, Schulman H, et al. A proteomic study of serum from children with autism showing differential expression of apolipoproteins and complement proteins. *Mol Psychiatry*. 2007;12(3):292–306.
11. Singh V. Elevation of serum C-reactive protein and S100 proteins for systemic inflammation in autistic children. *J Spec Educ Rehab*. 2005;6:117–125.
12. Zhu F, Zheng Y, Liu Y, Zhang X, Zhao J. Minocycline alleviates behavioral deficits and inhibits microglial activation in the offspring of pregnant mice after administration of polyriboinosinic-polyribocytidilic acid. *Psychiatry Res*. 2014;219(3):680–686.
13. Hui CW, St-Pierre A, El Hajj H, et al. Prenatal immune challenge in mice leads to partly sex-dependent behavioral, microglial, and molecular abnormalities associated with schizophrenia. *Front Mol Neurosci*. 2018;11:13.
14. Fernández de Cossío L, Guzmán A, van der Veldt S, Luheshi GN. Prenatal infection leads to ASD-like behavior and altered synaptic pruning in the mouse offspring. *Brain Behav Immun*. 2017;63:88–98.
15. Jawaid S, Kidd GJ, Wang J, Swetlik C, Dutta R, Trapp BD. Alterations in CA1 hippocampal synapses in a mouse model of fragile X syndrome. *Glia*. 2018;66(4):789–800.
16. Heo Y, Zhang Y, Gao D, Miller VM, Lawrence DA. Aberrant immune responses in a mouse with behavioral disorders. *PLoS One*. 2011;6(7): e20912.
17. Han S, Tai C, Jones CJ, Scheuer T, Catterall WA. Enhancement of inhibitory neurotransmission by GABAA receptors having α2,3-subunits ameliorates behavioral deficits in amouse model of autism. *Neuron*. 2014;81(6):1282–1289.
18. Tang G, Gudsnuk K, Kuo SH, et al. Loss of mTOR-dependent macroautophagy causes autistic-like synaptic pruning deficits. *Neuron*. 2014;83(5):1131–1143.
19. Kim HJ, Cho MH, Shim WH, et al. Deficient autophagy in microglia impairs synaptic pruning and causes social behavioral defects. *Mol Psychiatry*. 2017;22(11):1576–1584.
20. Zhan Y, Paolicelli RC, Sforazzini F, et al. Deficient neuron-microglia signaling results in impaired functional brain connectivity and social behavior. *Nat Neurosci*. 2014;17(3):400–406.
21. Filipello F, Morini R, Corradini I, et al. The microglial innate immune receptor TREM2 is required for synapse elimination and normal brain connectivity. *Immunity*. 2018;48(5):979–991.

22. Werling DM, Parikshak NN, Geschwind DH. Gene expression in human brain implicates sexually dimorphic pathways in autism spectrum disorders. *Nat Commun.* 2016;7:10717.
23. Weinhard L, Neniskyte U, Vadisiute A, et al. Sexual dimorphism of microglia and synapses during mouse postnatal development. *Dev Neurobiol.* 2018;78(6):618–626.
24. Petrelli F, Pucci L, Bezzi P. Astrocytes and microglia and their potential link with autism spectrum disorders. *Front Cell Neurosci.* 2016;10:21.
25. Petrelli F, Bezzi P. mGlu5-mediated signalling in developing astrocyte and the pathogenesis of autism spectrum disorders. *Curr Opin Neurobiol.* 2018;48:139–145.
26. Pacey LK, Guan S, Tharmalingam S, Thomsen C, Hampson DR. Persistent astrocyte activation in the fragile X mouse cerebellum. *Brain Behav.* 2015;5(10): e00400.
27. Lee TT, Skafidas E, Dottori M, et al. No preliminary evidence of differences in astrocyte density within the white matter of the dorsolateral prefrontal cortex in autism. *Mol Autism.* 2017;8:64.
28. Marchezan J, Winkler Dos Santos EGA, Deckmann I, Riesgo RDS. Immunological dysfunction in autism spectrum disorder: a potential target for therapy. *Neuroimmunomodulation.* 2018;25(5–6):300–319.
29. Kato T, Monji A, Hashioka S, Kanba S. Risperidone significantly inhibits interferon-gamma-induced microglial activation in vitro. *Schizophr Res.* 2007;92(1–3):108–115.
30. Sato-Kasai M, Kato TA, Ohgidani M, et al. Aripiprazole inhibits polyI:C-induced microglial activation possibly via TRPM7. *Schizophr Res.* 2016;178(1–3):35–43.
31. Masi A, Quintana DS, Glozier N, Lloyd AR, Hickie IB, Guastella AJ. Cytokine aberrations in autism spectrum disorder: a systematic review and meta-analysis. *Mol Psychiatry.* 2015;20(4):440–446.
32. Tsilioni I, Taliou A, Francis K, Theoharides TC. Children with autism spectrum disorders, who improved with a luteolin-containing dietary formulation, show reduced serum levels of TNF and IL-6. *Transl Psychiatry.* 2015;5: e647.
33. Theoharides TC, Asadi S, Panagiotidou S. A case series of a luteolin formulation (NeuroProtek®) in children with autism spectrum disorders. *Int J Immunopathol Pharmacol.* 2012;25(2):317–323.
34. Parker-Athill E, Luo D, Bailey A, et al. Flavonoids, a prenatal prophylaxis via targeting JAK2/STAT3 signaling to oppose IL-6/MIA associated autism. *J Neuroimmunol.* 2009;217(1–2):20–27.
35. Horváth G, Otrokocsi L, Beko K, et al. P2X7 receptors drive poly(I:C) induced autism-like behavior in mice. *J Neurosci.* 2019;39(13):2542–2561.
36. Altun H, Kurutaş EB, Şahin N, Güngör O, Fındıklı E. The levels of vitamin D, vitamin D receptor, homocysteine and complex B vitamin in children with autism Spectrum disorders. *Clin Psychopharmacol Neurosci.* 2018;16(4):383–390.
37. Kumar H, Sharma B. Minocycline ameliorates prenatal valproic acid induced autistic behaviour, biochemistry and blood brain barrier impairments in rats. *Brain Res.* 1630;2016:83–97.
38. Mattei D, Ivanov A, Ferrai C, et al. Maternal immune activation results in complex microglial transcriptome signature in the adult offspring that is reversed by minocycline treatment. *Transl Psychiatry.* 2017;7(5): e1120.
39. Naviaux RK, Curtis B, Li K, et al. Low-dose suramin in autism spectrum disorder: a small, phase I/II, randomized clinical trial. *Ann Clin Transl Neurol.* 2017;4(7):491–505.
40. Naviaux JC, Schuchbauer MA, Li K, et al. Reversal of autism-like behaviors and metabolism in adult mice with single-dose antipurinergic therapy. *Transl Psychiatry.* 2014;4: e400.
41. Saad K, Abdel-Rahman AA, Elserogy YM, et al. Randomized controlled trial of vitamin D supplementation in children with autism spectrum disorder. *J Child Psychol Psychiatry.* 2018;59(1):20–29.

42. Boontanrart M, Hall SD, Spanier JA, Hayes CE, Olson JK. Vitamin D3 alters microglia immune activation by an IL-10 dependent SOCS3 mechanism. *J Neuroimmunol*. 2016;292:126–136.
43. Kang DW, Adams JB, Gregory AC, et al. Microbiota transfer therapy alters gut ecosystem and improves gastrointestinal and autism symptoms: an open-label study. *Microbiome*. 2017;5(1):10.
44. Niederhofer H, Staffen W, Mair A. Immunoglobulins as an alternative strategy of psychopharmacological treatment of children with autistic disorder. *Neuropsychopharmacology*. 2003;28(5):1014–1015.
45. Magga J, Puli L, Pihlaja R, et al. Human intravenous immunoglobulin provides protection against Aβ toxicity by multiple mechanisms in a mouse model of Alzheimer's disease. *J Neuroinflammation*. 2010;7:90.
46. Ghaleiha A, Rasa SM, Nikoo M, Farokhnia M, Mohammadi MR, Akhondzadeh S. A pilot double-blind placebo-controlled trial of pioglitazone as adjunctive treatment to risperidone: effects on aberrant behavior in children with autism. *Psychiatry Res*. 2015;229(1–2):181–187.
47. Boris M, Kaiser CC, Goldblatt A, et al. Effect of pioglitazone treatment on behavioral symptoms in autistic children. *J Neuroinflammation*. 2007;4:3.
48. Yamanaka M, Ishikawa T, Griep A, Axt D, Kummer MP, Heneka MT. PPARγ/RXRα-induced and CD36-mediated microglial amyloid-β phagocytosis results in cognitive improvement in amyloid precursor protein/presenilin 1 mice. *J Neurosci*. 2012;32(48):17321–17331.
49. Ji H, Wang H, Zhang F, Li X, Xiang L, Aiguo S. PPARγ agonist pioglitazone inhibits microglia inflammation by blocking p38 mitogen-activated protein kinase signaling pathways. *Inflamm Res*. 2010;59(11):921–929.
50. Singh K, Connors SL, Macklin EA, et al. Sulforaphane treatment of autism spectrum disorder (ASD). *Proc Natl Acad Sci USA*. 2014;111(43):15550–15555.
51. Townsend BE, Johnson RW. Sulforaphane induces Nrf2 target genes and attenuates inflammatory gene expression in microglia from brain of young adult and aged mice. *Exp Gerontol*. 2016;73:42–48.
52. Eren E, Tufekci KU, Isci KB, Tastan B, Genc K, Genc S. Sulforaphane inhibits lipopolysaccharide-induced inflammation, cytotoxicity, oxidative stress, and miR-155 expression and switches to Mox phenotype through activating extracellular signal-regulated kinase 1/2-nuclear factor erythroid 2-related factor 2/antioxidant response element pathway in murine microglial cells. *Front Immunol*. 2018;9:36.
53. Asadabadi M, Mohammadi MR, Ghanizadeh A, et al. Celecoxib as adjunctive treatment to risperidone in children with autistic disorder: a randomized, double-blind, placebo-controlled trial. *Psychopharmacology (Berl)*. 2013;225(1):51–59.
54. Kaizaki A, Tien LT, Pang Y, et al. Celecoxib reduces brain dopaminergic neuronaldysfunction, and improves sensorimotor behavioral performance in neonatal rats exposed to systemic lipopolysaccharide. *J Neuroinflammation*. 2013;10:45.
55. Chez M, Low R, Parise C, Donnel T. Safety and observations in a pilot study of lenalidomide for treatment in autism. *Autism Res Treat*. 2012;2012:291601.
56. Valera E, Mante M, Anderson S, Rockenstein E, Masliah E. Lenalidomide reduces microglial activation and behavioral deficits in a transgenic model of Parkinson's disease. *J Neuroinflammation*. 2015;12:93.
57. Bradstreet JJ, Smith S, Granpeesheh D, El-Dahr JM, Rossignol D. Spironolactone might be a desirable immunologic and hormonal intervention in autism spectrum disorders. *Med Hypotheses*. 2007;68(5):979–987.
58. Sun YE, Peng L, Sun X, et al. Intrathecal injection of spironolactone attenuates radicular pain by inhibition of spinal microglia activation in a rat model. *PLoS One*. 2012;7(6): e39897.

59. Dirscherl K, Karlstetter M, Ebert S, et al. Luteolin triggers global changes in the microglial transcriptome leading to a unique anti-inflammatory and neuroprotective phenotype. *J Neuroinflammation*. 2010;7:3.
60. Ghaleiha A, Alikhani R, Kazemi MR, et al. Minocycline as adjunctive treatment to risperidone in children with autistic disorder: a randomized, double-blind placebo-controlled trial. *J Child Adolesc Psychopharmacol*. 2016;26(9):784–791.
61. Kealy J, Greene C, Campbell M. Blood-brain barrier regulation in psychiatric disorders. *Neurosci Lett*. 2018. https://doi.org/10.1016/j.neulet.2018.06.033.
62. Ju F, Ran Y, Zhu L, et al. Increased BBB permeability enhances activation of microglia and exacerbates loss of dendritic spines after transient global cerebral ischemia. *Front Cell Neurosci*. 2018;12:236.
63. Wu Y, Willcockson HH, Maixner W, Light AR. Suramin inhibits spinal cord microglia activation and long-term hyperalgesia induced by formalin injection. *J Pain*. 2004;5(1):48–55.
64. Shieh CH, Heinrich A, Serchov T, van Calker D, Biber K. P2X7-dependent, but differentially regulated release of IL-6, CCL2, and TNF-α in cultured mouse microglia. *Glia*. 2014;62(4):592–607.
65. Li Q, Han Y, Dy ABC, Hagerman RJ. The gut microbiota and autism Spectrum disorders. *Front Cell Neurosci*. 2017;11:120.
66. Erny D, Hrabě de Angelis AL, Jaitin D, et al. Host microbiota constantly control maturation and function of microglia in the CNS. *Nat Neurosci*. 2015;18(7):965–977.
67. Hsiao EY, McBride SW, Hsien S, et al. Microbiota modulate behavioral and physiological abnormalities associated with neurodevelopmental disorders. *Cell*. 2013;155(7):1451–1463.
68. Sharma A, Badhe P, Gokulchandran N, et al. An improved case of autism as revealed by PET CT scan in patient transplanted with autologous bone marrow derived mononuclear cells. *J Stem Cell Res Ther*. 2013;3(2):1000139.
69. Sharma A, Gokulchandran N, Sane H, et al. Autologous bone marrow mononuclear cell therapy for autism: an open label proof of concept study. *Stem Cells Int*. 2013;2013: 623875.
70. Bradstreet JJ, Sych N, Antonucci N, et al. Efficacy of fetal stem cell transplantation in autism spectrum disorders: an open-labeled pilot study. *Cell Transplant*. 2014;23(suppl 1):S105–S112.
71. Shroff G. Human embryonic stem cells in the treatment of autism: a case series. *Innov Clin Neurosci*. 2017;14(3–4):12–16.
72. Dawson G, Sun JM, Davlantis KS, et al. Autologous cord blood infusions are safe and feasible in young children with autism spectrum disorder: results of a single-center phase I open-label trial. *Stem Cells Transl Med*. 2017;6(5):1332–1339.
73. Lv YT, Zhang Y, Liu M, et al. Transplantation of human cord blood mononuclear cells and umbilical cord-derived mesenchymal stem cells in autism. *J Transl Med*. 2013;11:196.
74. Ashrafi G, Ryan TA. Glucose metabolism in nerve terminals. *Curr Opin Neurobiol*. 2017;45:156–161.
75. Bae JS, Han HS, Youn DH, et al. Bone marrow-derived mesenchymal stem cells promote neuronal networks with functional synaptic transmission after transplantation into mice with neurodegeneration. *Stem Cells*. 2007;25(5):1307–1316.
76. Abud EM, Ramirez RN, Martinez ES, et al. iPSC-derived human microglia-like cells to study neurological diseases. *Neuron*. 2017;94(2)278–93.e9.
77. Tsuchiya T, Park KC, Toyonaga S, et al. Characterization of microglia induced from mouse embryonic stem cells and their migration into the brain parenchyma. *J Neuroimmunol*. 2005;160(1–2):210–218.
78. Ohgidani M, Kato TA, Kanba S. Introducing directly induced microglia-like (iMG) cells from fresh human monocytes: a novel translational research tool for psychiatric disorders. *Front Cell Neurosci*. 2015;9:184.

79. Ryan KJ, White CC, Patel K, et al. A human microglia-like cellular model for assessing the effects of neurodegenerative disease gene variants. *Sci Transl Med*. 2017;9(421). https://doi.org/10.1126/scitranslmed.aai7635.
80. Rice D, Barone S. Critical periods of vulnerability for the developing nervous system: evidence from humans and animal models. *Environ Health Perspect*. 2000;108 (suppl 3):511–533.
81. Hensch TK. Critical period regulation. *Annu Rev Neurosci*. 2004;27:549–579.
82. Marco EJ, Hinkley LB, Hill SS, Nagarajan SS. Sensory processing in autism: a review of neurophysiologic findings. *Pediatr Res*. 2011;69(5 pt 2):48R–54R.
83. Doll CA, Broadie K. Impaired activity-dependent neural circuit assembly and refinement in autism spectrum disorder genetic models. *Front Cell Neurosci*. 2014;8:30.
84. Morrow EM, Yoo SY, Flavell SW, et al. Identifying autism loci and genes by tracing recent shared ancestry. *Science*. 2008;321(5886):218–223.
85. Schafer DP, Lehrman EK, Kautzman AG, et al. Microglia sculpt postnatal neural circuits in an activity and complement-dependent manner. *Neuron*. 2012;74(4):691–705.
86. Grayton HM, Missler M, Collier DA, Fernandes C. Altered social behaviours in neurexin 1α knockout mice resemble core symptoms in neurodevelopmental disorders. *PLoS One*. 2013;8(6): e67114.
87. Won H, Lee HR, Gee HY, et al. Autistic-like social behaviour in Shank2-mutant mice improved by restoring NMDA receptor function. *Nature*. 2012;486(7402):261–265.
88. Peça J, Feliciano C, Ting JT, et al. Shank3 mutant mice display autistic-like behaviours and striatal dysfunction. *Nature*. 2011;472(7344):437–442.
89. Harlow EG, Till SM, Russell TA, Wijetunge LS, Kind P, Contractor A. Critical period plasticity is disrupted in the barrel cortex of FMR1 knockout mice. *Neuron*. 2010;65(3):385–398.
90. Bhattacharya A, Kaphzan H, Alvarez-Dieppa AC, Murphy JP, Pierre P, Klann E. Genetic removal of p70 S6 kinase 1 corrects molecular, synaptic, and behavioral phenotypes in fragile X syndrome mice. *Neuron*. 2012;76(2):325–337.
91. Cohen S, Gabel HW, Hemberg M, et al. Genome-wide activity-dependent MeCP2 phosphorylation regulates nervous system development and function. *Neuron*. 2011;72(1):72–85.
92. Kwon CH, Luikart BW, Powell CM, et al. Pten regulates neuronal arborization and social interaction in mice. *Neuron*. 2006;50(3):377–388.
93. Auerbach BD, Osterweil EK, Bear MF. Mutations causing syndromic autism define an axis of synaptic pathophysiology. *Nature*. 2011;480(7375):63–68.
94. Greer PL, Hanayama R, Bloodgood BL, et al. The Angelman syndrome protein Ube3A regulates synapse development by ubiquitinating arc. *Cell*. 2010;140(5):704–716.
95. Ebert DH, Greenberg ME. Activity-dependent neuronal signalling and autism spectrum disorder. *Nature*. 2013;493(7432):327–337.
96. Chubykin AA, Atasoy D, Etherton MR, et al. Activity-dependent validation of excitatory versus inhibitory synapses by neuroligin-1 versus neuroligin-2. *Neuron*. 2007;54(6):919–931.
97. Choi YB, Li HL, Kassabov SR, et al. Neurexin-neuroligin transsynaptic interaction mediates learning-related synaptic remodeling and long-term facilitation in aplysia. *Neuron*. 2011;70(3):468–481.
98. Arida RM, Scorza CA, Scorza FA, Gomes da Silva S, da Graça Naffah-Mazzacoratti M, Cavalheiro EA. Effects of different types of physical exercise on the staining of parvalbumin-positive neurons in the hippocampal formation of rats with epilepsy. *Prog Neuropsychopharmacol Biol Psychiatry*. 2007;31(4):814–822.
99. Anderson-Hanley C, Tureck K, Schneiderman RL. Autism and exergaming: effects on repetitive behaviors and cognition. *Psychol Res Behav Manag*. 2011;4:129–137.
100. Dillon SR, Adams D, Goudy L, Bittner M, McNamara S. Evaluating exercise as evidence-based practice for individuals with autism spectrum disorder. *Front Public Health*. 2016;4:290.

101. Aberg MA, Pedersen NL, Torén K, et al. Cardiovascular fitness is associated with cognition in young adulthood. *Proc Natl Acad Sci USA*. 2009;106(49):20906–20911.
102. Gomes da Silva S, Unsain N, Mascó DH, et al. Early exercise promotes positive hippocampal plasticity and improves spatial memory in the adult life of rats. *Hippocampus*. 2012;22(2):347–358.
103. de Almeida AA, Gomes da Silva S, Fernandes J, et al. Differential effects of exercise intensities in hippocampal BDNF, inflammatory cytokines and cell proliferation in rats during the postnatal brain development. *Neurosci Lett*. 2013;553:1–6.
104. Pereira AC, Huddleston DE, Brickman AM, et al. An in vivo correlate of exercise-induced neurogenesis in the adult dentate gyrus. *Proc Natl Acad Sci USA*. 2007;104(13):5638–5643.
105. Amaral DG, Schumann CM, Nordahl CW. Neuroanatomy of autism. *Trends Neurosci*. 2008;31(3):137–145.
106. Kelly NA, Wood KH, Allendorfer JB, et al. High-intensity exercise acutely increases substantia nigra and prefrontal brain activity in Parkinson's disease. *Med Sci Monit*. 2017;23:6064–6071.
107. Clark PJ, Bhattacharya TK, Miller DS, Rhodes JS. Induction of c-Fos, Zif268, and Arc from acute bouts of voluntary wheel running in new and pre-existing adult mouse hippocampal granule neurons. *Neuroscience*. 2011;184:16–27.
108. Inoue T, Ninuma S, Hayashi M, Okuda A, Asaka T, Maejima H. Effects of long-term exercise and low-level inhibition of GABAergic synapses on motor control and the expression of BDNF in the motor related cortex. *Neurol Res*. 2018;40(1):18–25.
109. Uysal N, Agilkaya S, Sisman AR, et al. Exercise increases leptin levels correlated with IGF-1 in hippocampus and prefrontal cortex of adolescent male and female rats. *J Chem Neuroanat*. 2017;81:27–33.
110. Yoshii A, Constantine-Paton M. Postsynaptic BDNF-TrkB signaling in synapse maturation, plasticity, and disease. *Dev Neurobiol*. 2010;70(5):304–322.
111. Shcheglovitov A, Shcheglovitova O, Yazawa M, et al. SHANK3 and IGF1 restore synaptic deficits in neurons from 22q13 deletion syndrome patients. *Nature*. 2013;503(7475):267–271.
112. Griesi-Oliveira K, Acab A, Gupta AR, et al. Modeling non-syndromic autism and the impact of TRPC6 disruption in human neurons. *Mol Psychiatry*. 2015;20(11):1350–1365.
113. Danielson NB, Kaifosh P, Zaremba JD, et al. Distinct contribution of adult-born hippocampal granule cells to context encoding. *Neuron*. 2016;90(1):101–112.
114. Zhang Z, van Praag H. Maternal immune activation differentially impacts mature and adult-born hippocampal neurons in male mice. *Brain Behav Immun*. 2015;45:60–70.
115. Vukovic J, Colditz MJ, Blackmore DG, Ruitenberg MJ, Bartlett PF. Microglia modulate hippocampal neural precursor activity in response to exercise and aging. *J Neurosci*. 2012;32(19):6435–6443.
116. Nakayama H, Miyazaki T, Kitamura K, et al. GABAergic inhibition regulates developmental synapse elimination in the cerebellum. *Neuron*. 2012;74(2):384–396.
117. Peñagarikano O, Abrahams BS, Herman EI, et al. Absence of CNTNAP2 leads to epilepsy, neuronal migration abnormalities, and core autism-related deficits. *Cell*. 2011;147(1):235–246.
118. Han S, Tai C, Jones CJ, Scheuer T, Catterall WA. Enhancement of inhibitory neurotransmission by GABAA receptors having α2,3-subunits ameliorates behavioral deficits in a mouse model of autism. *Neuron*. 2014;81(6):1282–1289.
119. Mao W, Watanabe T, Cho S, et al. Shank 1 regulates excitatory synaptic transmission in mouse hippocampal parvalbumin-expressing inhibitory interneurons. *Eur J Neurosci*. 2015;41(8):1025–1035.
120. Fatemi SH, Halt AR, Stary JM, Kanodia R, Schulz SC, Realmuto GR. Glutamic acid decarboxylase 65 and 67 kDa proteins are reduced in autistic parietal and cerebellar cortices. *Biol Psychiatry*. 2002;52(8):805–810.

121. Fatemi SH, Reutiman TJ, Folsom TD, Thuras PD. GABA (A) receptor downregulation in brains of subjects with autism. *J Autism Dev Disord*. 2009;39(2):223–230.
122. Fatemi SH, Folsom TD, Reutiman TJ, Thuras PD. Expression of GABA (B) receptors is altered in brains of subjects with autism. *Cerebellum*. 2009;8(1):64–69.
123. Gomes da Silva S, Doná F, da Silva Fernandes MJ, Scorza FA, Cavalheiro EA, Arida RM. Physical exercise during the adolescent period of life increases hippocampal parvalbumin expression. *Brain Dev*. 2010;32(2):137–142.
124. Schoenfeld TJ, Rada P, Pieruzzini PR, Hsueh B, Gould E. Physical exercise prevents stress-induced activation of granule neurons and enhances local inhibitory mechanisms in the dentate gyrus. *J Neurosci*. 2013;33(18):7770–7777.
125. Wu Y, Dissing-Olesen L, Mac Vicar BA, Stevens B. Microglia: dynamic mediators of synapse development and plasticity. *Trends Immunol*. 2015;36(10):605–613.
126. Jiang T, Zhang L, Pan X, et al. Physical exercise improves cognitive function together with microglia phenotype modulation and remyelination in chronic cerebral hypoperfusion. *Front Cell Neurosci*. 2017;11:404.
127. Kohman RA, Bhattacharya TK, Wojcik E, Rhodes JS. Exercise reduces activation of microglia isolated from hippocampus and brain of aged mice. *J Neuroinflammation*. 2013;10:114.

Index

Note: Page numbers followed by "*f*" indicate figures, and "*t*" indicate tables.

I

K

L

M

N

O

P

FSC
www.fsc.org
MIX
Papier | Fördert gute Waldnutzung
FSC® C083411

Kolibri 360 GmbH
Ferdinand-Jühlke-Straße 7
99095 Erfurt, Deutschland
produktsicherheit@kolibri360.de